AMS SHORT COURSE LECTURE NOTES

Introductory Survey Lectures

Published as a subseries of

Proceedings of Symposia in Applied Mathematics

Proceedings of Symposia in
APPLIED MATHEMATHICS

Volume 73

Finite Frame Theory
A Complete Introduction to Overcompleteness

AMS Short Course
Finite Frame Theory: A Complete Introduction
to Overcompleteness
January 8–9, 2015
San Antonio, Texas

Kasso A. Okoudjou
Editor

American Mathematical Society
Providence, Rhode Island

LECTURE NOTES PREPARED FOR THE AMERICAN MATHEMATICAL SOCIETY SHORT COURSE ON FINITE FRAME THEORY: A COMPLETE INTRODUCTION TO OVERCOMPLETENESS HELD IN SAN ANTONIO, TEXAS, JANUARY 8–9, 2015

The AMS Short Course Series is sponsored by the Society's Program Committee for National Meetings. The series is under the direction of the Short Course Subcommittee of the Program Committee for National Meetings.

2010 *Mathematics Subject Classification.* Primary 15A29, 41A45, 42A10, 42C15, 47B99, 52A20, 52B11, 65H10, 90C26.

Library of Congress Cataloging-in-Publication Data

Library of Congress Cataloging-in-Publication Data
Names: Okoudjou, Kasso A., 1973- editor.
Title: Finite frame theory : a complete introduction to overcompleteness : AMS short course, January 8-9, 2015, San Antonio, TX / Kasso A. Okoudjou, editor.
Description: Providence, Rhode Island : American Mathematical Society, [2016] — Series: Proceedings of symposia in applied mathematics ; volume 73 — Includes bibliographical references and index. Identifiers: LCCN 2016002469 — ISBN 9781470420192 (alk. paper)
Subjects: LCSH: Frames (Vector analysis)–Congresses. — Hilbert space–Congresses. — AMS: Linear and multilinear algebra; matrix theory – Basic linear algebra – Inverse problems. msc — Approximations and expansions – Approximations and expansions – Approximation by arbitrary linear expressions. msc — Harmonic analysis on Euclidean spaces – Harmonic analysis in one variable – Trigonometric approximation. msc — Harmonic analysis on Euclidean spaces – Nontrigonometric harmonic analysis – General harmonic expansions, frames. msc — Operator theory – Special classes of linear operators – None of the above, but in this section. msc — Convex and discrete geometry – General convexity – Convex sets in n dimensions (including convex hypersurfaces). msc — Convex and discrete geometry – Polytopes and polyhedra – n-dimensional polytopes. msc — Numerical analysis – Nonlinear algebraic or transcendental equations – Systems of equations. msc — Operations research, mathematical programming – Mathematical programming – Nonconvex programming, global optimization. msc — Information and communication, circuits – Communication, information – Signal theory (characterization, reconstruction, filtering, etc.). msc Classification: LCC QA433 .F48 2016 — DDC 512/.5–dc23 LC record available at http://lccn.loc.gov/2016002469

DOI: http://dx.doi.org/10.1090/psapm/073

∞ The paper used in this book is acid-free and falls within the guidelines
established to ensure permanence and durability.
Visit the AMS home page at http://www.ams.org/

10 9 8 7 6 5 4 3 2 1   16 15 14 13 12 11

Contents

Preface vii

Introduction ix

A Brief Introduction to Hilbert Space Frame Theory and its Applications
PETER G. CASAZZA and RICHARD G. LYNCH 1

Unit norm tight frames in finite-dimensional spaces
DUSTIN G. MIXON 53

Algebro-Geometric Techniques and Geometric Insights for Finite Frames
NATE STRAWN 79

Preconditioning techniques in frame theory and probabilistic frames
KASSO A. OKOUDJOU 105

Quantization, finite frames, and error diffusion
ALEXANDER DUNKEL, ALEXANDER M. POWELL, ANNELIESE H. SPAETH, and ÖZGUR YILMAZ 143

Frames and Phaseless Reconstruction
RADU BALAN 175

Compressed sensing and dictionary learning
GUANGLIANG CHEN and DEANNA NEEDELL 201

Preface

The formula used to define Fourier frames is written explicitly by Paley and Wiener, [**10**] page 115, in the context of one of their fundamental theorems dealing with non-harmonic Fourier series. Fledgling forms of Fourier frames go back to Dini (1880) and then G. D. Birkhoff (1917), and leap to the profound results of Beurling and H. J. Landau in the 1960s, [**2**], [**9**]. Fourier frames lead naturally to non-uniform sampling formulas, and this is far from a complete story, e.g., [**1**].

Notwithstanding the importance of Fourier frames, and research such as that by Paley, Wiener, Beurling, and Landau, frames are with us today because of the celebrated work of Duffin and Schaeffer [**5**] in 1952. They explicitly found and featured the mathematical power of Fourier frames, and did the *right thing* mathematically by formulating such frames for Hilbert spaces, extracting central features of frames such as the decomposition of functions in terms of frames and understanding the role of overcomplete systems such as frames as opposed to orthonormal bases. It should also be pointed out that parallel to this development, a major analysis of bases was under way by the likes of Bari and Köthe; and their results could be rewritten in terms of frames, see, e.g., [**11**].

And then wavelet theory came along! More precisely, with regard to frames, there was the important work of Daubechies, Grossmann, and Meyer (1986) [**4**]; and there was the subsequent wonderful mix of mathematics and engineering and physics providing new insights as regards the value of frames. We now understood a basic role for frames with regard to noise reduction, stable decompositions, and robust representations of signals. In retrospect, frame research in the 1990s, besides its emerging prominence in wavelet theory and time-frequency analysis and their applications, was an *analytic* incubator, with all the accompanying excitement, that led to *finite* frames!

By the late 1990s and continuing today as an expanding mysterious universe, finite frame theory has become a dominant, intricate, relevant, and vital field. There were specific topics such as frame potential energy theory, $\Sigma-\Delta$ quantization, quantum detection, and periodic approximants in ambiguity function behavior, all with important applications. This has brought to bear a whole new vista of advanced technologies to understand frames and to unify ideas. The power of harmonic analysis and engineering brilliance are still part and parcel of frames, whether finite or not, but now we also use geometry and algebraic geometry, combinatorics, number theory, representation theory, and advanced linear and abstract algebra. There are major influences from compressive sampling, graph theory, and finite uncertainty principle inequalities.

The time was right just a few years ago *to stop and smell the roses*, and the volume on finite frames, edited by Casazza and Kutyniok [**3**] appeared (2013).

Amazingly and not surprisingly, given the talent pool of researchers, the intrigue and intricacies of the problems, and the applicability of the subject, the time is *still* right. Kasso Okoudjou's 2015 AMS Short Course on Finite Frame Theory was perfectly conceived. He assembled the leading experts in the field, not least of whom in my opinion was Okoudjou himself, to explain the latest and deepest results. This book is the best step possible towards the future. Enjoy!

John J. Benedetto
Norbert Wiener Center
Department of Mathematics
University of Maryland
College Park, MD 20742
USA

Introduction

The present volume is the proceedings of the AMS Short Course on Finite Frame Theory: A Complete Introduction to Overcompleteness organized in San-Antonio, January 8 & 9, 2015, prior to the Joint Mathematical Meetings.

Hilbert space frames have traditionally been used to decompose, process and reconstruct signals/images. In this volume we shall focus on frames for finite dimensional Euclidean spaces. In this setting, a set of vectors $\{f_k\}_{k=1}^M \subset \mathbb{K}^N$ is a frame for $\mathbb{K}^N$ if and only if there exist $0 < A \leq B < \infty$ such that

$$A\|x\|^2 \leq \sum_{k=1}^{M} |\langle x, f_k\rangle|^2 \leq B\|x\|^2$$

for each $x \in \mathbb{K}^N$, where $\mathbb{K} = \mathbb{R}$, or $\mathbb{K} = \mathbb{C}$. We refer to Section 4 of Chapter 1 for more details on frames. Today, frame theory is an exciting, dynamic subject with applications to pure mathematics, applied mathematics, engineering, medicine, computer science, and quantum computing. From a mathematical perspective, frame theory is at the intersection of many fields such as functional and harmonic analysis, numerical analysis, matrix theory, numerical linear algebra, algebraic and differential geometry, probability, statistics, and convex geometry. Problems in frame design arising in applications often present fundamental, completely new challenges never before encountered in mathematics. We refer to [**3**, **6**–**8**] for more on frame theory and its applications.

Finite unit norm tight frames (FUNTFs) are one of the most fundamental objects in frame theory with several applications. For example, in communication theory, FUNTFs are the optimal frames amongst unit norm frames to defeat additive white Gaussian noise, and amongst tight frames to defeat one erasure. Additionally, FUNTFs with minimal worst-case coherence are often optimal codes for synchronous multiple access communication channels, and equiangular FUNTFs can be used to embed fingerprints into media content to defeat piracy and common forgery attacks. Finally, compressed sensing matrices tend to (nearly) have FUNTF structure, and conversely, FUNTFs with small worst-case coherence typically perform well as compressed sensing matrices.

Perhaps the most beautiful theory behind FUNTFs concerns the *frame potential.* This quantifies the potential energy in a collection of particles on the sphere corresponding to a force between the particles which encourages pairwise orthogonality; while this *frame force* doesn't naturally arise from physics, the formulation still leads to a satisfying physical intuition. It's easy to show that FUNTFs are precisely the configurations which minimize the frame potential, and it turns out that all local minimizers are also global, meaning FUNTFs can be visualized as the steady state of a dynamical system governed by the frame force. In three dimensions, minimizers of the frame potential form the vertices of recognizable figures such as Platonic solids and the soccer ball.

To illustrate the many faces of frame theory, we observe that from a geometric point of view, the FUNTFs lie in the intersection of a product of spheres and a scaled Stiefel manifold. These restrictions can be viewed as a system of multivariate quadratic equations, making this space a quadratic algebraic variety. Though several basic and long standing problems have been tackled using algebro-geometric techniques, our understanding of these spaces remains incomplete. For example, the classification of the singular points of the FUNTF varieties remains one of the big open questions in this area. By contrast, the nonsingular points can be characterized as FUNTFs which admit a nontrivial decomposition into two mutually orthogonal collections. These "orthodecomposable" frames frequently present obstructions to nice theoretical arguments in finite frame theory. For example, the characterization of the local geometry of the FUNTF space near these points remains also an open and challenging question in the area. Nonetheless, algebro-geometric methods are increasingly being used to better understand frame theoretical questions.

Despite the promises and the power of the recently developed algebro-geometric methods for constructing FUNTFs, many questions in the area remain unsolved. For example, no effective method is known to construct FUNTFs when additional constraints on the frame vectors are imposed, e.g., the construction of equiangular FUNTFs (though some convex optimization techniques have been proposed). Furthermore, it seems desirable to have generic methods that would allow one to ideally transform a frame into a tight (or "nearly tight") one. These methods will be analogs of preconditioning methods prevalent in numerical linear algebra. Recently, techniques from convex geometry have been used to describe a class of frames called *scalable frames* which have the property that their frame vectors can be rescaled to result in tight frames.

Frames are intrinsically defined through their spanning properties. However, in real euclidean spaces, they can also be viewed as distributions of point masses. This point of view is also partially justified by the frame potential described above. In this context, the notion of *probabilistic frames* was introduced as a class of probability measures with finite second moment and whose support spans the entire space. This notion is a special case of continuous frames for Hilbert spaces that has applications in quantum computing. In this framework, probabilistic tools related to the Wasserstein metric can be appealed to in order to investigate questions in frame theory.

One of the fundamental application areas of frame theory remains modern signal processing. In this context, frame expansions and dual frames allow one to reconstruct a signal from its frame coefficients — the use of redundant sets of vectors ensures that this process is robust against noise and other forms of data loss. Although frame expansions provide discrete signal decompositions, the frame coefficients generally take on a continuous range of values and must also undergo a lossy step to discretize their amplitudes so that they become amenable to digital processing and storage. This analog-to-digital conversion step is known as quantization. A very well-developed theory of quantization based on finite frame theory is now part of the applied mathematics infrastructure.

An emerging topic in applied harmonic analysis is the question of nonlinear (signal) reconstruction. For example, frame design for phaseless reconstruction belongs to this class of problems. The problem of phaseless reconstruction can be simply stated as follows. Given the magnitudes of the coefficients of an output of a linear redundant system (frame), we want to reconstruct the unknown input. This problem has first occurred in X-ray crystallography starting from the early $20th$ century. In 1985 the Nobel prize in chemistry was awarded to Herbert Hauptman (a mathematician) for his contributions to the development of X-ray crystallography. The same nonlinear reconstruction problem shows up in speech processing, particularly in speech recognition. An age-old problem is about the importance of phase in signal processing and whether the magnitude of short-time Fourier transform encodes enough information to allow essentially unique reconstruction of the input signal. Generically, frame theory provides a unifying language to state and solve the problem.

The broader question of nonlinear signal analysis has also been investigated in the context of the new field of compressed sensing that arose as a response to inefficient traditional signal acquisition schemes. For example, assuming that the signal of interest is sparse (with respect to some fixed orthonormal basis), a typical problem is to algorithmically reconstruct this signal from a small number of linear measurements. During the last decade some deep results have been obtained in compressed sensing. However, in a number of applications, the signal of interest is sparse with respect to a (redundant) tight frame and many of the traditional compressed sensing methods are not applicable. A general theory of compressed sensing for signals that are sparse in tight frames has now emerged. Another problem related to compress sensing is the dictionary learning problem which consists of finding sparse and interpretable signal representations which are then used in applications such as compression, denoising, and super-resolution. In this context a trade-off is often made between analytic dictionaries which are backed by a very rich theoretical foundation, and the data-dependent dictionaries which are more flexible and are application based.

The seven chapters in this volume will cover most of the topics discussed above. In particular, in Chapter 1 Casazza and Lynch give an introduction to (finite dimensional) Hilbert space frame theory and its applications, and list a number of open problems. In chapter 2 Mixon gives a motivated account of the study of FUNTFs in finite dimensional spaces focusing on topics such as the frame potential, eigensteps, and equiangular tight frames. In Chapter 3, Strawn starts from a series of examples and explores algebraic varieties of FUNTFs leading to an exposition of the algebraic and differential geometry roots of certain of the known methods for constructing FUNTFs. In Chapter 4, Okoudjou through a series of motivating examples surveys certain preconditioning techniques recently introduced in frame theory and gives an account of a probabilistic interpretation of finite frame theory. In Chapter 5, Dunkel, Powell, Spaeth, and Yilmaz provide an expository introduction to the theory and practice of quantization for finite frame coefficients. In particular, the chapter focuses on memoryless scalar quantization (MSQ), the uniform noise model, first order Sigma-Delta ($\Sigma\Delta$) quantization, and the role of error diffusion in quantization. In Chapter 6, Balan reviews existing analytical results as well as algorithms for signal recovery for the phaseless reconstruction problem. In Chapter 7, Chen and Needell give an overview of recent results extending the theory of compressed sensing to the case of sparsity in tight frames, and describe another application of frame theory in the context of dictionary learning.

We would like to thank Sivaram Narayan for soliciting a proposal to organize the AMS 2015 Short Course. We also thank all the lecturers not only for the outstanding talks they gave, but also for their contributions to the present volume. Our gratitudes also go to the AMS staff for their help with the organizational aspects of the course. Finally, we would like to thank the reviewers for their useful comments which helped improve this volume.

Kasso A. Okoudjou
College Park, MD

Bibliography

[1] Enrico Au-Yeung and John J. Benedetto, *Generalized Fourier frames in terms of balayage*, J. Fourier Anal. Appl. **21** (2015), no. 3, 472–508, DOI 10.1007/s00041-014-9369-7. MR3345364

[2] Arne Beurling, *The collected works of Arne Beurling. Vol. 2*, Contemporary Mathematicians, Birkhäuser Boston, Inc., Boston, MA, 1989. Harmonic analysis; Edited by L. Carleson, P. Malliavin, J. Neuberger and J. Wermer. MR1057614 (92k:01046b)

[3] P. G. Casazza and G. Kutyniok, "Finite Frame Theory," Eds., Birkhaüser, Boston, 2012.

[4] Ingrid Daubechies, A. Grossmann, and Y. Meyer, *Painless nonorthogonal expansions*, J. Math. Phys. **27** (1986), no. 5, 1271–1283, DOI 10.1063/1.527388. MR836025 (87e:81089)

[5] R. J. Duffin and A. C. Schaeffer, *A class of nonharmonic Fourier series*, Trans. Amer. Math. Soc. **72** (1952), 341–366. MR0047179 (13,839a)

[6] Christopher Heil, *What is ... a frame?*, Notices Amer. Math. Soc. **60** (2013), no. 6, 748–750, DOI 10.1090/noti1011. MR3076247

[7] J. Kovacevic and A. Chebira, *Life Beyond Bases: The Advent of Frames (Part I),* Signal Processing Magazine, IEEE Volume **24**, Issue 4, July 2007, 86–104

[8] J. Kovacevic and A. Chebira, *Life Beyond Bases: The Advent of Frames (Part II),* Signal Processing Magazine, IEEE Volume **24**, Issue 5, Sept. 2007, 115–125.

[9] H. J. Landau, *Necessary density conditions for sampling and interpolation of certain entire functions*, Acta Math. **117** (1967), 37–52. MR0222554 (36 #5604)

[10] R. E. A. C. Paley and N Wiener, "Fourier Transforms in the Complex Domain," Amer. Math. Society Colloquium Publications, **XIX**, American Mathematical Society, Providence, RI, 1934.

[11] R Young, "An Introduction to Nonharmonic Fourier Series," Academic Press, New York, 1980, revised edition 2001.

Proceedings of Symposia in Applied Mathematics
Volume **73**, 2016
http://dx.doi.org/10.1090/psapm/073/00627

A Brief Introduction to Hilbert Space Frame Theory and its Applications

Peter G. Casazza and Richard G. Lynch

ABSTRACT. This is a short introduction to Hilbert space frame theory and its applications for those outside the area who want to enter the subject. We will emphasize finite frame theory since it is the easiest way to get into the subject.

CONTENTS

1. Reading List
2. The Basics of Hilbert Space Theory
3. The Basics of Operator Theory
4. Hilbert Space Frames
5. Constants Related to Frames
6. Constructing Finite Frames
7. Gramian Operators
8. Fusion Frames
9. Infinite Dimensional Hilbert Spaces
10. Major Open Problems in Frame Theory
References

1. Reading List

For a more complete treatment of frame theory we recommend the books of Han, Kornelson, Larson, and Weber [**59**], Christensen [**41**], the book of Casazza and Kutyniok [**32**], the tutorials of Casazza [**23, 24**] and the memoir of Han and Larson [**60**]. For a complete treatment of frame theory in time-frequency analysis we recommend the book of Gröchenig [**55**]. For an introduction to frame theory and filter banks plus applications to engineering we recommend Kovačević and Chebira [**64**]. Also, a wealth of information can be found at the Frame Research Center's website [**53**].

2010 *Mathematics Subject Classification.* Primary 42C15.

Key words and phrases. Hilbert space frames, operators related to frames, frame operator, Gramian operator, Hilbert space theory, operator theory, dual frames, frame constants, tight frames, construction of frames, fusion frames, applications of frames.

The authors were supported by NSF DMS 1307685; and NSF ATD 1042701 and 1321779; AFOSR DGE51: FA9550-11-1-0245.

2. The Basics of Hilbert Space Theory

Given a positive integer N, we denote by $\mathbb{H}^N$ the real or complex Hilbert space of dimension N. This is either $\mathbb{R}^N$ or $\mathbb{C}^N$ with the **inner product** given by

$$\langle x, y\rangle = \sum_{i=1}^{N} a_i \overline{b_i}$$

for $x = (a_1, a_2, \cdots, a_N)$ and $y = (b_1, b_2, \cdots, b_N)$ and the **norm** of a vector x is

$$\|x\|^2 = \langle x, x\rangle.$$

For $x, y \in \mathbb{H}^N$, $\|x - y\|$ is the **distance** from the vector x to the vector y. For future reference, note that in the real case,

$$\begin{aligned}\|x-y\|^2 &= \langle x-y, x-y\rangle \\ &= \langle x, x\rangle - \langle x, y\rangle - \langle y, x\rangle + \langle y, y\rangle \\ &= \|x\|^2 - 2\langle x, y\rangle + \|y\|^2\end{aligned}$$

and in the complex case we have

$$\langle x, y\rangle + \langle y, x\rangle = \langle x, y\rangle + \overline{\langle x, y\rangle} = 2Re\, \langle x, y\rangle,$$

where $Re\, c$ denotes the real part of the complex number c. We will concentrate on finite dimensional Hilbert spaces since it is the easiest way to get started on the subject of frames. Most of these results hold for infinite dimensional Hilbert spaces and at the end we will look at the infinite dimensional case.

The next lemma contains a **standard trick** for calculations.

LEMMA 2.1. *If $x \in \mathbb{H}^N$ and $\langle x, y\rangle = 0$ for all $y \in \mathbb{H}^N$ then $x = 0$.*

PROOF. Letting $y = x$ we have

$$0 = \langle x, y\rangle = \langle x, x\rangle = \|x\|^2,$$

and so $x = 0$. □

DEFINITION 2.2. A set of vectors $\{e_i\}_{i=1}^M$ in $\mathbb{H}^N$ is called:

(1) **linearly independent** if for any scalars $\{a_i\}_{i=1}^M$,

$$\sum_{i=1}^{M} a_i e_i = 0 \quad \Rightarrow \quad a_i = 0, \text{ for all } i = 1, 2, \cdots, M.$$

Note that this requires $e_i \neq 0$ for all $i = 1, 2, \cdots, M$.

(2) **complete** (or a **spanning set**) if

$$\overline{\text{span}}\{e_i\}_{i=1}^M = \mathbb{H}^N.$$

(3) **orthogonal** if for all $i \neq j$, $\langle e_i, e_j\rangle = 0$.

(4) **orthornormal** if it is orthogonal and unit norm.

(5) an **orthonormal basis** if it is complete and orthonormal.

The following is immediate from the definitions.

PROPOSITION 2.3. *If $\{e_i\}_{i=1}^N$ is an orthonormal basis for $\mathbb{H}^N$, then for every $x \in \mathbb{H}$ we have*

$$x = \sum_{i=1}^{N} \langle x, e_i\rangle e_i.$$

From the previous proposition, we can immediately deduce an essential identity called Parseval's Identity.

PROPOSITION 2.4 (Parseval's Identity). *If* $\{e_i\}_{i=1}^N$ *is an orthonormal basis for* $\mathbb{H}^N$*, then for every* $x \in \mathbb{H}^N$*, we have*

$$\|x\|^2 = \sum_{i=1}^N |\langle x, e_i\rangle|^2.$$

Some more basic identities and inequalities for Hilbert space that are frequently used are contained in the next proposition.

PROPOSITION 2.5. *Let* $x, y \in \mathbb{H}^N$.

(1) **Cauchy-Schwarz Inequality**:

$$|\langle x, y\rangle| \le \|x\|\|y\|,$$

with equality if and only if $x = cy$ *for some constant* c.

(2) **Triangle Inequality**:

$$\|x + y\| \le \|x\| + \|y\|.$$

(3) **Polarization Identity**: *Assuming* $\mathbb{H}^N$ *is real,*

$$\langle x, y\rangle = \frac{1}{4}\left[\|x + y\|^2 - \|x - y\|^2\right].$$

If $\mathbb{H}^N$ *is complex, then*

$$\langle x, y\rangle = \frac{1}{4}\left[\|x + y\|^2 - \|x - y\|^2 + i\|x + iy\|^2 - i\|x - iy\|^2\right].$$

(4) **Pythagorean Theorem**: *Given pairwise orthogonal vectors* $\{x_i\}_{i=1}^M$,

$$\left\|\sum_{i=1}^M x_i\right\|^2 = \sum_{i=1}^M \|x_i\|^2.$$

PROOF. (1) The inequality is trivial if $y = 0$. If $y \neq 0$, we may assume $\|y\| = 1$ by dividing through the inequality with $\|y\|$. Now we compute:

$$\begin{aligned} 0 &< \|x - \langle x, y\rangle y\|^2 \\ &= \|x\|^2 - 2\langle x, y\rangle\overline{\langle x, y\rangle} + |\langle x, y\rangle|^2\|y\|^2 \\ &= \|x\|^2 - |\langle x, y\rangle|^2 \\ &= \|x\|^2\|y\|^2 - |\langle x, y\rangle|^2. \end{aligned}$$

Note that the strict inequality would be equality if $x = cy$.

(2) Applying (1) to obtain the second inequality:

$$\begin{aligned} \|x + y\|^2 &= \|x\|^2 + 2Re\,\langle x, y\rangle + \|y\|^2 \\ &\le \|x\|^2 + 2|\langle x, y\rangle| + \|y\|^2 \\ &\le \|x\|^2 + 2\|x\|\|y\| + \|y\|^2 \\ &= (\|x\| + \|y\|)^2. \end{aligned}$$

(3) We compute assuming $\mathbb{H}^N$ is a real Hilbert space:

$$\begin{aligned}\|x+y\|^2 - \|x-y\|^2 &= \|x\|^2 + 2\langle x,y\rangle + \|y\|^2 - (\|x\|^2 - 2\langle x,y\rangle + \|y\|^2)\\ &= 4\langle x,y\rangle.\end{aligned}$$

The proof in the complex case is similar.

(4) Since $\langle x_i, x_j\rangle = 0$ for all $i \neq j$, we have

$$\begin{aligned}\left\|\sum_{i=1}^M x_i\right\|^2 &= \left\langle \sum_{i=1}^M x_i, \sum_{j=1}^M x_j \right\rangle\\ &= \sum_{i,j=1}^M \langle x_i, x_j\rangle\\ &= \sum_{i=1}^M \langle x_i, x_i\rangle\\ &= \sum_{i=1}^M \|x_i\|^2.\end{aligned}$$

□

We now look at subspaces of the Hilbert space.

DEFINITION 2.6. Let W, V be subspaces of $\mathbb{H}^N$.

(1) A vector $x \in \mathbb{H}^N$ is **orthogonal to a subspace** W, denoted $x \perp W$ if

$$\langle x, y\rangle = 0 \text{ for all } y \in W.$$

The **orthogonal complement** of W is

$$W^\perp = \{x \in \mathbb{H} : x \perp W\}.$$

(2) The subspaces W, V are **orthogonal subspaces**, denoted $W \perp V$ if $W \subset V^\perp$, that is,

$$\langle x, y\rangle = 0 \text{ for all } x \in W, \quad y \in V.$$

A simple calculation shows that $W^\perp$ is always closed and so if W is closed then $W^{\perp\perp} = W$. Fundamental to Hilbert space theory are **orthogonal projections** as defined next.

DEFINITION 2.7. An operator $P : \mathbb{H}^N \to \mathbb{H}^N$ is called a **projection** if $P^2 = P$. It is an **orthogonal projection** if P is also self-adjoint (See definition 3.3).

For any subspace $W \subset \mathbb{H}^N$, there is an orthogonal projection of $\mathbb{H}$ onto W called the **nearest point projection**. One way to define it is to pick any orthonormal basis $\{e_i\}_{i=1}^K$ for W and define

$$Px = \sum_{i=1}^K \langle x, e_i\rangle e_i.$$

Note that for all $j = 1, 2, \cdots, K$ we have

$$\langle Px, e_j\rangle = \left\langle \sum_{i=1}^K \langle x, e_i\rangle e_i, e_j \right\rangle = \sum_{i=1}^K \langle x, e_i\rangle\langle e_i, e_j\rangle = \langle x, e_j\rangle.$$

We need to check that this operator is well defined, that is, we must show it is independent of the choice of basis.

LEMMA 2.8. *If* $\{e_i\}_{i=1}^K$ *and* $\{g_i\}_{i=1}^K$ *are orthonormal bases for* W, *then*

$$\sum_{i=1}^K \langle x, e_i\rangle e_i = \sum_{i=1}^K \langle x, g_i\rangle g_i.$$

PROOF. We compute:

$$\begin{aligned}
\sum_{i=1}^K \langle x, e_i\rangle e_i &= \sum_{i=1}^K \left\langle x, \sum_{j=1}^K \langle e_i, g_j\rangle g_j \right\rangle e_i \\
&= \left(\sum_{i,j=1}^K \overline{\langle e_i, g_j\rangle} \langle x, g_j\rangle \right) e_i \\
&= \sum_{i=1}^K \left\langle \sum_{j=1}^K \langle x, g_j\rangle g_j, e_i \right\rangle e_i \\
&= \sum_{j=1}^K \langle x, g_j\rangle g_j.
\end{aligned}$$ □

In this case, $Id - P$ is also an orthogonal projection onto $W^\perp$. This projection maps each vector in $\mathbb{H}^N$ onto the unique nearest vector to x in W. In particular, $\|Px\| \le \|x\|$ for all $x \in \mathbb{H}$.

PROPOSITION 2.9. *Let* P *be an orthogonal projection onto a subspace* W. *Then*

$$\|x - Px\| \le \|x - y\|, \text{ for all } y \in W.$$

PROOF. Choose an orthonormal basis $\{e_i\}_{i=1}^K$ for W. Then

$$\begin{aligned}
Re\, \langle x, y\rangle &\le |\langle x, y\rangle| \\
&= \left| \sum_{i=1}^K \langle x, e_i\rangle \langle y, e_i\rangle \right| \\
&= \left| \sum_{i=1}^K \langle Px, e_i\rangle \langle y, e_i\rangle \right| \\
&\le \left(\sum_{i=1}^K |\langle Px, e_i\rangle|^2 \right)^{1/2} \left(\sum_{i=1}^K |\langle y, e_i\rangle|^2 \right)^{1/2} \\
&\le \|Px\| \|y\| \\
&\le \frac{1}{2}(\|Px\|^2 + \|y\|^2)
\end{aligned}$$

Therefore,

$$\begin{aligned}\|x-y\|^2 &= \|x\|^2 + \|y\|^2 - 2Re\,\langle x,y\rangle \\ &\geq \|x\|^2 + \|y\|^2 - (\|Px\|^2 + \|y\|^2) \\ &= \|x\|^2 - \|Px\|^2 \\ &= \|x\|^2 + \|Px\|^2 - 2\|Px\|^2 \\ &= \|x\|^2 + \|Px\|^2 - 2\langle Px, Px\rangle \\ &= \|x\|^2 + \|Px\|^2 - 2\langle Px, x\rangle \\ &= \|x - Px\|^2. \end{aligned}$$

□

Another important concept is taking orthogonal sums of subspaces of a Hilbert space.

DEFINITION 2.10. If $\{W_i\}_{i\in I}$ (the index set I is allowed to be infinite) are subspaces of a Hilbert space $\mathbb{H}^N$, their **orthogonal direct sum** is

$$\left(\sum_{i\in I}\oplus W_i\right)_{\ell_2} = \left\{(x_i)_{i\in I} : x_i \in W_i, \text{ and } \sum_{i\in I}\|x_i\|^2 < \infty\right\}$$

and inner product defined by

$$\big\langle (x_i)_{i\in I}, (y_i)_{i\in I}\big\rangle = \sum_{i\in I}\langle x_i, y_i\rangle.$$

It follows that

$$\big\|(x_i)_{i\in I}\big\|^2 = \sum_{i\in I}\|x_i\|^2.$$

3. The Basics of Operator Theory

DEFINITION 3.1. A **linear operator** $T : \mathbb{H}^N \to \mathbb{H}^K$ between Hilbert spaces $\mathbb{H}^N$ and $\mathbb{H}^K$ satisfies:

$$T(ax+by) = aT(x) + bT(y) \text{ for all } x, y \in \mathbb{H}^N \text{ and scalars a,b.}$$

The **operator norm** is

$$\begin{aligned}\|T\| &= \sup_{\|x\|=1}\|Tx\| \\ &= \sup_{\|x\|\leq 1}\|Tx\| \\ &= \sup_{x\neq 0}\frac{\|Tx\|}{\|x\|}.\end{aligned}$$

From the definition, for all $x \in \mathbb{H}^N$ we have $\|Tx\| \leq \|T\|\|x\|$. Furthermore, if $T : \mathbb{H}^{N_1} \to \mathbb{H}^{N_2}$ and $S : \mathbb{H}^{N_2} \to \mathbb{H}^{N_3}$ then

$$\|STx\| \leq \|S\|\|Tx\| \leq \|S\|\|T\|\|x\|$$

showing that

$$\|ST\| \leq \|S\|\|T\|.$$

Linear operators on finite spaces can always be represented as a matrix, so we can work with linear operators and their matrices interchangeably.

DEFINITION 3.2. Let $T : \mathbb{H}^N \to \mathbb{H}^K$ be a linear operator, let $\{e_i\}_{i=1}^N$ be an orthonormal basis for $\mathbb{H}^N$ and let $\{g_j\}_{j=1}^K$ be an orthonormal basis for $\mathbb{H}^K$. The **matrix representation** of T (with respect to these orthonormal bases) is $T = [a_{ij}]_{1 \le i \le N, 1 \le j \le K}$ where

$$a_{ij} = \langle Te_i, g_j \rangle.$$

The following definition holds some fundamental concepts that we often work with when dealing with linear operators.

DEFINITION 3.3. Let $T : \mathbb{H}^N \to \mathbb{H}^K$ be a linear operator.

(1) The **kernel of T** is

$$\ker T = \{x \in \mathbb{H}^N : Tx = 0\}.$$

The **range of T** (or **image of T**) is

$$\operatorname{ran} T = \{Tx : x \in \mathbb{H}^N\}.$$

The **rank of T**, denoted rank T is the dimension of the $\operatorname{ran} T$. A standard result from linear algebra known as the **rank-nullity theorem** states

$$N = \dim \ker T + \operatorname{rank} T.$$

(2) T is **injective** if $\ker T = \{0\}$, and **surjective** if $\operatorname{ran} T = \mathbb{H}^K$. It is **bijective** if it is both injective and surjective.

(3) The **adjoint operator** $T^* : \mathbb{H}^K \to \mathbb{H}^N$ is defined by:

$$\langle Tx, y \rangle = \langle x, T^* y \rangle \text{ for all } x \in \mathbb{H}^N,\ y \in \mathbb{H}^K.$$

Note that $T^{**} = T$ and $(S + T)^* = S^* + T^*$.

(4) T is **bounded** if $\|T\| < \infty$.

(5) If $\mathbb{H}^N = \mathbb{H}^K$, then T is **invertible** if it bounded and there is a bounded linear map $S : \mathbb{H}^N \to \mathbb{H}^N$ so that $TS = ST = Id$. If such an S exists, it is unique and we call it the **inverse operator** of T and denote it by $S = T^{-1}$. In the finite setting, the existence of an inverse is equivalent to T being any of bounded, injective, or surjective.

A useful alternative method that we will use later for calculating $\|T\|$ is given in the following proposition.

PROPOSITION 3.4. *If $T : \mathbb{H}^N \to \mathbb{H}^K$ is a linear operator, then*

$$\|T\| = \sup\{|\langle Tx, y \rangle| : \|x\| = \|y\| = 1\}.$$

PROOF. For $\|x\| = 1 = \|y\|$, Cauchy-Schwarz gives

$$|\langle Tx, y \rangle| \le \|Tx\| \|y\| \le \|T\| \|x\| = \|T\|.$$

Hence,

$$\|T\| \ge \sup\{|\langle Tx, y \rangle| : \|x\| = \|y\| = 1\}.$$

Conversely,

$$\begin{aligned}\sup\{|\langle Tx, y\rangle| : \|x\| = \|y\| = 1\} \\ &\geq \sup\left\{\left|\left\langle Tx, \frac{Tx}{\|Tx\|}\right\rangle\right| : \|x\| = 1 \text{ and } Tx \neq 0\right\} \\ &= \sup_{\|x\|=1} \frac{\|Tx\|^2}{\|Tx\|} \\ &= \sup_{\|x\|=1} \|Tx\| = \|T\|. \end{aligned}$$ □

The following gives a way to identify operators.

PROPOSITION 3.5. *If $S, T : \mathbb{H}^N \to \mathbb{H}^K$ are operators satisfying*

$$\langle Tx, y\rangle = 0, \textit{ for all } x, y \in \mathbb{H}^N,$$

then $T = 0$. Hence, if

$$\langle Tx, y\rangle = \langle Sx, y\rangle \textit{ for all } x, y \in \mathbb{H}^N,$$

then $S = T$.

PROOF. Given $x \in \mathbb{H}^N$, by letting $y = Tx$ we obtain:

$$0 = \langle Tx, Tx\rangle = \|Tx\|^2 = 0,$$

and so $Tx = 0$ and $T = 0$. □

There are important relationships between the kernel of T and the range of T^* and vice versa.

PROPOSITION 3.6. *Let $T : \mathbb{H}^N \to \mathbb{H}^K$ be a linear operator. Then*

(1) $\ker T = [\operatorname{ran} T^*]^\perp$.
(2) $[\ker T]^\perp = \operatorname{ran} T^*$.
(3) $\ker T^* = [\operatorname{ran}\ T]^\perp$.
(4) $[\ker T^*]^\perp = \operatorname{ran} T$.

PROOF. (1) We have $x \in \ker T$ if and only if $Tx = 0$ then for all $y \in \mathbb{H}^K$ if and only if $\langle Tx, y\rangle = 0 = \langle x, T^*y\rangle$ for all $y \in H^K$ if and only if $x \in [\operatorname{ran} T^*]^\perp$.

(2) Observe from (1) that:

$$[\ker T]^\perp = [T^*(\mathbb{H}^K)]^{\perp\perp} = T^*(\mathbb{H}^K) = \operatorname{ran} T^*.$$

The relations (3) and (4) follow by replacing T by T^* in (1) and (2). □

We also have the following relationships for adjoint operators.

PROPOSITION 3.7. *For linear operators $T : \mathbb{H}^{N_1} \to H^{N_2}$ and $S : \mathbb{H}^{N_2} \to \mathbb{H}^{N_3}$, we have:*

(1) $(ST)^* = T^*S^*$.
(2) $\|T^*\| = \|T\|$.
(3) $\|T^*T\| = \|T\|^2$.

PROOF. (1) We compute:

$$\begin{aligned}\langle x, (ST)^*y\rangle &= \langle STx, y\rangle \\ &= \langle Tx, S^*y\rangle \\ &= \langle x, T^*S^*y\rangle.\end{aligned}$$

(2) We use Proposition 3.4 to get:

$$\begin{aligned}\|T\| &= \sup\{|\langle Tx, y\rangle| : \|x\| = \|y\| = 1\}\\ &= \sup\{|\langle x, T^*y\rangle| : \|x\| = \|y\| = 1\}\\ &= \|T^*\|.\end{aligned}$$

(3) We know that $\|T^*T\| \leq \|T^*\|\|T\| = \|T\|^2$. On the other hand, for all $\|x\| = 1$ via Cauchy-Schwarz,

$$\|Tx\|^2 = \langle Tx, Tx\rangle = \langle T^*Tx, x\rangle \leq \|T^*Tx\|\|x\| \leq \|T^*T\|\|x\| = \|T^*T\|.$$

Hence, $\|T\|^2 \leq \|T^*T\|$. □

DEFINITION 3.8. Let $T : \mathbb{H}^N \to \mathbb{H}^K$ be an injective linear operator. Then $(T^*T)^{-1}$ exists and the **Moore-Penrose inverse of T**, denoted by $T^\dagger$, is defined by

$$T^\dagger := (T^*T)^{-1}T^*.$$

The map $T^\dagger$ is a left inverse, that is, $T^\dagger T = I$.

DEFINITION 3.9. A linear operator $T : \mathbb{H}^N \to \mathbb{H}^K$ is called:

(1) **self-adjoint**, if $\mathbb{H}^N = \mathbb{H}^K$ and $T = T^*$.
(2) **normal**, if $\mathbb{H}^N = \mathbb{H}^K$ and $T^*T = TT^*$.
(3) an **isometry**, if $\|Tx\| = \|x\|$ for all $x \in \mathbb{H}^N$.
(4) a **partial isometry** if T restricted to $[\ker T]^\perp$ is an isometry.
(5) **positive** , if $\mathbb{H}^N = \mathbb{H}^K$, T is self-adjoint, and

$$\langle Tx, x\rangle \geq 0 \text{ for all } x \in \mathbb{H}.$$

In this case we write $T \geq 0$.
(6) **unitary**, if $TT^* = T^*T = Id$.

It follows that TT^* and T^*T are self-adjoint for any operator T.

EXAMPLE 3.10. A fundamental example of a positive, self-adjoint operator important in frame theory is to take vectors $\Phi = \{\varphi_i\}_{i=1}^M$ in $\mathbb{H}^N$ and define the operator:

$$Sx = \sum_{i=1}^{M} \langle x, \varphi_i\rangle\varphi_i.$$

It follows that

$$\langle Sx, x\rangle = \sum_{i=1}^{M}\langle x, \varphi_i\rangle\langle\varphi_i, x\rangle = \sum_{i=1}^{M} |\langle x, \varphi_i\rangle|^2 = \langle x, Sx\rangle$$

showing that it is positive and self-adjoint. This operator is called the **frame operator** of the sequence Φ.

Note that given two positive operators S and T, the sum $S + T$ is a positive operator but ST may not be as the next example shows.

EXAMPLE 3.11. Take $S : \mathbb{R}^2 \to \mathbb{R}^2$ to be the operator defined by

$$S\begin{bmatrix} x_1 \\ x_2 \end{bmatrix} = \begin{bmatrix} 0 & -1 \\ 1 & 0 \end{bmatrix}\begin{bmatrix} x_1 \\ x_2 \end{bmatrix} = \begin{bmatrix} -x_2 \\ x_1 \end{bmatrix}$$

Then $\langle Sx, x\rangle = 0$ for all $x \in \mathbb{R}^2$ so that S is positive. However,

$$S^2 \begin{bmatrix} x_1 \\ x_2 \end{bmatrix} = \begin{bmatrix} -1 & 0 \\ 0 & -1 \end{bmatrix} \begin{bmatrix} x_1 \\ x_2 \end{bmatrix} = \begin{bmatrix} -x_1 \\ -x_2 \end{bmatrix}$$

so that $\langle S^2x, x\rangle = -\|x\|^2$ for all $x \in \mathbb{R}^2$ and hence S^2 is not positive.

Nevertheless, we can define inequalities for positive operators:

DEFINITION 3.12. If S, T are positive operators on $\mathbb{H}^N$, we write $S \leq T$ if $T - S \geq 0$.

PROPOSITION 3.13. *Let $T : \mathbb{H}^N \to \mathbb{H}^K$ be a linear operator.*

(1) If $\mathbb{H}^N = \mathbb{H}^K$, the following are equivalent:
  (a) T is self-adjoint.
  (b) $\langle Tx, y\rangle = \langle x, Ty\rangle$, for all $x, y \in \mathbb{H}^N$.

(2) The following are equivalent:
  (a) T is an isometry.
  (b) $T^*T = Id$.
  (c) $\langle Tx, Ty\rangle = \langle x, y\rangle$, for all $x, y \in \mathbb{H}^N$.

(3) The following are equivalent:
  (a) T is unitary.
  (b) T and T^* are isometries.
  (c) T is an isometry and T^* is injective.
  (d) T is a surjective isometry.
  (e) T is bijective and $T^{-1} = T^*$.

(4) If U is any unitary operator, then $\|T\| = \|TU\| = \|UT\|$.

PROOF. (1) Applying Proposition 3.5, we have $T^* = T$ if and only if $\langle Tx, y\rangle = \langle T^*x, y\rangle$ for all x, y if and only if $\langle Tx, y\rangle = \langle x, Ty\rangle$ for all x, y.

(2) $(a) \Rightarrow (b)$: We prove the real case, but the complex case is similar using the complex version of the Polarization Identity. We have by the (real) Polarization Identity that for any $x, y \in \mathbb{H}^N$,

$$\begin{aligned}
\langle x, y\rangle &= \frac{1}{4}\left(\|x+y\|^2 - \|x-y\|^2\right) \\
&= \frac{1}{4}\left(\|T(x+y)\|^2 - \|T(x-y)\|^2\right) \\
&= \frac{1}{4}\left(\|Tx\|^2 + \|Ty\|^2 + 2\langle Tx, Ty\rangle - [\|Tx\|^2 + \|Ty\|^2 - 2\langle Tx, Ty\rangle]\right) \\
&= \langle Tx, Ty\rangle \\
&= \langle T^*Tx, y\rangle.
\end{aligned}$$

So, $T^*T = Id$ by Proposition 3.5.

$(b) \Rightarrow (c)$: For any $x, y \in \mathbb{H}^N$ we have

$$\langle Tx, Ty\rangle = \langle T^*Tx, y\rangle = \langle x, y\rangle.$$

$(c) \Rightarrow (a)$: (c) implies,

$$\begin{aligned}
\|Tx\|^2 &= \langle Tx, Tx\rangle \\
&= \langle x, x\rangle \\
&= \|x\|^2.
\end{aligned}$$

(3) $(a) \Leftrightarrow (b)$: This is immediate by the definitions and (2).

$(b) \Rightarrow (c)$: By (2, b) T^* is injective.
$(c) \Rightarrow (d)$: We observe:

$$\mathbb{H}^K = \{0\}^\perp = [\ker T^*]^\perp = T(\mathbb{H}^N) = \operatorname{ran} T.$$

$(d) \Rightarrow (e)$: Since T is bijective by assumption, $S = T^{-1}$ exists. We compute using (2, b):

$$T^* = T^*I = T^*(TS) = (T^*T)S = IS = S.$$

$(e) \Rightarrow (a)$: Immediate by the definitions.
(4) Since U is an isometry by (3), the result is now immediate from Lemma 3.4. □

DEFINITION 3.14. A linear operator $T : \mathbb{H}^N \to \mathbb{H}^K$ is a **Hilbert space isomorphism** if for all $x, y \in \mathbb{H}^N$ we have

$$\langle Tx, Ty\rangle = \langle x, y\rangle.$$

Two Hilbert spaces $\mathbb{H}^N$ and $\mathbb{H}^M$ are **isomorphic** if there is a Hilbert space isomorphism $T : \mathbb{H}^N \to \mathbb{H}^M$.

Proposition 3.13(2) implies that T is an isometry and $T^*T = Id$ for any Hilbert space isomorphism T. Thus, it is automatically injective. We see in the next proposition that every two Hilbert spaces of the same dimension are isomorphic.

PROPOSITION 3.15. *Every two N-dimensional Hilbert spaces are Hilbert space isomorphic. Thus, any N-dimensional Hilbert space is isomorphic to $\mathbb{C}^N$.*

PROOF. Let $\{e_i\}_{i=1}^N$ be an orthonormal basis of $\mathbb{H}_1^N$ and let $\{g_i\}_{i=1}^N$ be an orthonormal basis of $\mathbb{H}_2^N$. The operator T given by: $Te_i = g_i$ for all $i = 1, 2, \cdots, N$ is clearly a Hilbert space isomorphism. □

In the complex case, an operator T is self-adjoint if and only if $\langle Tx, x\rangle \in \mathbb{R}$ for any x. We need a lemma to prove this.

LEMMA 3.16. *If $\mathbb{H}^N$ is a complex Hilbert space and $T : \mathbb{H}^N \to \mathbb{H}^N$ is a linear operator satisfying $\langle Tx, x\rangle = 0$ for all $x \in \mathbb{H}^N$, then $T = 0$, the zero operator.*

PROOF. We prove this by showing that $\langle Tx, y\rangle$ for all $x, y \in \mathbb{H}^N$ and then apply Proposition 3.5 to conclude that $T = 0$. Note by assumption that we have for all x, y,

$$0 = \langle T(x+y), x+y\rangle = \langle Tx, y\rangle + \langle y, Tx\rangle$$

and

$$0 = \langle T(x+iy), x+iy\rangle = i\langle Tx, y\rangle - i\langle y, Tx\rangle.$$

Thus, $\langle Tx, y\rangle = 0$ for all x, y and therefore $T = 0$. □

Now we formally state the theorem and then prove it.

THEOREM 3.17. *An operator $T : \mathbb{H}^N \to \mathbb{H}^N$ on a complex Hilbert space $\mathbb{H}^N$ is self-adjoint if and only if $\langle Tx, x\rangle \in \mathbb{R}$ for any $x \in \mathbb{H}^N$.*

PROOF. We have that $\langle Tx, x\rangle \in \mathbb{R}$ for all x if and only if

$$\langle Tx, x\rangle = \overline{\langle Tx, x\rangle} = \langle x, Tx\rangle = \langle T^*x, x\rangle \quad \text{for all } x$$

if and only if $\langle (T - T^*)x, x\rangle = 0$ for all x, which by Lemma 3.16 happens if and only if $T = T^*$. □

COROLLARY 3.18. *If $T : \mathbb{H}^N \to \mathbb{H}^N$ is a positive operator on a complex Hilbert space $\mathbb{H}^N$, then T is self-adjoint.*

REMARK 3.19. Notice that Theorem 3.17 fails when the Hilbert space is real. This is mainly due to the fact that Lemma 3.16 fails, in particular, the operator given in Example 3.11 is a specific counterexample.

Next we introduce the concept of diagonalization, which is fundamental to studying the behavior of operators.

DEFINITION 3.20. Let $T : \mathbb{H}^N \to \mathbb{H}^N$ be a linear operator. A nonzero vector $x \in \mathbb{H}^N$ is an **eigenvector** of T with **eigenvalue** λ if

$$Tx = \lambda x.$$

The operator T is **diagonalizable** if there exists an orthonormal basis for $\mathbb{H}^N$ consisting of eigenvectors for T.

The above definition is not the *standard one* which states that an operator is diagonalizable if there is some basis consisting of eigenvectors for T.

As an example, if P is a projection from $\mathbb{H}^N$, we have

$$\langle Px, x\rangle = \langle P^2x, x\rangle = \langle Px, Px\rangle = \|Px\|^2,$$

so that P is a positive operator. If $P : \mathbb{H}^N \to W$ and dim $W = m$, then P has eigenvalue 1 with multiplicity m and eigenvalue 0 of multiplicity $N - m$.

DEFINITION 3.21. Let $T : \mathbb{H}^N \to \mathbb{H}^N$ be an invertible positive operator with eigenvalues $\lambda_1 \geq \lambda_2 \geq \cdots \geq \lambda_N > 0$. The **condition number** is $\frac{\lambda_1}{\lambda_N}$.

The condition number is an important concept to frame theorists because its gives a way to measure **redundancy**, as we will see later.

The following relationship holds for T^*T and TT^*.

PROPOSITION 3.22. *If $T : \mathbb{H}^N \to \mathbb{H}^K$ is any linear operator, then T^*T and TT^* have the same nonzero eigenvalues, including multiplicity.*

PROOF. If $T^*Tx = \lambda x$ with $\lambda \neq 0$ and $x \neq 0$, then $Tx \neq 0$ and

$$TT^*(Tx) = T(T^*Tx) = T(\lambda x) = \lambda Tx.$$ □

We will see in the next section that T^*T may have zero eigenvalues even if TT^* does not have zero eigenvalues.

Further restrictions on T gives more information on the eigenvalues. The following result is a corollary of Proposition 3.13.

COROLLARY 3.23. *Let $T : \mathbb{H}^N \to \mathbb{H}^N$ be a linear operator.*
(1) If T is unitary, then its eigenvalues have modulus one.
(2) If T is self-adjoint, then its eigenvalues are real.
(3) If T is positive, then its eigenvalues are nonnegative.

Whenever T is diagonalizable, it has a nice representation using its eigenvalues and eigenvectors. It also gives an easy way to compute its norm.

THEOREM 3.24. *If T is diagonalizable, so that there exists an orthonormal basis $\{e_i\}_{i=1}^N$ of eigenvectors for T with respective eigenvalues $\{\lambda_i\}_{i=1}^N$, then*

$$Tx = \sum_{i=1}^N \lambda_i \langle x, e_i\rangle e_i \text{ for all } x \in \mathbb{H}^N,$$

and

$$\|T\| = \max_{1\le i\le N} |\lambda_i|.$$

PROOF. Since $\{e_i\}_{i=1}^N$ is orthonormal, we can write any x as

$$x = \sum_{i=1}^N \langle x, e_i\rangle e_i$$

and therefore

$$Tx = T\Big(\sum_{i=1}^N \langle x, e_i\rangle e_i\Big) = \sum_{i=1}^N \langle x, e_i\rangle Te_i = \sum_{i=1}^N \langle x, e_i\rangle \lambda_i e_i = \sum_{i=1}^N \lambda_i\langle x, e_i\rangle e_i.$$

From this combined with Parseval's identity, we get

$$\|Tx\|^2 = \sum_{i=1}^N |\lambda_i|^2|\langle x, e_i\rangle|^2 \le \max_{1\le i\le N} |\lambda_i|^2 \sum_{i=1}^N |\langle x, e_i\rangle|^2 = \max_{1\le i\le N}|\lambda_i|^2\|x\|^2.$$

To see that there is an x for which $\|Tx\|$ obtains this maximum, let j be so that $|\lambda_j|$ is the maximum across all $|\lambda_i|$ and then take $x = e_j$ above. $\square$

We can classify the diagonalizable operators. In the infinite dimensional setting this is called the **Spectral Theorem**. We will do a special case which is more instructive than the general case. Specifically, we will show that all self-adjoint operators are diagonalizable. We need a series of lemmas.

LEMMA 3.25. *If T is a self-adjoint operator on a Hilbert space $\mathbb{H}^N$, then*

$$\|T\| = \sup_{\|x\|=1} |\langle Tx, x\rangle|.$$

PROOF. We once again give the proof in the complex setting, but the same general idea works in the real case. To begin, set

$$M := \sup_{\|x\|=1} \{|\langle Tx, x\rangle|\}.$$

We will first show that $M \le \|T\|$: For any $y \in \mathbb{H}^N$, we have

$$|\langle Tx, y\rangle| \le \|Tx\|\|y\| \le \|T\|\|x\|\|y\|,$$

and thus if $\|x\| = 1$, then $|\langle Tx, x\rangle| \le \|T\|$. Taking the supremum over all $x \in \mathbb{H}^N$ such that $\|x\| = 1$, we get $M \le \|T\|$.

Next we will show that $\|T\| \le M$: For any $x, y \in \mathbb{H}^N$, we have

$$\begin{aligned} 4\mathrm{Re}\,\langle Tx, y\rangle &= \big(\langle Tx, x\rangle + 2\mathrm{Re}\,\langle Tx, y\rangle + \langle Ty, y\rangle\big) \\ &\quad - \big(\langle Tx, x\rangle - 2\mathrm{Re}\,\langle Tx, y\rangle + \langle Ty, y\rangle\big) \\ &= \langle T(x+y), x+y\rangle - \langle T(x-y), x-y\rangle \end{aligned}$$

where the fact that T is self-adjoint was used in the second equality. Hence,

$$\begin{aligned} 4\mathrm{Re}\,\langle Tx, y\rangle &= \|x+y\|^2\Big\langle T\Big(\frac{x+y}{\|x+y\|}\Big), \frac{x+y}{\|x+y\|}\Big\rangle \\ &\quad - \|x-y\|^2\Big\langle T\Big(\frac{x-y}{\|x-y\|}\Big), \frac{x-y}{\|x-y\|}\Big\rangle \\ &\le M(\|x+y\|^2 + \|x-y\|^2) = 2M(\|x\|^2 + \|y\|^2). \end{aligned}$$

Note that there exists a $\theta \in [0, 2\pi)$ such that $e^{i\theta}\langle Tx, y\rangle = |\langle Tx, y\rangle|$. Now, replace y with $e^{-i\theta}y$ to obtain:

$$4|\langle Tx, y\rangle| \le 2M(\|x\|^2 + \|y\|^2).$$

Finally, if $\|x\| = 1$ and $y = Tx/\|Tx\|$, then

$$\|Tx\| = \left\langle Tx, \frac{Tx}{\|Tx\|} \right\rangle \le M.$$

Combining both steps gives $\|T\| = M$ as desired. □

LEMMA 3.26. *If T is normal then $\|Tx\| = \|T^*x\|$ for all $x \in \mathbb{H}^N$.*

PROOF. We compute:

$$\begin{aligned} \|Tx\|^2 &= \langle Tx, Tx\rangle \\ &= \langle T^*Tx, x\rangle \\ &= \langle TT^*x, x\rangle \\ &= \langle T^*x, T^*x\rangle \\ &= \|T^*x\|^2. \end{aligned}$$ □

LEMMA 3.27. *If T is normal and*

$$Tx = \lambda x$$

for some $x \in \mathbb{H}^N$ and scalar λ, then

$$T^*x = \overline{\lambda}x.$$

PROOF. If T is normal, then $T - \lambda \cdot Id$ is normal. Applying Lemma 3.26 gives

$$\begin{aligned} Tx = \lambda x &\Leftrightarrow \|(T - \lambda \cdot Id)x\| = 0 \\ &\Leftrightarrow \|(T - \lambda \cdot Id)^*x\| = 0 \\ &\Leftrightarrow \|T^*x - \overline{\lambda}x\| = 0 \\ &\Leftrightarrow T^*x = \overline{\lambda}x. \end{aligned}$$ □

Finally, we will need:

LEMMA 3.28. *If T is normal and $Tx = \lambda x$, then*

$$T\left([x]^\perp\right) \subset [x]^\perp.$$

PROOF. We have $y \perp x$ if and only if

$$\langle Ty, x\rangle = \langle y, T^*x\rangle = \langle y, \overline{\lambda}x\rangle = 0$$

if and only if $Ty \perp x$. □

Now we are ready to prove the main theorem on diagonalization.

THEOREM 3.29. *If T is a self-adjoint, then T is diagonalizable.*

PROOF. By Lemma 3.25, we can choose $x \in \mathbb{H}^N$ with $\|x\| = 1$ and $\|T\| = |\langle Tx, x\rangle|$. By Theorem 3.17 we know that $\langle Tx, x\rangle \in \mathbb{R}$, even if the Hilbert space is complex. Therefore, $\|T\| = \langle Tx, x\rangle$ or $\|T\| = -\langle Tx, x\rangle$. If $\|T\| = \langle Tx, x\rangle$, then

$$\begin{aligned}\|(T - \|T\| \cdot Id)x\|^2 &= \langle Tx - \|T\|x, Tx - \|T\|x\rangle \\ &= \langle Tx, Tx\rangle + \|T\|^2 - 2\|T\|\langle Tx, x\rangle \\ &\leq 2\|T\|(\|T\| - \langle Tx, x\rangle) \\ &= 0\end{aligned}$$

and so $\lambda_1 = \|T\|$ is an eigenvalue for T with eigenvector $e_1 = x$. On the other hand, if $\|T\| = -\langle Tx, x\rangle$, then $\|(T + \|T\| \cdot Id)x\| = 0$ by a similar argument, so that $\lambda_1 = -\|T\|$.

Now, if we restrict the operator T to $[e_1]^\perp$, we have by Lemma 3.28 that

$$T|_{[e_1]^\perp} : [e_1]^\perp \to [e_1]^\perp,$$

and is still self-adjoint. So we can repeat the above argument to get a second orthogonal eigenvector of norm one and eigenvalue of the form:

$$\lambda_2 = \left\|T|_{[e_1]^\perp}\right\| \quad \text{or} \quad \lambda_2 = -\left\|T|_{[e_1]^\perp}\right\|, \quad e_2 \in [e_1]^\perp.$$

Continuing, we get a sequence of eigenvalues $\{\lambda_i\}_{i=1}^N$ so that

$$\lambda_j = \left\|T|_{[\text{span}\,\{e_i\}_{i=1}^{j-1}]^\perp}\right\| \quad \text{or} \quad \lambda_j = -\left\|T|_{[\text{span}\,\{e_i\}_{i=1}^{j-1}]^\perp}\right\|$$

with corresponding orthonormal eigenvectors $\{e_i\}_{i=1}^N$. □

It is possible to take well-defined **powers of an operator** when it is positive, invertible, and diagonalizable.

COROLLARY 3.30. *Let T be an positive, invertible, diagonalizable operator on $\mathbb{H}^N$ with eigenvectors $\{e_i\}_{i=1}^N$ and respective eigenvalues $\{\lambda_i\}_{i=1}^N$. For any nonnegative $a \in \mathbb{R}$ we define the operator T^a by*

$$T^a x = \sum_{i=1}^N \lambda_i^a \langle x, e_i\rangle e_i, \ \textit{for all } x \in \mathbb{H}^N.$$

Thus, T^a is also positive operator and $T^aT^b = T^{a+b}$ for all $a, b \in \mathbb{R}$. In particular, T^{-1} and $T^{-1/2}$ are positive operators. We also note that the definition makes sense for nonnegative powers when T is not invertible.

Notice that Corollary 3.18 combined with Theorem 3.29 gives that all positive operators on a complex Hilbert space are diagonalizable. However, it is possible for a positive operator on a real Hilbert space to not be diagonalizable and hence we cannot obtain powers as defined above. One again, the operator S as given in Example 3.11 provides a counterexample since this operator has no eigenvalues over the reals.

We conclude this section with the concept of trace. In order for it to be well-defined, we need the following proposition.

PROPOSITION 3.31. *Let T be a linear operator on $\mathbb{H}^N$ and let $\{e_i\}_{i=1}^N$ and $\{g_i\}_{i=1}^N$ be orthonormal bases for $\mathbb{H}^N$. Then*

$$\sum_{i=1}^N \langle Te_i, e_i\rangle = \sum_{i=1}^N \langle Tg_i, g_i\rangle.$$

PROOF. We compute:

$$\begin{aligned}
\sum_{i=1}^{N}\langle Te_i, e_i\rangle &= \sum_{i=1}^{N}\left\langle \sum_{j=1}^{N}\langle Te_i, g_j\rangle g_j, e_i\right\rangle \\
&= \sum_{i=1}^{N}\sum_{j=1}^{N}\langle Te_i, g_j\rangle\langle g_j, e_i\rangle \\
&= \sum_{i=1}^{N}\sum_{j=1}^{N}\langle e_i, T^*g_j\rangle\langle g_j e_i\rangle \\
&= \sum_{j=1}^{N}\left\langle \sum_{i=1}^{N}\langle g_j, e_i\rangle e_i, T^*g_j\right\rangle \\
&= \sum_{j=1}^{N}\langle g_j, T^*g_j\rangle \\
&= \sum_{j=1}^{N}\langle Tg_j, g_j\rangle. \qquad \square
\end{aligned}$$

DEFINITION 3.32. Let $T : \mathbb{H}^N \to \mathbb{H}^N$ be an operator. The **trace of T** is

$$Tr\ T = \sum_{i=1}^{N}\langle Te_i, e_i\rangle,$$

where $\{e_i\}_{i=1}^N$ is any orthonormal basis of $\mathbb{H}^N$. The previous proposition shows that this quantity is independent of the choice of orthonormal basis and therefore the trace is well-defined.

When T is diagonalizable, the trace can be easily computed using its eigenvalues.

COROLLARY 3.33. *If $T : \mathbb{H}^N \to \mathbb{H}^N$ is a diagonalizable operator with eigenvalues $\{\lambda_i\}_{i=1}^N$, then*

$$Tr\ T = \sum_{i=1}^{N}\lambda_i.$$

PROOF. If $\{e_i\}_{i=1}^N$ is an orthonormal basis of eigenvectors associated to $\{\lambda_i\}_{i=1}^N$, then

$$Tr\ T = \sum_{i=1}^{N}\langle Te_i, e_i\rangle = \sum_{i=1}^{M}\langle \lambda_i e_i, e_i\rangle = \sum_{i=1}^{N}\lambda_i. \qquad \square$$

4. Hilbert Space Frames

Hilbert space frames were introduced by Duffin and Schaeffer in 1952 [**44**] to address some deep questions in non-harmonic Fourier series. The idea was to weaken Parseval's Identity as given in Proposition 2.4. We do not need to have an orthonormal sequence to have equality in Parseval's identity. For example, if

$\{e_i\}_{i=1}^N$ and $\{g_i\}_{i=1}^N$ are orthonormal bases for a Hilbert space $\mathbb{H}^N$ then

$$\left\{\frac{1}{\sqrt{2}}e_1, \frac{1}{\sqrt{2}}g_1, \frac{1}{\sqrt{2}}e_2, \frac{1}{\sqrt{2}}g_2, \ldots, \frac{1}{\sqrt{2}}e_N, \frac{1}{\sqrt{2}}g_N\right\}_{i=1}^N$$

satisfies Parseval's identity.

DEFINITION 4.1. A family of vectors $\{\varphi_i\}_{i=1}^M$ is a **frame** for a Hilbert space $\mathbb{H}^N$ if there are constants $0 < A \leq B < \infty$ so that for all $x \in \mathbb{H}^N$

$$A\|x\|^2 \leq \sum_{i=1}^M |\langle x, \varphi_i\rangle|^2 \leq B\|x\|^2. \tag{4.1}$$

We include some common, often used terms:

- The constants A and B are called **lower and upper frame bounds**, respectively, for the frame. The largest lower frame bound and the smallest upper frame bound are called the **optimal frame bounds**.
- If $A = B$ this is an **A-tight frame** and if $A = B = 1$ this is a **Parseval frame**.
- If $\|\varphi_i\| = \|\varphi_j\|$ for all $i, j \in I$, this is an **equal norm frame** and if $\|\varphi_i\| = 1$ for all $i \in I$ this is a **unit norm frame**.
- If $\|\varphi_i\| = 1$ for $i \in I$ and there exists a constant c so that $|\langle\varphi_i, \varphi_j\rangle| = c$ for all $i \neq j$, then the frame is called an **equiangular frame**.
- The values $\{\langle x, \varphi_i\rangle\}_{i=1}^M$ are called the **frame coefficients** of the vector x with respect to the frame.
- The frame is a **bounded frame** if $\min_{1\leq i\leq M}\|\varphi_i\| > 0$.
- If only the right hand inequality holds in (4.1) we call $\{\varphi_i\}_{i\in I}$ a **B-Bessel** sequence or simply **Bessel** if explicit reference to the constant is not needed.

It follows from the left-hand-side of inequality (4.1), that the closed linear span of a frame must equal the Hilbert space and so $M \geq N$. In the finite dimensional case, spanning is equivalent to being a frame.

PROPOSITION 4.2. *$\Phi = \{\varphi_i\}_{i=1}^M$ is a frame for $\mathbb{H}^N$ if and only if* span $\Phi = \mathbb{H}^N$.

PROOF. We only need to prove the *if* part. For the right hand inequality,

$$\sum_{i=1}^M |\langle x, \varphi_i\rangle|^2 \leq \sum_{i=1}^M \|x\|^2\|\varphi_i\|^2 \leq B\|x\|^2,$$

where

$$B = \sum_{i=1}^M \|\varphi_i\|^2.$$

For the left hand inequality, we proceed by contradiction. Suppose we can find a sequence $\{x_n\}_{n=1}^\infty$ with $\|x_n\| = 1$ (by scaling) so that

$$\sum_{i=1}^M |\langle x_n, \varphi_i\rangle|^2 \leq \frac{1}{n},$$

and thus there is a norm convergent subsequence $\{x_{n_j}\}_{j=1}^\infty$ of $\{x_n\}_{n=1}^\infty$, say $x_{n_j} \to x$.

Then

$$\sum_{i=1}^{M} |\langle x, \varphi_i \rangle|^2 = \lim_{j\to\infty} \sum_{i=1}^{M} \left|\langle x_{n_j}, \varphi_i \rangle\right|^2 = 0.$$

That is, $x \perp \varphi_i$ for all $i = 1, 2, \cdots, M$ and so Φ does not span $\mathbb{H}^N$. □

Spanning does not necessarily imply that a sequence is a frame when the space is infinite dimensional. For example, suppose $\{e_i\}_{i=1}^{\infty}$ is an orthonormal basis for an infinite dimensional Hilbert space $\mathbb{H}$. Then the sequence $\{e_i/i\}_{i=1}^{\infty}$ spans the space, but is not frame since a lower frame bound does not exist.

It is important to note that there are no restrictions put on the frame vectors. For example, if $\{e_i\}_{i=1}^{N}$ is an orthonormal basis for $\mathbb{H}^N$, then

$$\{e_1, 0, e_2, 0, e_3, 0, \cdots, e_N, 0\}$$

and

$$\left\{e_1, \frac{e_2}{\sqrt{2}}, \frac{e_2}{\sqrt{2}}, \frac{e_3}{\sqrt{3}}, \frac{e_3}{\sqrt{3}}, \frac{e_3}{\sqrt{3}}, \cdots, \frac{e_N}{\sqrt{N}}, \cdots, \frac{e_N}{\sqrt{N}}\right\}$$

are both Parseval frames for $\mathbb{H}^N$. That is, zeros and repetitions are allowed.

The smallest redundant family in $\mathbb{R}^2$ has three vectors and can be chosen to be a unit norm, tight, and equiangular frame called the **Mercedes Benz Frame**, given by

$$\left\{\sqrt{\frac{2}{3}}\begin{pmatrix} 0 \\ 1 \end{pmatrix}, \sqrt{\frac{2}{3}}\begin{pmatrix} \frac{\sqrt{3}}{2} \\ -\frac{1}{2} \end{pmatrix}, \sqrt{\frac{2}{3}}\begin{pmatrix} -\frac{\sqrt{3}}{2} \\ -\frac{1}{2} \end{pmatrix}\right\}.$$

Drawing the vectors might illuminate where it got its name.

4.1. Frame Operators. If $\{\varphi_i\}_{i=1}^{M}$ is a frame for $\mathbb{H}^N$ with frame bounds A, B, define the **analysis operator** of the frame $T : \mathbb{H}^N \to \ell_2^M$ to be

$$Tx = \sum_{i=1}^{M} \langle x, \varphi_i \rangle e_i = \left\{\langle x, \varphi_i \rangle\right\}_{i=1}^{M}, \quad \text{for all } x \in \mathbb{H}^N,$$

where $\{e_i\}_{i=1}^{M}$ is the natural orthonormal basis of ℓ_2^M. It follows that

$$\|Tx\|^2 = \sum_{i=1}^{M} |\langle x, \varphi_i \rangle|^2,$$

so $\|T\|^2$ is the optimal Bessel bound of the frame.

The adjoint of the analysis operator is the **synthesis operator** which is given by

$$T^* e_i = \varphi_i.$$

Note that the matrix representation of the synthesis operator of a frame $\{\varphi_i\}_{i=1}^{M}$ is the $N \times M$ matrix with the frame vectors as its columns.

$$T^* = \begin{bmatrix} | & | & \cdots & | \\ \varphi_1 & \varphi_2 & \cdots & \varphi_M \\ | & | & \cdots & | \end{bmatrix}$$

In practice, we often work with the matrix representation of the synthesis operator with respect to the eigenbasis of the frame operator (as defined below). It will be shown later that the rows and columns of this matrix representation must satisfy very specific properties that proves useful in constructing frames. See Proposition 4.28 and Proposition 4.29.

THEOREM 4.3. *Let $\{\varphi_i\}_{i=1}^M$ be a family of vectors in a Hilbert space $\mathbb{H}^N$. The following are equivalent:*

(1) $\{\varphi_i\}_{i=1}^M$ is a frame for $\mathbb{H}$.
(2) The operator T^* is bounded, linear, and surjective.
(3) The operator T bounded, linear, and injective.

Moreover, $\{\varphi_i\}_{i=1}^M$ is a Parseval frame if and only if the synthesis operator is a quotient map (that is, a partial isometry) if and only if $T^*T = Id$ if and only if T is an isometry.

PROOF. (1) $\Leftrightarrow$ (2) is immediate by Proposition 4.2, and (2) $\Leftrightarrow$ (3) is immediate by Proposition 3.6. Proposition 3.13 gives the moreover part. $\square$

The **frame operator** for the frame is $S = T^*T : \mathbb{H}^N \to \mathbb{H}^N$ given by

$$Sx = T^*Tx = T^*\left(\sum_{i=1}^M \langle x, \varphi_i\rangle e_i\right) = \sum_{i=1}^M \langle x, \varphi_i\rangle T^*e_i = \sum_{i=1}^M \langle x, \varphi_i\rangle \varphi_i.$$

A direct calculation, as given in Example 3.10, yields

$$\langle Sx, x\rangle = \sum_{i=1}^M |\langle x, \varphi_i\rangle|^2. \tag{4.2}$$

PROPOSITION 4.4. *The frame operator of a frame is a* **positive, self-adjoint, and invertible operator** *on $\mathbb{H}^N$. Moreover, if A and B are frame bounds, then S satisfies the* **operator inequality**

$$A \cdot Id \le S \le B \cdot Id.$$

PROOF. Example 3.10 shows that it is positive and self-adjoint. We check the operator inequality:

$$\langle Ax, x\rangle = A\|x\|^2 \le \sum_{i=1}^M |\langle x, \varphi_i\rangle|^2 = \langle Sx, x\rangle \le B\|x\|^2 = \langle Bx, x\rangle$$

for all $x \in \mathbb{H}^N$. Note that this also shows that S is invertible. $\square$

The frame operator can be used to **reconstruct** vectors in the space using the computation

$$\begin{aligned} x &= S^{-1}Sx \\ &= SS^{-1}x \\ &= \sum_{i=1}^M \langle S^{-1}x, \varphi_i\rangle \varphi_i \\ &= \sum_{i=1}^M \langle x, S^{-1}\varphi_i\rangle \varphi_i \end{aligned}$$

Also,

$$\begin{aligned} \sum_{i=1}^M \langle x, S^{-1/2}\varphi_i\rangle S^{-1/2}\varphi_i &= S^{-1/2}\left(\sum_{i=1}^M \langle S^{-1/2}x, \varphi_i\rangle \varphi_i\right) \\ &= S^{-1/2}(S(S^{-1/2}x) \\ &= x. \end{aligned}$$

It follows that $\{S^{-1/2}\varphi_i\}_{i=1}^M$ is a Parseval frame. Since S is invertible, the family $\{S^{-1}\varphi_i\}_{i=1}^M$ is also a frame for $\mathbb{H}^N$ called the **canonical dual frame**. See Subsection 4.2 for the definition and properties of dual frames.

The following result is basically just a restatement of the definitions and known facts.

PROPOSITION 4.5. *Let $\Phi = \{\varphi_i\}_{i=1}^M$ be a frame for $\mathbb{H}^N$ with analysis operator T and frame operator S. The following are equivalent:*

(1) $\{\varphi_i\}_{i=1}^M$ is an A-tight frame for $\mathbb{H}^N$.
(2) $S = A \cdot Id$.
(3) For every $x \in \mathbb{H}^N$,

$$x = \frac{1}{A}\sum_{i=1}^M \langle x, \varphi_i\rangle \varphi_i.$$

(4) For every $x \in \mathbb{H}^N$,

$$A\|x\|^2 = \sum_{i=1}^M |\langle x, \varphi_i\rangle|^2.$$

(5) $T/\sqrt{A}$ is an isometry.

Moreover, if Φ is a Parseval frame then $A = 1$ above.

Tight frames have the property that all of the eigenvalues of the frame operator coincide and are equal the frame bound. For an arbitrary frame, it turns out the smallest and largest eigenvalues of the frame operator are the optimal lower and upper frame bounds, respectively.

THEOREM 4.6. *Suppose $\{\varphi_i\}_{i=1}^M$ is a frame for $\mathbb{H}^N$ with frame operator S with eigenvalues $\lambda_1 \geq \lambda_2 \geq \cdots \geq \lambda_N$. Then λ_1 is the optimal upper frame bound and λ_N is the optimal lower frame bound.*

PROOF. Let $\{e_j\}_{j=1}^N$ be an orthonormal eigenbasis of the frame operator S with associated eigenvalues $\{\lambda_j\}_{j=1}^N$ given in decreasing order. Now, given an $x \in \mathbb{H}^N$, write

$$x = \sum_{j=1}^N \langle x, e_j\rangle e_j$$

to obtain

$$Sx = \sum_{j=1}^N \lambda_j \langle x, e_j\rangle e_j.$$

By Equation (4.2), this gives that

$$\begin{aligned}\sum_{i=1}^M |\langle x, \varphi_i\rangle|^2 = \langle Sx, x\rangle &= \sum_{i,j=1}^N \left\langle \lambda_i \langle x, e_i\rangle e_i, \langle x, e_j\rangle e_j\right\rangle \\ &= \sum_{j=1}^N \lambda_j |\langle x, e_j\rangle|^2 \leq \lambda_1 \|x\|^2\end{aligned}$$

proving that λ_1 is an upper frame bound. To see that it is optimal, note

$$\sum_{i=1}^{M} |\langle e_1, \varphi_i \rangle|^2 = \langle Se_1, e_1 \rangle = \lambda_1.$$

The lower bound is proven similarly. □

Another type of sequence that we often deal with are *Riesz bases*, which rids of the orthonormality assumption, but retains a unique composition.

DEFINITION 4.7. A family of vectors $\{\varphi_i\}_{i=1}^N$ in a Hilbert space $\mathbb{H}^N$ is a **Riesz basis** if there are constants $A, B > 0$ so that for all families of scalars $\{a_i\}_{i=1}^N$ we have

$$A \sum_{i=1}^{N} |a_i|^2 \leq \left\| \sum_{i=1}^{N} a_i \varphi_i \right\|^2 \leq B \sum_{i=1}^{N} |a_i|^2.$$

The following gives an equivalent formulation, in particular, Riesz bases are precisely sequences of vectors that are images of orthonormal bases under an invertible map. It follows directly from the definitions.

PROPOSITION 4.8. *Let $\Phi = \{\varphi_i\}_{i=1}^N$ be a family vectors in $\mathbb{H}^N$. Then the following are equivalent.*

(1) Φ is a Riesz basis for $\mathbb{H}^N$ with Riesz bounds A and B.
(2) For any orthonormal basis $\{e_i\}_{i=1}^N$ for $\mathbb{H}^N$, the operator F on $\mathbb{H}^N$ given by $Fe_i = \varphi_i$ for all $i = 1, 2, \cdots, N$ is an invertible operator with $\|F\|^2 \leq B$ and $\|F^{-1}\|^{-2} \geq A$.

It follows that if Φ is a Riesz basis with bounds A and B, then it is a frame with these same bounds.

Next we see that applying an invertible operator to a frame still gives a frame.

PROPOSITION 4.9. *Let $\Phi = \{\varphi_i\}_{i=1}^M$ be a sequence of vectors in $\mathbb{H}^N$ with analysis operator T and let F be a linear operator on $\mathbb{H}^N$. Then the analysis operator of the sequence $F\Phi = \{F\varphi_i\}_{i=1}^M$ is given by*

$$T_{F\Phi} = TF^*.$$

Moreover, if Φ is a frame for $\mathbb{H}^N$ and F is invertible, then $F\Phi$ is a also a frame for $\mathbb{H}^N$.

PROOF. For $x \in \mathbb{H}^N$,

$$T_{F\Phi}x = \left\{\langle x, F\varphi_i \rangle\right\}_{i=1}^M = \left\{\langle F^*x, \varphi_i \rangle\right\}_{i=1}^M = TF^*x.$$

The moreover part follows from Theorem 4.3. □

Furthermore, if we apply an invertible operator to a frame, there is a specific form that the frame operator of the new frame must take.

PROPOSITION 4.10. *Suppose $\Phi = \{\varphi_i\}_{i=1}^M$ is a frame for $\mathbb{H}^N$ with frame operator S and F is an invertible operator on $\mathbb{H}^N$. Then the frame operator of the frame $F\Phi = \{F\varphi_i\}_{i=1}^M$ is the operator FSF^*.*

PROOF. This follows immediately from Proposition 4.9 and the definition. □

Corollary 3.33 concerning the trace formula for diagonalizable operators, has a corresponding result for Parseval frames.

PROPOSITION 4.11 (Trace formula for Parseval frames). *Let $\Phi = \{\varphi_i\}_{i=1}^M$ be a parseval frame on $\mathbb{H}^N$ and let F be a linear operator on $\mathbb{H}^N$. Then*

$$Tr(F) = \sum_{i=1}^{M} \langle F\varphi_i, \varphi_i \rangle.$$

PROOF. If $\{e_i\}_{i=1}^N$ is an orthonormal basis for $\mathbb{H}^N$ then by definition

$$Tr(F) = \sum_{j=1}^{N} \langle Fe_i, e_i \rangle.$$

This along with the fact that Φ is Parseval gives that

$$\begin{aligned} Tr(F) &= \sum_{j=1}^{N} \left\langle \sum_{i=1}^{M} \langle Fe_j, \varphi_i \rangle \varphi_i, e_j \right\rangle \\ &= \sum_{j=1}^{N} \sum_{i=1}^{M} \langle e_j, F^*\varphi_i \rangle \langle \varphi_i, e_j \rangle \\ &= \sum_{i=1}^{M} \left\langle \sum_{j=1}^{N} \langle \varphi_i, e_j \rangle e_j, F^*\varphi_i \right\rangle \\ &= \sum_{i=1}^{M} \langle \varphi_i, F^*\varphi_i \rangle \\ &= \sum_{i=1}^{M} \langle F\varphi_i, \varphi_i \rangle. \end{aligned}$$

□

DEFINITION 4.12. Two frames $\{\varphi_i\}_{i=1}^M$ and $\{\psi_i\}_{i=1}^M$ in a Hilbert space $\mathbb{H}^N$ are **isomorphic frames** if there exists a bounded, invertible operator $L : \mathbb{H}^N \to \mathbb{H}^N$ so that $L\varphi_i = \psi_i$ for all $1 \le i \le M$. We say they are **unitarily isomorphic frames** if L is a unitary operator.

PROPOSITION 4.13. *Let $\Phi = \{\varphi_i\}_{i=1}^M$ and $\Psi = \{\psi_i\}_{i=1}^M$ be frames for $\mathbb{H}^N$ with analysis operators T_1 and T_2 respectively. The following are equivalent:*

(1) Φ *and* Ψ *are isomorphic.*
(2) $\operatorname{ran} T_1 = \operatorname{ran} T_2$.
(3) $\ker T_1^* = \ker T_2^*$.

Moreover, in this case $F = T_2^*(T_1^*|_{\operatorname{ran} T_1})^{-1}$ satisfies $F\varphi_i = \psi_i$ for all $i = 1, 2, \cdots, M$.

PROOF. The equivalence of (2) and (3) follows by Proposition 3.6.

(1) $\Rightarrow$ (3): Let $F\varphi_i = \psi_i$ be a well-defined invertible operator on $\mathbb{H}^N$. Then by Proposition 4.9 we have that $T_2 = T_1F^*$ and hence $FT_1^* = T_2^*$. Since F is invertible, (3) follows.

(2) $\Rightarrow$ (1): Let P be the orthogonal projection onto $W = \operatorname{ran} T_1 = \operatorname{ran} T_2$. Then $(Id - P)$ is an orthogonal projection onto $W^\perp = \ker T_1^* = \ker T_2^*$ so that

$$\varphi_i = T_1^* e_i = T_1^* P e_i + T_1^*(Id - P)e_i = T_1^* P e_i$$

and similarly $\psi_i = T_2^* P e_i$. The operators T_1^* and T_2^* both map W bijectively onto $\mathbb{H}^N$. Therefore, the operator $F := T_2^*(T_1^*|_W)^{-1}$ maps $\mathbb{H}^N$ bijectively onto itself.

Consequently, for all $i = 1, 2, \cdots, M$ we have

$$F\varphi_i = T_2^*(T_1^*|_W)^{-1}T_1^*Pe_i = T_2^*Pe_i = \psi_i. \qquad \square$$

As a consequence of Proposition 4.10 we have:

THEOREM 4.14. *Every frame $\{\varphi_i\}_{i=1}^M$ (with frame operator S) is isomorphic to the Parseval frame $\{S^{-1/2}\varphi_i\}_{i\in I}$.*

PROOF. The frame operator of $\{S^{-1/2}\varphi_i\}_{i=1}^N$ is $S^{-1/2}S(S^{-1/2})^* = Id$. $\square$

As a consequence, only unitary operators can map Parseval frames to Parseval frames.

COROLLARY 4.15. *If two Parseval frames $\Phi = \{\varphi_i\}_{i=1}^M$ and $\Psi = \{\psi_i\}_{i=1}^M$ are isomorphic, then they are unitarily isomorphic.*

PROOF. Since both frames have the identity as their frame operator, if F maps one Parseval frame to another and is invertible, then by Propositon 4.10,

$$Id = F(Id)F^* = FF^*.$$

Since F^*F is injective, F is a unitary operator by Proposition 3.13. $\square$

We can always "move" one frame operator to another.

PROPOSITION 4.16. *Let $\Phi = \{\varphi_i\}_{i=1}^M$ and $\Psi = \{\psi_i\}_{i=1}^M$ be frames for $\mathbb{H}^N$ with frame operators S_1 and S_2 respectively. Then there exists an invertible operator F on $\mathbb{H}^N$ so that S_1 is the frame operator of the frame $F\Psi = \{F\psi_i\}_{i=1}^M$.*

PROOF. If S is the frame operator of $F\Psi$, letting $F = S_1^{1/2}S_2^{-1/2}$ we have

$$S = FS_2F^* = (S_1^{1/2}S_2^{-1/2})S_2(S_1^{1/2}S_2^{-1/2})^* = S_1. \qquad \square$$

4.2. Dual Frames. We begin with the definition.

DEFINITION 4.17. If $\Phi = \{\varphi_i\}_{i=1}^M$ is a frame for $\mathbb{H}^N$, a frame $\{\psi_i\}_{i=1}^M$ for $\mathbb{H}^N$ is called a **dual frame** for Φ if

$$\sum_{i=1}^M \langle x, \psi_i\rangle \varphi_i = x, \text{ for all } x \in \mathbb{H}^N.$$

It follows that the canonical dual frame $\{S^{-1}\varphi_i\}_{i=1}^M$, where S is the frame operator of Φ, is a dual frame. But there are many other dual frames in general.

PROPOSITION 4.18. *Let $\Phi = \{\varphi_i\}_{i=1}^M$ and $\Psi = \{\psi_i\}_{i=1}^M$ be frames for $\mathbb{H}^N$ with analysis operators T_1, T_2 respectively. The following are equivalent:*

(1) Ψ is a dual frame of Φ.
(2) $T_1^*T_2 = Id$.

PROOF. Note that for any $x \in \mathbb{H}^N$,

$$T_1^*T_2x = T_1^*\left[(\langle x, \psi_i\rangle)_{i=1}^M\right] = \sum_{i=1}^M \langle x, \psi_i\rangle \varphi_i.$$

The result is now immediate. $\square$

THEOREM 4.19. *Suppose $\Phi = \{\varphi_i\}_{i=1}^M$ is a frame with analysis operator T_1 and frame operator S. The class of all dual frames of Φ are frames of the form $\{\eta_i\}_{i=1}^M := \{S^{-1}\varphi_i + \psi_i\}_{i=1}^M$, where if T_2 is the analysis operator of $\Psi = \{\psi_i\}_{i=1}^M$, then $T_1^*T_2 = 0$. That is,* $\operatorname{ran} T_1 \perp \operatorname{ran} T_2$.

PROOF. Note that the analysis operator of $\{\eta_i\}_{i=1}^M$ is $(T_1^*)^{-1} + T_2$. Now,

$$T_1^*\left((T_1^*)^{-1} + T_2\right) = T_1^*(T_1^*)^{-1} + T_1^*T_2 = Id + 0 = Id.$$

By Proposition 4.18, $\{\eta_i\}_{i=1}^M$ is a dual frame of Φ.

Conversely, if $\{\eta_i\}_{i=1}^M$ is a dual frame for Φ, let $\psi_i = \eta_i - S^{-1}\varphi_i$, for all $i = 1, 2, \cdots, M$. Then for all $x \in \mathbb{H}^N$,

$$\begin{aligned}
T_1^*T_2x &= \sum_{i=1}^M \langle x, \psi_i\rangle \varphi_i \\
&= \sum_{i=1}^M \langle x, \eta_i - S^{-1}\varphi_i\rangle \varphi_i \\
&= \sum_{i=1}^M \langle x, \eta_i\rangle \varphi_i - \sum_{i=1}^M \langle x, S^{-1}\varphi_i\rangle \varphi_i \\
&= x - x = 0.
\end{aligned}$$

This implies that for all $x, y \in \mathbb{H}^N$,

$$\langle T_1x, T_2y\rangle = \langle x, T_1^*T_2y\rangle = 0,$$

which is precisely $\operatorname{ran} T_1 \perp \operatorname{ran} T_2$. □

PROPOSITION 4.20. *Let $\Phi = \{\varphi_i\}_{i=1}^M$ be a frame for $\mathbb{H}^N$ with frame operator S. Then the only dual frame of Φ which is isomorphic to Φ is $\{S^{-1}\varphi_i\}_{i=1}^M$.*

PROOF. Let $\{\psi_i\}_{i=1}^M$ be a dual frame for Φ and assume there is an invertible operator F so that $\psi_i = FS^{-1}\varphi_i$ for all $i = 1, 2, \cdots, M$. Then, for every $x \in \mathbb{H}^N$ we have

$$\begin{aligned}
F^*x &= \sum_{i=1}^M \langle F^*x, S^{-1}\varphi_i\rangle \varphi_i \\
&= \sum_{i=1}^M \langle x, FS^{-1}\varphi_i\rangle \varphi_i \\
&= \sum_{i=1}^M \langle x, \psi_i\rangle \varphi_i \\
&= x.
\end{aligned}$$

It follows that $F^* = Id$ and so $F = Id$. □

4.3. Redundancy. The main property of frames which makes them so useful in applied problems is their **redundancy**. That is, each vector in the space has infinitely many representations with respect to the frame but it also has one natural representation given by the frame coefficients. The role played by redundancy varies with specific applications. One important role is its **robustness**. That is, by spreading our information over a wider range of vectors, we are better able to sustain

losses (called **erasures** in this setting) and still have accurate reconstruction. This shows up in internet coding (for transmission losses), distributed processing (where "sensors" are constantly fading out), modeling the brain (where memory cells are constantly dying out) and a host of other applications. Another advantage of spreading our information over a wider range is to mitigate the effects of noise in our signal or to make it prominent enough so it can be removed as in signal/image processing. A further upside of redundancy is in areas such as quantum tomography where we need classes of orthonormal bases which have "constant" interactions with one another or we need vectors to form a Parseval frame but have the absolute values of their inner products with all other vectors the same. In speech recognition, we need a vector to be determined by the absolute value of its frame coefficients. This is a very natural frame theory problem since this is impossible for a linearly independent set to achieve. Redundancy is a fundamental issue in this setting.

Our next proposition shows the relationship between the frame elements and the frame bounds.

PROPOSITION 4.21. *Let $\{\varphi_i\}_{i=1}^M$ be a frame for $\mathbb{H}^N$ with frame bounds A, B. Then we have $\|\varphi_i\|^2 \leq B$ for all $1 \leq i \leq M$, and if $\|\varphi_i\|^2 = B$ holds for some i, then $\varphi_i \perp \operatorname{span}\{\varphi_j\}_{j\neq i}$. If $\|\varphi_i\|^2 < A$, then $\varphi_i \in \overline{\operatorname{span}}\,\{\varphi_j\}_{j\neq i}$.*

PROOF. If we replace x in the frame definition by φ_i we see that

$$A\|\varphi_i\|^2 \leq \|\varphi_i\|^4 + \sum_{j\neq i} |\langle \varphi_i, \varphi_j\rangle|^2 \leq B\|\varphi_i\|^2.$$

The first part of the result is now immediate. For the second part, assume to the contrary that $E = \overline{\operatorname{span}}\,\{\varphi_j\}_{j\neq i}$ is a proper subspace of $\mathbb{H}^N$. Replacing φ_i in the above inequality by $P_{E^\perp}\varphi_i$ and using the left hand side of the inequality yields an immediate contradiction. □

As a particular case of Proposition 4.21 we have for a Parseval frame $\{\varphi_i\}_{i=1}^M$ that $\|\varphi_i\|^2 \leq 1$ for all i, and $\|\varphi_i\| = 1$ for some i if and only if $\varphi_i \perp \overline{\operatorname{span}}\,\{\varphi_j\}_{j\neq i}$. We call $\{\varphi_i\}_{i=1}^M$ an **exact frame** if it ceases to be a frame when any one of its vectors is removed. If $\{\varphi_i\}_{i=1}^M$ is an exact frame then $\langle S^{-1}\varphi_i, \varphi_j\rangle = \langle S^{-1/2}\varphi_i, S^{-1/2}\varphi_j\rangle = \delta_{ij}$ (where δ_{ij} is the Kronecker delta) since $\{S^{-1/2}\varphi_i\}_{i=1}^M$ is now an orthonormal basis for $\mathbb{H}^N$. That is, $\{S^{-1}\varphi_i\}_{i=1}^M$ and $\{\varphi_i\}_{i=1}^M$ form a **biorthogonal system**. Also, it follows that $\{e_i\}_{i=1}^N$ is an orthonormal basis for $\mathbb{H}^N$ if and only if it is an exact, Parseval frame. Another consequence of Proposition 4.21 is the following.

PROPOSITION 4.22. *The removal of a vector from a frame leaves either a frame or an incomplete set.*

PROOF. By Theorem 4.14, we may assume that $\{\varphi_i\}_{i=1}^M$ is a Parseval frame. Now by Proposition 4.21, for any i, either $\|\varphi_i\| = 1$ and $\varphi_i \perp \operatorname{span}\{\varphi_j\}_{j\neq i}$, or $\|\varphi_i\| < 1$ and $\varphi_i \in \overline{\operatorname{span}}\,\{\varphi_j\}_{j\neq i}$. □

4.4. Minimal Moments. Since a frame is not independent (unless it is a Riesz basis) a vector in the space will have many representations relative to the frame besides the natural one given by the frame coefficients. However, the natural representation of a vector is the unique representation of minimal ℓ_2-norm as the following result of Duffin and Schaeffer [**44**] shows.

THEOREM 4.23. *Let $\Phi = \{\varphi_i\}_{i=1}^M$ be a frame for a Hilbert space $\mathbb{H}^N$ with frame operator S and $x \in \mathbb{H}^N$. If $\{b_i\}_{i=1}^M$ is any sequence of scalars such that*

$$x = \sum_{i=1}^{M} b_i \varphi_i,$$

then

$$\sum_{i=1}^{M} |b_i|^2 = \sum_{i=1}^{M} |\langle S^{-1}x, \varphi_i\rangle|^2 + \sum_{i=1}^{M} |\langle S^{-1}x, \varphi_i\rangle - b_i|^2. \tag{4.3}$$

PROOF. We have by assumption

$$\sum_{i=1}^{M} \langle S^{-1}x, \varphi_i\rangle \varphi_i = x = \sum_{i=1}^{M} b_i \varphi_i.$$

Therefore, the sequence $\{\langle S^{-1}x, \varphi_i\rangle - b_i\}_{i=1}^M \in \ker T^* = [\operatorname{ran} T]^\perp$, where T is the analysis operator of Φ. Now, writing

$$b_i = \langle S^{-1}x, \varphi_i\rangle - (\langle S^{-1}x, \varphi_i\rangle - b_i)$$

and noting that the sequence $\{\langle S^{-1}x, \varphi_i\rangle\}_{i=1}^M \in \operatorname{ran} T$ and therefore perpendicular to $\{\langle S^{-1}x, \varphi_i\rangle - b_i\}_{i=1}^M$ gives (4.3). □

4.5. Orthogonal Projections and Naimark's Theorem. A major advantage of frames over wavelets is that orthogonal projections take frames to frames but do not map wavelets to wavelets.

PROPOSITION 4.24. *Let $\{\varphi_i\}_{i=1}^M$ be a frame for $\mathbb{H}^N$ with frame bounds A, B, and let P be an orthogonal projection on H. Then $\{P\varphi_i\}_{i=1}^M$ is a frame for $P(H)$ with frame bounds A, B. In particular, an orthogonal projection of a orthonormal basis (or a Parseval frame) is a Parseval frame.*

PROOF. For any $x \in P(H)$ we have

$$\sum_{i=1}^{M} |\langle x, P\varphi_i\rangle|^2 = \sum_{i=1}^{M} |\langle Px, \varphi_i\rangle|^2 = \sum_{i=1}^{M} |\langle x, \varphi_i\rangle|^2.$$

The result is now immediate. □

Proposition 4.24 gives that an orthogonal projection P applied an orthonormal basis $\{e_i\}_{i=1}^M$ for $\mathbb{H}^M$ leaves a Parseval frame $\{Pe_i\}_{i=1}^M$ for $P(\mathbb{H}^M)$. The converse of this is also true and is a result of Naimark (see [**30**] and Han and Larson [**60**]).

THEOREM 4.25 (Naimark's Theorem). *A sequence $\{\varphi_i\}_{i=1}^M$ is a Parseval frame for a Hilbert space $\mathbb{H}^N$ if and only if there is a larger Hilbert space $\mathbb{H}^M \supset \mathbb{H}^N$ and an orthonormal basis $\{e_i\}_{i=1}^M$ for $\mathbb{H}^M$ so that the orthogonal projection P of $\mathbb{H}^M$ onto $\mathbb{H}^N$ satisfies $Pe_i = \varphi_i$ for all $i = 1, 2, \cdots, M$.*

PROOF. The "if" part follows from Proposition 4.24. For the "only if" part, note that if $\{\varphi_i\}_{i=1}^M$ is a Parseval for $\mathbb{H}^N$, then the synthesis operator $T^* : \ell_2^M \to \mathbb{H}^N$ is a partial isometry. Let $\{e_i\}_{i=1}^M$ be an orthonormal basis for ℓ_2^M for which $T^* e_i = \varphi_i$ for all $i = 1, 2, \cdots, M$. Since the analysis operator T is an isometry we can identify $\mathbb{H}^N$ with $T(\mathbb{H}^N)$. Now let $\mathbb{H}^M = \ell_2^M$ and let P be the orthogonal projection of $\mathbb{H}^M$ onto $T(\mathbb{H}^N)$. Then for all $i = 1, 2, \cdots, M$ and all $y = Tx \in T(\mathbb{H}^N)$ we have

$$\langle Tx, Pe_i\rangle = \langle PTx, e_i\rangle = \langle Tx, e_i\rangle = \langle x, T^* e_i\rangle = \langle x, \varphi_i\rangle = \langle Tx, T\varphi_i\rangle.$$

It follows that $Pe_i = T\varphi_i$, and the result follows from the association of $\mathbb{H}^N$ with $T(\mathbb{H}^N)$. □

DEFINITION 4.26. The **standard N-simplex** (or **regular N-gon**) is the subset of $\mathbb{R}^{N-1}$ given by unit norm $\{\varphi_i\}_{i=1}^N$ which are equiangular.

It follows from Proposition 4.24:

COROLLARY 4.27. *A regular simplex is an equiangular tight frame.*

PROOF. Given a regular N-simplex, it can be realized by letting $\{e_i\}_{i=1}^N$ be an orthonormal basis for $\mathbb{R}^N$ and letting

$$x = \sum_{i=1}^N \varepsilon_i e_i, \quad \text{for any } \varepsilon_i = 1, \text{ for } i = 1, 2, \cdots, N.$$

Now let P be the orthogonal projection onto the span of x, which is given by:

$$Py = \left[\frac{1}{N}\sum_{i=1}^N \varepsilon_i \langle y, e_i\rangle\right] x$$

Then the N-simplex is a scaling by $1/\|(Id - P)e_i\|$ of $\{(Id - P)e_i\}_{i=1}^N$, which is N-vectors in an $N-1$-dimensional Hilbert space and is a Parseval frame by Proposition 4.24. □

4.6. Frame Representations. Another important property of frames is given in the next proposition.

PROPOSITION 4.28. *Let $\Phi = \{\varphi_i\}_{i=1}^M$ be a sequence of vectors in $\mathbb{H}^N$, let $\{e_j\}_{j=1}^N$ be an orthonormal basis for $\mathbb{H}^N$, and let $\{\lambda_j\}_{j=1}^N$ be positive real numbers. The following are equivalent:*

(1) Φ is a frame for $\mathbb{H}^N$ with frame operator S having eigenvectors $\{e_j\}_{j=1}^N$ and respective eigenvalues $\{\lambda_j\}_{j=1}^N$.
(2) The following hold:
  (a) If for all $j = 1, 2, \cdots, N$,

$$\psi_j := (\langle\varphi_1, e_j\rangle, \langle\varphi_2, e_j\rangle, \cdots, \langle\varphi_M, e_j\rangle),$$

  then for all $1 \le j \ne k \le N$ we have

$$\langle\psi_j, \psi_k\rangle = 0.$$

  (b) For all $j = 1, 2, \cdots, N$ we have $\|\psi_j\|_2^2 = \lambda_j$.

PROOF. If $\{\varphi_i\}_{i=1}^M$ is a frame for $\mathbb{H}^N$ with frame operator S having eigenvectors $\{e_j\}_{j=1}^N$ and respective eigenvalues $\{\lambda_j\}_{j=1}^N$, then for all $j = 1, 2, \cdots, N$ we have

$$\sum_{i=1}^M \langle e_j, \varphi_i\rangle\varphi_i = \lambda_j e_j.$$

Hence, for $1 \leq j \neq k \leq N$ we have

$$\langle \psi_j, \psi_k \rangle = \sum_{i=1}^{M} \overline{\langle \varphi_i, e_j \rangle} \langle \varphi_i, e_k \rangle = \sum_{i=1}^{M} \langle e_j, \varphi_i \rangle \langle \varphi_i, e_k \rangle = \langle \lambda_j e_j, e_k \rangle = 0.$$

Similarly,

$$\|\psi_j\|_2^2 = \langle \lambda_j e_j, e_j \rangle = \lambda_j. \qquad \square$$

This gives some important properties of the matrix representation of the synthesis operator. As stated before, these criterion are often used in constructing frames with a given synthesis operator.

PROPOSITION 4.29. *If $\{\varphi_i\}_{i=1}^M$ is a frame for $\mathbb{H}^N$ and $T^* = [a_{ij}]_{i=1,j=1}^{N\ ,M}$ is the synthesis matrix with respect to the eigenvectors of the frame operator, then the following hold.*

(1) The rows of T^* are orthogonal.
(2) The square sum of the columns are the square norm of the frame vectors.
(3) The square sum of the rows are the eigenvalues of the frame operator.

5. Constants Related to Frames

Let $\{\varphi_i\}_{i=1}^M$ be a frame for $\mathbb{H}^N$ with frame operator S.

General Frame: The sum of the eigenvalues of S equals the sum of the squares of the lengths of the frame vectors:

$$\sum_{j=1}^{N} \lambda_j = \sum_{i=1}^{M} \|\varphi_i\|^2.$$

Equal Norm Frame: For an equal norm frame in which $\|\varphi_i\| = c$ holds for all $i = 1, 2, \cdots, M$ we have

$$\sum_{j=1}^{N} \lambda_j = \sum_{i=1}^{M} \|\varphi_i\|^2 = M \cdot c^2.$$

Tight Frame: Since tightness means $A = B$, we have

$$\sum_{i=1}^{M} |\langle x, \varphi_i \rangle|^2 = A\|x\|^2, \quad \text{for all } x \in \mathbb{H}^N.$$

We have that $S = A \cdot I_N$ and thus the sum of the eigenvalues becomes:

$$N \cdot A = \sum_{j=1}^{N} \lambda_j = \sum_{i=1}^{M} \|\varphi_i\|^2.$$

Parseval Frame: If the frame is Parseval, then $A = B = 1$ and so

$$\sum_{i=1}^{M} |\langle x, \varphi_i \rangle|^2 = \|x\|^2, \quad \text{for all } x \in \mathbb{H}_N.$$

We have that $S = Id$ and

$$N = \sum_{j=1}^{N} \lambda_j = \sum_{i=1}^{M} \|\varphi_i\|^2.$$

Equal Norm Tight Frame: For an equal norm A-tight frame in which $\|\varphi_i\| = c$ holds for all $i = 1, 2, \cdots, M$ we have

$$N \cdot A = \sum_{j=1}^{N} \lambda_j = \sum_{i=1}^{M} \|\varphi_i\|^2 = M \cdot c^2.$$

Hence $A = M \cdot c^2/N$ and thus

$$\sum_{i=1}^{M} |\langle x, \varphi_i \rangle|^2 = \frac{M}{N} c^2 \|x\|^2, \quad \text{for all } x \in \mathbb{H}^N.$$

Equal Norm Parseval Frame: For an equal norm Parseval frame we have

$$N = \sum_{j=1}^{M} \lambda_j = \sum_{i=1}^{M} \|\varphi_i\|^2 = c^2 M.$$

6. Constructing Finite Frames

For applications, we need to construct finite frames with extra properties such as:

(1) Prescribing in advance the norms of the frame vectors. (See for example [**21**, **25**, **36**]).

(2) Constructing equiangular frames. That is, frames $\{\varphi_i\}_{i=1}^M$ for which there is a constant $c > 0$ and

$$|\langle \varphi_i, \varphi_j \rangle| = c, \quad \text{for all } i \neq j.$$

(See for example [**37**, **61**, **76**, **77**]).

(3) Frames for which the operator

$$\pm x \mapsto \{|\langle x, \varphi_i \rangle|\}_{i=1}^M, \quad \text{is one-to-one.}$$

(See for example [**3**, **11**]).

For a good introduction to constructive methods for frames see [**25**]

6.1. Finding Parseval Frames. There is a unique way to get Parseval frames of M vectors in $\mathbb{H}^N$. Take a $M \times M$ unitary matrix $U = (a_{ij})_{i,j=1}^M$. Take the submatrix

$$V = (a_{ij})_{i=1,j=1}^{M\ ,N}.$$

The rows of V form a parseval frame for $\mathbb{H}^N$, since these vectors are just the rows of the matrix U (which are an orthonormal basis for $\mathbb{H}^M$) projected onto $\mathbb{H}^N$ and so form a Parseval frame. The converse to this is also true - and follows directly from **Naimark's Theorem** (Theorem 4.25).

6.2. Adding vectors to a frame to make it tight. Every finite frame for $\mathbb{H}_N$ can be turned into a tight frame with the addition of at most $N-1$-vectors.

PROPOSITION 6.1. *If $\{\varphi_i\}_{i=1}^M$ is a frame for $\mathbb{H}^N$, then there are vectors $\{\psi_j\}_{j=2}^N$ so that $\{\varphi_i\}_{i=1}^M \cup \{\psi_j\}_{j=2}^N$ is a tight frame.*

PROOF. Let S be the frame operator for the frame with eigenvectors $\{e_j\}_{j=1}^N$ and respective eigenvalues $\{\lambda_j\}_{j=1}^N$ that satisfy $\lambda_1 \geq \lambda_2 \geq \cdots \geq \lambda_N$. We define ψ_j for $j = 2, 3, \cdots, N$ by

$$\psi_j = \sqrt{\lambda_1 - \lambda_j}\, e_j.$$

This family $\{\varphi_i\}_{i=1}^M \cup \{\psi_j\}_{j=2}^N$ is a λ_1-tight frame. □

6.3. Majorization. One of the main constructive methods for frames is due to Casazza and Leon [**21**, **36**]. They gave a construction for the important results of Benedetto and Fickus [**8**] and Casazza, Fickus, Kovačević, Leon, Tremain [**29**].

THEOREM 6.2. [**29**] *Fix $N \leq M$ and $a_1 \geq a_2 \geq \cdots \geq a_M > 0$. The following are equivalent:*

(1) There is a tight frame $\{\varphi_i\}_{i=1}^M$ for $\mathbb{H}^N$ satisfying $\|\varphi_i\| = a_i$, for all $i = 1, 2, \cdots, M$.
(2) For all $1 \leq n < N$ we have

$$a_n^2 \leq \frac{\sum_{i=n+1}^M a_i^2}{N-n}.$$

(3) We have

$$\sum_{i=1}^M a_i^2 \geq N a_1^2.$$

(4) If

$$\lambda = \sqrt{\frac{N}{\sum_{i=1}^M a_i^2}},$$

then

$$\lambda a_i \leq 1, \quad \textit{for all } i = 1, 2, \cdots, M.$$

This result was generalized by Casazza and Leon [**21**] to:

THEOREM 6.3. *Let S be a positive self-adjoint operator on $\mathbb{H}^N$ and let $\lambda_1 \geq \lambda_2 \geq \cdots \geq \lambda_N > 0$ be the eigenvalues of S. Fix $M \geq N$ and real numbers $a_1 \geq a_2 \geq \cdots \geq a_M > 0$. The following are equivalent:*

(1) There is a frame $\{\varphi_i\}_{i=1}^M$ for $\mathbb{H}^N$ with frame operator S satisfying $\|\varphi_i\| = a_i$ for all $i = 1, 2, \cdots, M$.
(2) For every $1 \leq k \leq N$ we have

$$\sum_{j=1}^k a_j^2 \leq \sum_{j=1}^k \lambda_j,$$

and

$$\sum_{i=1}^M a_i^2 = \sum_{j=1}^N \lambda_j.$$

Theorem 6.3(2) is the so called **majorization** of one sequence over another. The next result follows readily from the above results.

COROLLARY 6.4. *Let S be a positive self-adjoint operator on $\mathbb{H}^N$. For any $M \geq N$ there is an equal norm sequence $\{\varphi_i\}_{i=1}^M$ in $\mathbb{H}^N$ which has S as its frame operator.*

PROOF. Let $\lambda_1 \geq \lambda_2 \geq \ldots \lambda_N > 0$ be the eigenvalues of S and let

$$a^2 = \frac{1}{M}\sum_{j=1}^{N} \lambda_j. \tag{6.1}$$

Now we check condition (2) of Theorem 6.3 to see that there is a sequence $\{\varphi_i\}_{i=1}^M$ in $\mathbb{H}^N$ with $\|\varphi_i\| = a$ for all $i = 1, 2, \cdots, M$. That is, we check the condition with $a_1 = a_2 = \cdots a_M = a$. To check the equality in Theorem 6.3(2), note that by Equation (6.1) we have

$$\sum_{i=1}^{M} a_i^2 = Ma^2 = \sum_{i=1}^{N} \lambda_i. \tag{6.2}$$

For the first inequality with $k = 1$ in Theorem 6.3(2), we note that by Equation (6.1) we have that

$$a_1^2 = a^2 = \frac{1}{M}\sum_{i=1}^{N} \lambda_i \leq \frac{1}{N}\sum_{i=1}^{N} \lambda_i \leq \lambda_1.$$

So our inequality holds for $k = 1$. Suppose there is an $1 < k \leq N$ for which this inequality fails and k is the first time this fails. So,

$$\sum_{i=1}^{k-1} a_i^2 = (k-1)a^2 \leq \sum_{i=1}^{k-1} \lambda_i,$$

while

$$\sum_{i=1}^{k} a_i^2 = ka^2 > \sum_{i=1}^{k} \lambda_i.$$

It follows that

$$a_{k+1}^2 = \cdots = a_N^2 = a^2 > \lambda_k \geq \lambda_{k+1} \geq \cdots \geq \lambda_N.$$

Hence,

$$\begin{aligned} Ma^2 = \sum_{i=1}^{M} a_i^2 &\geq \sum_{i=1}^{k} a_i^2 + \sum_{i=k+1}^{N} a_i^2 \\ &> \sum_{i=1}^{k} \lambda_i + \sum_{i=k+1}^{N} a_i^2 \\ &> \sum_{i=1}^{k} \lambda_i + \sum_{i=k+1}^{N} \lambda_i \\ &= \sum_{i=1}^{N} \lambda_i. \end{aligned}$$

But this contradicts Equation (6.2). □

There was a recent **significant advance** on this subject due to due to Cahill/Fickus/Mixon/Poteet/Strawn [**18, 19**] where they give an algorithm for constructing all self-adjoint matrices with prescribed spectrum and diagonal and all finite frames with prescribed spectrum and diagonal. This work technically contains the solution to many of our frame questions. That is, if we could carry out their construction with an additional restriction on it (e.g. requiring "equiangular"—see Subsection 10.1) then we could construct equiangular tight frames.

6.4. Spectral Tetris. Another significant advance for frame theory came when **Spectral Tetris** was introduced by Casazza/Fickus/Mixon/Wang/Zhou [**27**]. This is now a massive subject and we refer the reader to a comprehensive survey of Casazza/Woodland [**39**]. We will just give an illustrative example here.

Before we begin our example, let us go over a few necessary facts for construction. Recall, that in order to construct an M-element unit norm tight frame (UNTF) in $\mathbb{H}^N$, we will construct an $N \times M$ synthesis matrix having the following properties:

(1) The columns square sum to one, to obtain unit norm vectors.
(2) The rows are orthogonal, which is equivalent to the frame operator, S, being a diagonal $N \times N$ matrix.
(3) The rows have constant norm, to obtain tightness, meaning that $S = A \cdot Id$ for some constant A.

REMARK 6.5. Since we will be constructing M-element UNTFs in $\mathbb{H}^N$, recall that the frame bound will be $A = \frac{M}{N}$.

Also, before the construction of a frame is possible, we must first ensure that such a frame exists by checking that the spectrum of the frame majorizes the square vector norms of the frame. However, this is not the only constraint. For Spectral Tetris to work, we also require that the frame has redundancy of at least 2, that is $M \geq 2N$, where M is the number of frame elements and N is the dimension of the Hilbert space. For a UNTF, since our unique eigenvalue is $\frac{M}{N}$, we see that this is equivalent to the requirement that the eigenvalue of the frame is greater than or equal to 2.

The main idea of Spectral Tetris is to iteratively construct a synthesis matrix, T^*, for a UNTF one to two vectors at a time, which satisfies properties (1) and (2) at each step and gets closer to and eventually satisfies property (3) when complete. When it is necessary to build two vectors at a time throughout the Spectral Tetris process, we will utilize the following key 2×2 matrix as a building block for our construction.

Spectral Tetris relies on the existence of 2×2 matrices $A(x)$, for given $0 \leq x \leq 2$, such that:

(1) the columns of $A(x)$ square sum to 1,
(2) $A(x)$ has orthogonal rows,
(3) the square sum of the first row is x.

These properties combined are equivalent to

$$A(x)\, A^*(x) = \begin{bmatrix} x & 0 \\ 0 & 2-x \end{bmatrix}.$$

A matrix which satisfies these properties and which is used as a building block in Spectral Tetris is:

$$A(x) = \begin{bmatrix} \sqrt{\frac{x}{2}} & \sqrt{\frac{x}{2}} \\ \sqrt{1-\frac{x}{2}} & -\sqrt{1-\frac{x}{2}} \end{bmatrix}.$$

We are now ready to give the example.

EXAMPLE 6.6. We would like to use Spectral Tetris to construct a sparse, unit norm, tight frame with 11 elements in $\mathbb{H}^4$, so our tight frame bound will be $\frac{11}{4}$.

To do this we will create a 4×11 matrix T^*, which satisfies the following conditions:

(1) The columns square sum to 1.
(2) T^* has orthogonal rows.
(3) The rows square sum to $\frac{11}{4}$.
(4) $S = T^*T = \frac{11}{4} \cdot Id$.

Note that (4) follows if (1), (2) and (3) are all satisfied. Also notice that the sequence of eigenvalues $\{\lambda_j\}_{j=1}^4 = \{\frac{11}{4}, \frac{11}{4}, \frac{11}{4}, \frac{11}{4}\}$ majorizes the sequence of square norms $\{a_i^2\}_{i=1}^{11} = \{1,1,1,1,1,1,1,1,1,1,1\}$, which, in general, is necessary for such a frame to exist.

Define $t_{i,j}$ to be the entry in the ith row and j^{th} column of T^*. With an empty 4×11 matrix, we start at $t_{1,1}$ and work our way left to right to fill out the matrix. By requirement (1), we need the square sum of column one to be 1 and by requirement (2) we need the square sum of row one to be $\frac{11}{4} \geq 1$. Hence, we will start by being greedy and put the maximum weight of 1 in $t_{1,1}$. This forces the rest of the entries in column 1 to be zero, from requirement (1). We get:

$$T^* = \begin{bmatrix} 1 & \cdot & \cdot & \cdot & \cdot & \cdot & \cdot & \cdot & \cdot & \cdot & \cdot \\ 0 & \cdot & \cdot & \cdot & \cdot & \cdot & \cdot & \cdot & \cdot & \cdot & \cdot \\ 0 & \cdot & \cdot & \cdot & \cdot & \cdot & \cdot & \cdot & \cdot & \cdot & \cdot \\ 0 & \cdot & \cdot & \cdot & \cdot & \cdot & \cdot & \cdot & \cdot & \cdot & \cdot \end{bmatrix}.$$

Next, since row one needs to square sum to $\frac{11}{4}$, by (3), and we only have a total weight of 1 in row one, then we need to add $\frac{11}{4} - 1 = \frac{7}{4} = 1 + \frac{3}{4} \geq 1$ more weight to row one. So we will again be greedy and add another 1 in $t_{1,2}$. This forces the rest of the entries in column 2 to be zero, by (1). Also note that we have a total square sum of 2 in row one. We get:

$$T^* = \begin{bmatrix} 1 & 1 & \cdot & \cdot & \cdot & \cdot & \cdot & \cdot & \cdot & \cdot & \cdot \\ 0 & 0 & \cdot & \cdot & \cdot & \cdot & \cdot & \cdot & \cdot & \cdot & \cdot \\ 0 & 0 & \cdot & \cdot & \cdot & \cdot & \cdot & \cdot & \cdot & \cdot & \cdot \\ 0 & 0 & \cdot & \cdot & \cdot & \cdot & \cdot & \cdot & \cdot & \cdot & \cdot \end{bmatrix}.$$

In order to have a total square sum of $\frac{11}{4}$ in the first row, we need to add a total of $\frac{11}{4} - 2 = \frac{3}{4} < 1$ more weight. If the remaining unknown entries are chosen so that T^* has orthogonal rows, then S will be a diagonal matrix. Currently, the diagonal entries of S are mostly unknowns, having the form $\{2+?, \cdot, \cdot, \cdot\}$. Therefore we need a way to add $\frac{3}{4}$ more weight in the first row without compromising the orthogonality of the rows of T^* nor the normality of its columns. That is, if we get "greedy" and try to add $\sqrt{\frac{3}{4}}$ to position $t_{1,3}$ then the rest of row one must be zero, yielding:

$$T^* = \begin{bmatrix} 1 & 1 & \sqrt{\frac{3}{4}} & 0 & 0 & 0 & 0 & 0 & 0 & 0 & 0 \\ 0 & 0 & \cdot & \cdot & \cdot & \cdot & \cdot & \cdot & \cdot & \cdot & \cdot \\ 0 & 0 & \cdot & \cdot & \cdot & \cdot & \cdot & \cdot & \cdot & \cdot & \cdot \\ 0 & 0 & \cdot & \cdot & \cdot & \cdot & \cdot & \cdot & \cdot & \cdot & \cdot \end{bmatrix}.$$

In order for column three to square sum to one, at least one of the entries $t_{2,3}, t_{3,3}$ or $t_{4,3}$ is non-zero. But then, it is impossible for the rows to be orthogonal and thus we cannot proceed. Hence, we need to instead add two columns of information in attempts to satisfy these conditions. The key idea is to utilize our 2×2 building block, $A(x)$, as defined at (a).

We define the third and fourth columns of T^* according to such a matrix $A(x)$, where $x = \frac{11}{4} - 2 = \frac{3}{4}$. Notice that by doing this, column three and column four now square sum to one within the first two rows, hence the rest of the unknown entries in these two columns will be zero. We get:

$$T^* = \begin{bmatrix} 1 & 1 & \sqrt{\frac{3}{8}} & \sqrt{\frac{3}{8}} & \cdot & \cdot & \cdot & \cdot & \cdot & \cdot & \cdot \\ 0 & 0 & \sqrt{\frac{5}{8}} & -\sqrt{\frac{5}{8}} & \cdot & \cdot & \cdot & \cdot & \cdot & \cdot & \cdot \\ 0 & 0 & 0 & 0 & \cdot & \cdot & \cdot & \cdot & \cdot & \cdot & \cdot \\ 0 & 0 & 0 & 0 & \cdot & \cdot & \cdot & \cdot & \cdot & \cdot & \cdot \end{bmatrix}.$$

The diagonal entries of S are now $\{\frac{11}{4}, \frac{5}{4}+?, \cdot, \cdot\}$. The first row of T^*, and equivalently the first diagonal entry of S, now have sufficient weight and so its remaining entries are set to zero. The second row, however, is currently falling short by $\frac{11}{4} - \left(\left(\sqrt{\frac{5}{8}}\right)^2 + \left(-\sqrt{\frac{5}{8}}\right)^2\right) = \frac{6}{4} = 1 + \frac{2}{4}$. Since $1 + \frac{2}{4} \geq 1$, we can be greedy and add a weight of 1 in $t_{2,5}$. Hence, column five becomes e_2. Next, with a weight of $\frac{2}{4} < 1$ left to add to row two we utilize our 2×2 building block $A(x)$, with $x = \frac{2}{4}$. Adding this 2×2 block in columns six and seven yields sufficient weight in these columns and hence we finish these two columns with zeros. We get:

$$T^* = \begin{bmatrix} 1 & 1 & \sqrt{\frac{3}{8}} & \sqrt{\frac{3}{8}} & 0 & 0 & 0 & 0 & 0 & 0 & 0 \\ 0 & 0 & \sqrt{\frac{5}{8}} & -\sqrt{\frac{5}{8}} & 1 & \sqrt{\frac{2}{8}} & \sqrt{\frac{2}{8}} & 0 & 0 & 0 & 0 \\ 0 & 0 & 0 & 0 & 0 & \sqrt{\frac{6}{8}} & -\sqrt{\frac{2}{8}} & \cdot & \cdot & \cdot & \cdot \\ 0 & 0 & 0 & 0 & 0 & 0 & 0 & \cdot & \cdot & \cdot & \cdot \end{bmatrix}.$$

The diagonal entries of T^* are now $\{\frac{11}{4}, \frac{11}{4}, \frac{6}{4}+?, \cdot\}$, where the third diagonal entry, and equivalently the third row, are falling short by $\frac{11}{4} - \frac{6}{4} = \frac{5}{4} = 1 + \frac{1}{4}$. Since $1 + \frac{1}{4} \geq 1$, then we take the eighth column of T^* to be e_3. We will complete our matrix following these same strategies, by letting the ninth and tenth columns arise from $A\left(\frac{1}{4}\right)$, and making the final column e_4, yielding the desired UNTF:

$$T^* = \begin{bmatrix} 1 & 1 & \sqrt{\frac{3}{8}} & \sqrt{\frac{3}{8}} & 0 & 0 & 0 & 0 & 0 & 0 & 0 \\ 0 & 0 & \sqrt{\frac{5}{8}} & -\sqrt{\frac{5}{8}} & 1 & \sqrt{\frac{2}{8}} & \sqrt{\frac{2}{8}} & 0 & 0 & 0 & 0 \\ 0 & 0 & 0 & 0 & 0 & \sqrt{\frac{6}{8}} & -\sqrt{\frac{2}{8}} & 1 & \sqrt{\frac{7}{8}} & \sqrt{\frac{7}{8}} & 0 \\ 0 & 0 & 0 & 0 & 0 & 0 & 0 & 0 & \sqrt{\frac{7}{8}} & -\sqrt{\frac{7}{8}} & 1 \end{bmatrix}.$$

In this construction, column vectors are either introduced one at a time, such as columns $1, 2, 5, 8$, and 11, or in pairs, such as columns $\{3, 4\}, \{6, 7\}$, and $\{9, 10\}$. Each singleton contributes a value of 1 to a particular diagonal entry of S, while each pair spreads two units of weight over two entries. Overall, we have formed a flat spectrum, $\{\frac{11}{4}, \frac{11}{4}, \frac{11}{4}, \frac{11}{4}\}$, from blocks of area one or two. This construction is reminiscent of the game Tetris, as we fill in blocks of mixed area to obtain a flat spectrum.

7. Gramian Operators

If $\{\varphi_i\}_{i=1}^M$ is a frame for $\mathbb{H}^N$ with analysis operator T, the **Gramian operator** is defined as

$$G := TT^*,$$

which has matrix representation

$$G = [\langle \varphi_j, \varphi_i \rangle]_{i,j=1}^M ,$$

called the **Gramian matrix**. Since we know T^*T and TT^* have the same non-zero eigenvalues, we have:

PROPOSITION 7.1. *Let $\{\varphi_i\}_{i=1}^M$ be a frame for $\mathbb{H}^N$ with frame bounds A, B and frame operator S. The Gramian operator has the same non-zero eigenvalues as S. That is, the largest eigenvalue of G is less than or equal to B and the smallest non-zero eigenvalue is greater than or equal to A.*

THEOREM 7.2. *If $\{\varphi_i\}_{i=1}^M$ is a Parseval frame with analysis operator T, then the Gramian operator is an orthogonal projection.*

PROOF. It is clear that TT^* is self-adjoint and

$$(TT^*)(TT^*) = T(T^*T)T^* = T(I)T^* = TT^*.$$ □

COROLLARY 7.3. *Let $\Phi = \{\varphi_i\}_{i=1}^M$ be vectors in $\mathbb{H}^N$. The Gramian of Φ is invertible if and only if Φ is a Riesz basis (that is, when $M = N$).*

PROOF. If $G = TT^*$ is invertible, by Proposition 7.1, T^*T is invertible. Hence, $\{\varphi_i\}_{i=1}^M$ is a frame for $\mathbb{H}^N$. Also, T^* is one-to-one, and further T^* is bounded, linear and onto. Hence, it is an isomorphism.

If $\{\varphi_i\}_{i=1}^M$ is a Riesz basis then T^* is an isomorphism and we have that T^* is invertible and so $G = TT^*$ is invertible. □

PROPOSITION 7.4. *Let $F = [a_{ij}]_{i,j=1}^M$ be a positive, self-adjoint matrix (operator) on $\mathbb{H}^N$ with $\dim\ker F = M - N$. Then $\{F^{1/2}e_i\}_{i=1}^M$ spans an N-dimensional space and*

$$\langle F^{1/2}e_i, F^{1/2}e_j \rangle = \langle Fe_i, e_j \rangle = a_{ij}.$$

Hence, F is the Gramian matrix for the vectors $\{F^{1/2}e_i\}_{i=1}^M$. Moreover,

$$\|F^{1/2}e_i\|^2 = a_{ii},$$

and so if $a_{ii} = 1$ for all $i = 1, 2, \cdots, M$ then $\{F^{1/2}e_i\}_{i=1}^M$ is a unit norm family.

8. Fusion Frames

A number of new applications have emerged which cannot be modeled naturally by one single frame system. Generally they share a common property that requires distributed processing. Furthermore, we are often overwhelmed by a deluge of data assigned to one single frame system, which becomes simply too large to be handled numerically. In these cases it would be highly beneficial to split a large frame system into a set of (overlapping) much smaller systems, and to process locally within each sub-system effectively.

A distributed frame theory for a set of local frame systems is therefore in demand. A variety of applications require distributed processing. Among them there are, for instance, wireless sensor networks [**62**], geophones in geophysics measurements and studies [**43**], and the physiological structure of visual and hearing systems [**73**]. To understand the nature, the constraints, and related problems of these applications, let us elaborate a bit further on the example of wireless sensor networks.

In wireless sensor networks, sensors of limited capacity and power are spread in an area sometimes as large as an entire forest to measure the temperature, sound, vibration, pressure, motion, and/or pollutants. In some applications, wireless sensors are placed in a geographical area to detect and characterize chemical, biological, radiological, and nuclear material. Such a sensor system is typically redundant, and there is no orthogonality among sensors, therefore each sensor functions as a frame element in the system. Due to practical and cost reasons, most sensors employed in such applications have severe constraints in their processing power and transmission bandwidth. They often have strictly metered power supply as well. Consequently, a typical large sensor network necessarily divides the network into redundant sub-networks – forming a set of subspaces. The primary goal is to have local measurements transmitted to a local sub-station within a subspace for a subspace combining. An entire sensor system in such applications could have a number of such local processing centers. They function as relay stations, and have the gathered information further submitted to a central processing station for final assembly.

In such applications, distributed/local processing is built in the problem formulation. A staged processing structure is prescribed. We will have to be able to process the information stage by stage from local information and to eventually fuse them together at the central station. We see therefore that a mechanism of coherently collecting sub-station/subspace information is required.

Also, due to the often unpredictable nature of geographical factors, certain local sensor systems are less reliable than others. While facing the task of combining local subspace information coherently, one has also to consider weighting the more reliable sets of substation information more than suspected less reliable ones. Consequently, the coherent combination mechanism we just saw as necessary often requires a weighted structure as well. This all leads naturally to what is called a fusion frame.

DEFINITION 8.1. Let I be a countable index set, let $\{W_i\}_{i=1}^M$ be a family of closed subspaces in $\mathbb{H}^N$, and let $\{v_i\}_{i=1}^M$ be a family of weights, i.e. $v_i > 0$ for all $i \in I$. Then $\{(W_i, v_i)\}_{i=1}^M$ is a **fusion frame**, if there exist constants $0 < C \leq D <$

∞ such that

$$C\|x\|^2 \leq \sum_{i=1}^{M} v_i^2 \|P_{W_i}(x)\|^2 \leq D\|x\|^2 \quad \text{for all } x \in \mathbb{H}^N, \tag{8.1}$$

where P_{W_i} is the orthogonal projection onto the subspace W_i. We call C and D the **fusion frame bounds**. The family $\{(W_i, v_i)\}_{i=1}^M$ is called a **C-tight fusion frame**, if in (8.1) the constants C and D can be chosen so that $C = D$, a **Parseval fusion frame** provided that $C = D = 1$, and an **orthonormal fusion basis** if $\mathbb{H}^N = \bigoplus_{i=1}^M W_i$. If $\{(W_i, v_i)\}_{i=1}^M$ possesses an upper fusion frame bound, but not necessarily a lower bound, we call it a **Bessel fusion sequence** with **Bessel fusion bound** D.

Often it will become essential to consider a fusion frame together with a set of local frames for its subspaces. In this case we speak of a fusion frame system.

DEFINITION 8.2. Let $\{(W_i, v_i)\}_{i=1}^M$ be a fusion frame for $\mathbb{H}^N$, and let the sequence of vectors $\{x_{ij}\}_{j=1}^{J_i}$ be a frame for W_i for each $1 \leq i \leq M$. Then we call $\{(W_i, v_i, \{\varphi_{ij}\}_{j=1}^{J_i})\}_{i=1}^M$ a **fusion frame system** for $\mathbb{H}^N$. The constants, C and D, are the associated **fusion frame bounds** if they are the fusion frame bounds for $\{(W_i, v_i)\}_{i=1}^M$, and A and B are the **local frame bounds** if these are the common frame bounds for the **local frames** $\{\varphi_{ij}\}_{j=1}^{J_i}$ for each $1 \leq i \leq M$. A collection of dual frames, $\{\psi_{ij}\}_{j=1}^{J_i}$ for each $1 \leq i \leq M$, associated with the local frames will be called **local dual frames**.

To provide a quick inside-look at some intriguing relations between properties of the associated fusion frame and the sequence consisting of all local frame vectors, we present the following theorem from [**31**] that provides a link between local and global properties.

THEOREM 8.3. [**31**, Thm. 3.2] *For each $1 \leq i \leq M$, let $v_i > 0$, let W_i be a closed subspace of $\mathbb{H}^N$, and let $\{\varphi_{ij}\}_{j=1}^{J_i}$ be a frame for W_i with frame bounds A_i and B_i. Suppose that $0 < A = \inf_{1 \leq i \leq M} A_i \leq \sup_{1 \leq i \leq M} B_i = B < \infty$. Then the following conditions are equivalent.*

(1) $\{(W_i, v_i)\}_{i=1}^M$ is a fusion frame for $\mathbb{H}^N$.
(2) $\{v_i\varphi_{ij}\}_{j=1,\, i=1}^{J_i,\, M}$ is a frame for $\mathbb{H}^N$.

In particular, if $\{(W_i, v_i, \{\varphi_{ij}\}_{j=1}^{J_i})\}_{i=1}^M$ is a fusion frame system for $\mathbb{H}^N$ with fusion frame bounds C and D, then $\{v_i\varphi_{ij}\}_{j=1,\, i=1}^{J_i,\, M}$ is a frame for $\mathbb{H}^N$ with frame bounds AC and BD. Conversely, if $\{v_i\varphi_{ij}\}_{j=1,\, i=1}^{J_i,\, M}$ is a frame for $\mathbb{H}^N$ with frame bounds C and D, then $\{(W_i, v_i, \{\varphi_{ij}\}_{j=1}^{J_i})\}_{i=1}^M$ is a fusion frame system for $\mathbb{H}^N$ with fusion frame bounds $\frac{C}{B}$ and $\frac{D}{A}$.

Tight frames play a vital role in frame theory due to the fact that they provide easy reconstruction formulas. Tight fusion frames will turn out to be particularly useful for distributed reconstruction as well. Notice, that the previous theorem also implies that $\{(W_i, v_i)\}_{i=1}^M$ is a C-tight fusion frame for $\mathbb{H}^N$ if and only if $\{v_i f_{ij}\}_{j=1,\, i=1}^{J_i,\, M}$ is a C-tight frame for $\mathbb{H}^N$.

The following result from [**33**] proves that the fusion frame bound C of a C-tight fusion frame can be interpreted as the **redundancy** of this fusion frame.

PROPOSITION 8.4. *If $\{(W_i, v_i)\}_{i=1}^M$ is a C-tight fusion frame for $\mathbb{H}^N$, then*

$$C = \frac{\sum_{i=1}^M v_i^2 \dim W_i}{N}.$$

Let $\mathcal{W} = \{(W_i, v_i)\}_{i=1}^M$ be a fusion frame for $\mathbb{H}^N$. In order to map a signal to the representation space, i.e., to analyze it, the **fusion analysis operator** $T_{\mathcal{W}}$ is employed, which is defined by

$$T_{\mathcal{W}} : \mathbb{H}^N \to \left(\sum_{i=1}^M \oplus W_i \right)_{\ell_2} \quad \text{with} \quad T_{\mathcal{W}}(x) = \left\{ v_i P_{W_i}(x) \right\}_{i=1}^M.$$

It can easily be shown that the **fusion synthesis operator** $T_{\mathcal{W}}^*$, which is defined to be the adjoint operator of the analysis operator, is given by

$$T_{\mathcal{W}}^* : \left(\sum_{i=1}^M \oplus W_i \right)_{\ell_2} \to \mathbb{H}^N$$

with

$$T_{\mathcal{W}}^*(x) = \sum_{i=1}^M v_i x_i, \text{ where } x = \{x_i\}_{i=1}^M \in \left(\sum_{i=1}^M \oplus W_i \right)_{\ell_2}.$$

The **fusion frame operator** $S_{\mathcal{W}}$ for $\mathcal{W}$ is defined by

$$S_{\mathcal{W}}(x) = T_{\mathcal{W}}^* T_{\mathcal{W}}(x) = \sum_{i \in I} v_i^2 P_{W_i}(x).$$

Interestingly, a fusion frame operator exhibits properties similar to a frame operator concerning invertibility. In fact, if $\{(W_i, v_i)\}_{i=1}^M$ is a fusion frame for $\mathbb{H}^N$ with fusion frame bounds C and D, then the associated fusion frame operator $S_{\mathcal{W}}$ is positive and invertible on $\mathbb{H}^N$, and

$$C \cdot Id \leq S_{\mathcal{W}} \leq D \cdot Id. \tag{8.2}$$

We refer the reader to [**31**, Prop. 3.16] for details.

There has been a significant amount of recent work on fusion frames. This topic now has its own website and we recommend visiting it for the latest developments on fusion frames, distributed processing and sensor networks.

http://www.fusionframe.org/

But also visit the Frame Research Center Website:

http://www.framerc.org/

9. Infinite Dimensional Hilbert Spaces

We work with two standard infinite dimensional Hilbert Spaces.

9.1. Hilbert Spaces of Sequences. We being with the definition.

DEFINITION 9.1. We define ℓ_2 by:

$$\{x = \{a_i\}_{i=1}^{\infty} : \|x\| := \sum_{i=1}^{\infty} |a_i|^2 < \infty.\}$$

The **inner product** of $x = \{a_i\}_{i=1}^{\infty}$ and $y = \{b_i\}_{i=1}^{\infty}$ is given by

$$\langle x, y \rangle = \sum_{i=1}^{\infty} a_i \overline{b_i}.$$

The space ℓ_2 has a natural orthonormal basis $\{e_i\}_{i=1}^{\infty}$ where

$$e_i = (0, 0, \cdots, 0, 1, 0, \cdots)$$

where the 1 is in the ith-coordinate. Most of the results on finite dimensional Hilbert spaces carry over here. The one major exception is that operators here may not have eigenvalues or be diagonalizable. We give an example in the next subsection. But all of the "identities" hold here. Another difference is the notion of **linear independence**.

DEFINITION 9.2. A family of vectors $\{x_i\}_{i=1}^{\infty}$ in ℓ_2 is **linearly independent** if for every finite subset $I \subset \mathbb{N}$ and any scalars $\{a_i\}_{i \in I}$ we have

$$\sum_{i \in I} a_i x_i = 0 \Rightarrow a_i = 0, \text{ for all } i \in I.$$

The family is ω-independent if for any family of scalars $\{a_i\}_{i=1}^{\infty}$, satisfying $\sum_{i=1}^{\infty} |a_i|^2 < \infty$, we have

$$\sum_{i=1}^{\infty} a_i x_i = 0 \Rightarrow a_i = 0 \text{ for all } i = 1, 2, \cdots.$$

An orthonormal basis $\{e_i\}_{i=1}^{\infty}$ is clearly ω-independent. But, linearly independent vectors may not be ω-independent. For example, if we let

$$x_i = e_i + \frac{1}{2^i} e_{i+1}, \quad i = 1, 2, \cdots,$$

it is easily checked that this family is finitely linearly independent. But,

$$\sum_{i=1}^{\infty} \frac{(-1)^{i-1}}{2^i} x_i = 0,$$

and so this family is not ω-independent.

9.2. Hilbert Spaces of Functions. We define a Hilbert space of functions:

DEFINITION 9.3. If $A \subset \mathbb{R}$ (or $\mathbb{C}$) we define

$$L^2(A) = \left\{ f : A \to \mathbb{R} \text{ (or } \mathbb{C}) : \|f\|^2 := \int_A |f(t)|^2 dt < \infty \right\}.$$

The inner product of $f, g \in L^2(A)$ is

$$\langle f, g \rangle = \int_I f(t) \overline{g(t)} dt.$$

The two cases we work with the most are $L^2([0,1])$ and $L^2(\mathbb{R})$.

The space $L^2([0,1])$ has a natural orthonormal basis given by the complex exponentials:

$$\{e^{2\pi int}\}_{n\in\mathbb{Z}}.$$

If we choose $A \subset [0,1]$, the orthogonal projection of $L^2([0,1])$ onto $L^2(A)$ satisfies:

$$P(e^{2\pi int}) = \chi_A e^{2\pi int}.$$

This family of vectors is a Parseval frame (since it is the image of an orthonormal basis under an orthogonal projection) called the **Fourier frame on** A. Recently, Marcus/Spielman/Srivastava [**69**] solved the **Feichtinger Conjecture** for this class of frames (See subsection 10.6).

To get a flavor of things that don't go so nicely in the infinite dimensional setting, we note that there are positive, self-adjoint, invertible operators on infinite dimensional Hilbert spaces which have no eigenvectors. For example, consider the operator $S : L^2([0,1]) \to L^2([0,1])$ defined by

$$S(f)(x) := (1+x)f(x).$$

For any $f \in L^2[0,1]$ we have:

$$\langle f, Sf\rangle = \int_0^1 (1+x)f^2(x)dx \geq \int_0^1 f^2(x)dx = \|f\|^2.$$

So S is a positive, self-adjoint, invertible operator. However, in order for $Sf = \lambda f$ we would need to have

$$(1+x)f(x) = \lambda f(x) \text{ almost everywhere on } [0,1].$$

This is clearly impossible unless $f = 0$, so S has no eigenvectors.

10. Major Open Problems in Frame Theory

In this section we look at some of the major open problems in Frame Theory.

10.1. Equiangular Frames. One of the simplest stated yet deepest problems in mathematics is the equiangular line problem.

PROBLEM 10.1. How many equiangular lines can be drawn through the origin in $\mathbb{R}^N$ or $\mathbb{C}^N$?

The easiest way to describe **equiangular lines** is to put a unit norm vector starting at the origin on each line, say $\{\varphi_i\}_{i=1}^M$, and the lines are **equiangular** if there is a constant $0 < c \leq 1$ so that

$$|\langle \varphi_i, \varphi_j\rangle| = c, \text{ for all } i \neq j.$$

These inner products represent the cosine of the acute angle between the lines. The problem of constructing any number (especially, the maximal number) of equiangular lines in $\mathbb{R}^N$ is one of the most elementary and at the same time one of the most difficult problems in mathematics. After sixty years of research, we do not know the answer for all dimensions ≤ 20 in either the real or complex case. This line of research was started in 1948 by Hanntjes [**58**] in the setting of elliptic geometry where he identified the maximal number of equiangular lines in $\mathbb{R}^N$ for $n = 2, 3$. Later, Van Lint and Seidel [**68**] classified the largest number of equiangular lines in $\mathbb{R}^N$ for dimensions $N \leq 7$ and at the same time emphasized the relations to discrete mathematics. In 1973, Lemmens and Seidel [**66**] made a comprehensive study of real equiangular line sets which is still today a fundamental piece of work. Gerzon **[66]** gave an upper bound for the maximal number of equiangular lines in $\mathbb{R}^N$:

THEOREM 10.2 (Gerzon). *If we have M equiangular lines in $\mathbb{R}^N$ then*

$$M \leq \frac{N(N+1)}{2}.$$

It is known that we cannot reach this maximum in most cases. It is also known that the maximal number of equiangular lines in $\mathbb{C}^N$ is less than or equal to N^2. It is believed that this number of lines exists in $\mathbb{C}^N$ for all N but until now a positive answer does not exist for all $N \leq 20$.

Also, P. Neumann **[66]** produced a fundamental result in the area:

THEOREM 10.3 (P. Neumann). *If $\mathbb{R}^N$ has M equiangular lines at angle $1/\alpha$ and $M > 2N$, then α is an odd integer.*

Finally, there is a lower bound on the angle formed by equiangular line sets.

THEOREM 10.4. *If $\{\varphi_m\}_{m=1}^M$ is a set of norm one vectors in $\mathbb{R}^N$, then*

$$\max_{i\neq j} |\langle \varphi_i, \varphi_j\rangle| \geq \sqrt{\frac{M-N}{N(M-1)}}. \tag{10.1}$$

Moreover, we have equality if and only if $\{\varphi_i\}_{i=1}^M$ is an equiangular tight frame and in this case the tight frame bound is $\frac{M}{N}$.

This inequality goes back to Welch **[79]**. Strohmer and Heath **[76]** and Holmes and Paulsen **[61]** give more direct arguments which also yields the "moreover" part. For some reason, in the literature there is a further assumption added to the "moreover" part of Theorem 10.4 that the vectors span $\mathbb{R}^N$. This assumption is not necessary. That is, equality in inequality 10.1 already implies that the vectors span the space **[37]**.

Equiangular Tight Frames:

Good references for real eqiuangular frames are **[37, 61, 76, 77]**. A unit norm frame with the property that there is a constant c so that

$$|\langle \varphi_i, \varphi_j\rangle| = c, \quad \text{for all } i \neq j,$$

is called an **equiangular frame** at angle c. Equiangular tight frames first appeared in discrete geometry **[68]** but today (especially the complex case) have applications in signal processing, communications, coding theory and more **[57, 76]**. A detailed study of this class of frames was initiated by Strohmer and Heath **[76]** and Holmes and Paulsen **[61]**. Holmes and Paulsen **[61]** showed that equiangular tight frames give error correction codes that are robust against two erasures. Bodmann and Paulsen **[13]** analyzed arbitrary numbers of erasures for equiangular tight frames. Recently, Bodmann, Casazza, Edidin and Balan **[11]** showed that equiangular tight frames are useful for signal reconstruction when all phase information is lost. Recently, Sustik, Tropp, Dhillon and Heath **[77]** made an important advance on this subject (and on the complex version). Other applications include the construction of capacity achieving signature sequences for multiuser communication systems in wireless communication theory **[79]**. The tightness condition allows equiangular tight frames to achiece the capacity of a Gaussian channel and their equiangularity allows them to satisfy an interference invariance property. Equiangular tight frames potentially have many more practical and theoretical applications. Unfortunately, we know very few of them and so their usefulness is largely untapped.

The main problem:

PROBLEM 10.5. Classify all equiangular tight frames, or find large classes of them.

Fickus/Jasper/Mixon [**50**] gave a large class of **Kirkman equiangular tight frames** and used them in coding theory.

THEOREM 10.6. *The following are equivalent:*

(1) *The space* $\mathbb{R}^N$ *has an equiangular tight frame with* M *elements at angle* $1/\alpha$.
(2) *We have*

$$M = \frac{(\alpha^2 - 1)N}{\alpha^2 - N},$$

and there exist M *equiangular lines in* $\mathbb{R}^N$ *at angle* $1/\alpha$.

Moreover, in this case we have:

(a) $\alpha \leq N \leq \alpha^2 - 2$.
(b) $N = \alpha$ *if and only if* $M = N + 1$.
(c) $N = \alpha^2 - 2$ *if and only if* $M = \frac{N(N+1)}{2}$.
(d) $M = 2N$ *if and only if*

$$\alpha^2 = 2N - 1 = a^2 + b^2, \quad a,b \text{ integers}.$$

If $M \neq N + 1, 2N$ *then:*

(e) α *is an odd integer.*
(f) M *is even.*
(g) α *divides M-1.*
(h) $\beta = \frac{M-1}{\alpha}$ *is the angle for the complementary equiangular tight frame.*

10.2. The Scaling Problem. The scaling problem is one of the deepest problems in frame theory.

PROBLEM 10.7 (Scaling Problem). Classify the frames $\{\varphi_i\}_{i=1}^M$ for $\mathbb{H}^N$ so that there are scalars $\{a_i\}_{i=1}^M$ for which $\{a_i\varphi_i\}_{i=1}^M$ is a Parseval frame? Give an algorithm for finding $\{a_i\}_{i=1}^M$.

This is really a special case of an even deeper problem.

PROBLEM 10.8. Given a frame $\{\varphi_i\}_{i=1}^M$ for $\mathbb{H}^N$, find the scalars $\{a_i\}_{i=1}^M$ so that $\{a_i\varphi_i\}_{i=1}^M$ has the minimal condition number with respect to all such scalings.

For recent results on the scaling problem see Chen/Kutyniok/Okoudjou/Philipp/Wang [**40**], Cahill/Chen [**15**] and Kutyniok/Okoudjou/Philipp/Tuley [**65**].

10.3. Sparse Orthonormal Bases for Subspaces. We will look at two questions concerning the construction of sparse orthonormal bases.

DEFINITION 10.9. Given a vector $x = (a_1, a_2, \cdots, a_N) \in \mathbb{H}^N$, we let

$$\|x\|_0 = |\{1 \leq i \leq N : a_i \neq 0\}.$$

A natural question in Hilbert space theory is:

PROBLEM 10.10. Given a Hilbert space $\mathbb{H}^N$ with orthonormal basis $\{e_i\}_{i=1}^N$ and a K dimensional subspace W, find the sparsest orthonormal basis for W with respect to $\{e_i\}_{i=1}^N$. That is, find a orthonormal basis $\{g_i\}_{i=1}^K$ for W so that

$$\sum_{i=1}^K \|g_i\|_0, \text{ is a minimum with respect to all orthonormal bases for W.}$$

Sparse Gram-Schmidt Orthogonalization:

There is a basic notion for turning a linearly independent set into an orthonormal set with the same partial spans: Gram-Schmidt Orthogonalization. Given a set $\{\varphi_i\}_{i=1}^M$ of linearly independent vectors in $\mathbb{H}^N$, first let

$$e_1 = \frac{\varphi_1}{\|\varphi_1\|}.$$

Assume we have constructed $\{e_i\}_{i=1}^K$ satisfying:

(1) $\{e_i\}_{i=1}^K$ is orthonormal.
(2) We have

$$\operatorname*{span}_{1\le i\le j} e_i = \operatorname*{span}_{1\le i\le j} \varphi_i, \text{ for all } j = 1, 2, \cdots, K.$$

We then let

$$\psi_{K+1} = \varphi_{K+1} - \sum_{i=1}^N \langle \varphi_{K+1}, e_i\rangle e_i$$

and let

$$e_{K+1} = \frac{\psi_{K+1}}{\|\psi_{K+1}\|}.$$

If we have a fixed basis $\{g_i\}_{i=1}^N$ for $\mathbb{H}^N$, we can compute

$$\sum_{i=1}^K \|e_i\|_0 \text{ with respect to the basis } \{g_i\}_{i=1}^N.$$

But this sum is different for different orderings of the original vectors $\{\varphi_i\}_{i=1}^M$. Related to Problem 10.10 we have:

PROBLEM 10.11. What is the correct ordering of $\{\varphi_i\}_{i=1}^M$ so that Gram-Schmidt Orthogonalization produces the sparsest orthonormal sequence with respect to all possible orderings?

10.4. The Paulsen Problem. To state the Paulsen Problem, we need some definitions.

DEFINITION 10.12. A frame $\{\varphi_i\}_{i=1}^M$ for $\mathbb{H}^N$ with frame operator S is said to be **ε-nearly equal norm** if

$$(1-\varepsilon)\frac{N}{M} \le \|\varphi_i\|^2 \le (1+\varepsilon)\frac{N}{M}, \text{ for all } i = 1, 2, \cdots, M,$$

and it is **ε-nearly Parseval** if

$$(1-\varepsilon)\cdot Id \le S \le (1+\varepsilon)\cdot Id.$$

DEFINITION 10.13. The **distance between two frames** $\Phi = \{\varphi_i\}_{i=1}^M$ and $\Psi = \{\psi_i\}_{i=1}^M$ is given by:

$$d(\Phi, \Psi) = \sum_{i=1}^{M} \|\varphi_i - \psi_i\|^2.$$

Because we did not take the square-root on the right-hand-side of the above inequality, the function d is not really a distance function.

The **Paulsen Problem** now states:

PROBLEM 10.14. How close in terms of d is an ε-nearly equal norm and ε-nearly Parseval frame to an equal norm Parseval frame?

The importance of this problem is that we have algorithms for finding frames which are equal norm and nearly Parseval. But we do not know that these are actually close to any equal norm Parseval frame.

The closest equal norm frame to a frame is known and the closest Parseval frame to a frame is known:

The closest equal norm frame to a frame $\{\varphi_i\}_{i=1}^M$ is

$$\left\{C\frac{\varphi_i}{\|\varphi_i\|}\right\}_{i=1}^M \quad \text{where } C := \frac{\sum_{i=1}^M \|\varphi_i\|}{M}.$$

If $\Phi = \{\varphi_i\}_{i=1}^M$ is a frame with frame operator S, the closest Parseval frame to Φ is the canonical Parseval frame $\{S^{-1/2}\varphi_i\}_{i=1}^M$ [**34**] (see [**10**] for a better calculation).

Also, there is an algorithm for turning a frame into an equal norm frame without changing the frame operator [**25**].

Casazza/Cahill [**16**] showed that the Paulsen Problem is equivalent to an old deep problem in Operator Theory.

PROBLEM 10.15. Given a projection P on $\mathbb{H}^N$ with ε-nearly equal diagonal elements of its matrix, what is the closest constant diagonal projection to P?

10.5. Concrete Construction of RIP Matrices. Compressive Sensing is one of the most active areas of research today. See the book [**56**] for an exhaustive coverage of this subject. A fundamental tool in this area matrices with the **Restricted Isometry Property**, denoted **RIP**. Compressive sensing is a method for solving **underdetermined systems** if we have some form of **sparsity** of the incoming signal.

DEFINITION 10.16. A vector $x = (a_1, a_2, \cdots, a_N) \in \mathbb{H}^N$ is **K-sparse** if

$$|\{1 \leq i \leq N : a_i \neq 0\}| \leq K.$$

The fundamental tool in compressive sensing is the class of **Restricted Isometry Property** (**RIP**) matrices.

DEFINITION 10.17. A matrix Φ has the (K, δ)- **Restricted Isometry Property**, **RIP** if

$$(1 - \delta)\|x\|^2 \leq \|\Phi x\|^2 \leq (1_+\delta)\|x\|^2,$$

for every K-sparse vector x. The smallest δ for which Φ is (K, δ)-RIP is the **restricted isometry constant (RIC)** δ_K.

The main result here is (see [**56**]):

THEOREM 10.18. *Given $\delta < 1$, there exist $N \times M$ matrices with restricted isometry constant $\delta_K \leq \delta$ for*

$$K \leq c\frac{N}{\ln(N/K)},$$

for a universal constant c.

This means, for example, in an N-dimensional Hilbert space, we can find a set of $100N$ norm one vectors $\{\varphi_i\}_{i=1}^{100N}$ for which every subset of size $N/100$ is a δ-Riesz basic sequence. This is a quite amazing result. We know that in an N-dimensional Hilbert space, any orthogonal set must have $\leq N$ elements. This result says that if we relax this requirement just a little, we can find huge sets of vectors for which every subset of size a proportion of the dimension of the space is nearly orthogonal.

In the language of frame theory, we are looking for a family of unit norm vectors $\Phi = \{\varphi_i\}_{i=1}^{M}$ in $\mathbb{H}^N$ so that every subset of Φ of size a proportion of N is nearly orthogonal and M is much larger than N. The existence of such matrices has been carried out by random matrix theory. Which means we do not know concretely a single such matrix, despite the fact that these are essential for compressive sensing. For years, the closest thing to **concrete** here was a result of DeVore [**45**] which constructed $N \times M$ matrices for which subsets of size $\sqrt{N}$ were δ-Riesz. But this is far from what we know is true which is subsets of size cN for $0 < c < 1$ independent of N. Bandira/Fickus/Mixon/Wong [**5**] investigated various methods for constructing RIP matrices. Bourgain [**14**] then **broke the square root barrier** by showing we can concretely construct RIP matrices with subsets of size $N^{1/2+\varepsilon}$ being δ-Riesz. There is also an important result of Rudelson/Vershynin [**74**] which says that if we take a random selection of rows from the Discrete Fourier Transform Matrix, then this submatrix will be a RIP matrix. Since these matrices are fundamental to compressive sensing, a longstanding, important and fundamental problem here is:

PROBLEM 10.19. Give a concrete construction of RIP matrices.

10.6. An Algorithm for the Feichtinger Conjecture. For nearly 50 years the Kadison-Singer problem [**63**] has defied the best efforts of some of the most talented mathematicians of our time. It was just recently solved by Marcus/Spielman/Srivastava [**69**]. For a good summary of the history of this problem and consequences of this achievement see [**26**].

In his work on time-frequency analysis, Feichtinger [**38, 54**] noted that all of the Gabor frames he was using had the property that they could be divided into a finite number of subsets which were Riesz basic sequences. This led to a conjecture known as the **Feichtinger Conjecture** [**38**]. There is a significant body of work on this conjecture and we refer the reader to [**26**] for the best reference.

First we need:

DEFINITION 10.20. A family of vectors $\{\varphi_i\}_{i\in I}$ is an ε-Riesz basic sequence for $0 < \varepsilon < 1$ if for every family of scalars $\{a_i\}_{i\in I}$ we have

$$(1-\varepsilon)\sum_{i\in I}|a_i|^2 \leq \left\|\sum_{i\in I}a_i\varphi_i\right\|^2 \leq (1+\varepsilon)\sum_{i\in I}|a_i|^2.$$

The following theorem gives the best quantative solution to the **Feichtinger Conjecture** from the results of [**69**].

THEOREM 10.21 (Marcus/Spielman/Srivastave). *Every unit norm B-Bessel sequence can be partitioned into r-subsets each of which is a ε-Riesz basic sequence, where*

$$r = \left(\frac{6(B+1)}{\varepsilon}\right)^2 \quad \textit{in the real case},$$

and

$$r = \left(\frac{6(B+1)}{\varepsilon}\right)^4 \quad \textit{in the complex case}.$$

This theorem could be quite useful, except that it is an **existence proof**. Now what we really need is:

PROBLEM 10.22. Find an implementable algorithm for proving the Feichtinger Conjecture.

10.7. Classifying Gabor Frames. Gabor frames form the basis for time-frequency analysis which is the mathematics behind signal processing. This is a huge subject which cannot be covered here except for a few remarks. We recommend the excellent book of Gröchenig [**55**] for a comprehensive coverage of this subject. We first define **translation** and **modulation**:

DEFINITION 10.23. Fix $a, b > 0$. For a function $f \in L^2(\mathbb{R})$ we define

$$\textit{Translation by } a: \quad T_a f(x) = f(x-a),$$

$$\textit{Modulation by } b: \quad M_b f(x) = e^{2\pi i b x} f(x).$$

In 1946, Gabor [**54**] formulated a fundamental approach to signal decomposition in terms of elementary signals. Gabor's approach quickly became a paradigm for the spectral analysis associated with time-frequency methods, such as the short-time Fourier transform and the Wigner transform. For Gabor's method, we need to fix a **window function** $g \in L^\infty(\mathbb{R})$ and $a, b \in \mathbb{R}^+$. If the family

$$\mathcal{G}(g, a, b) = \{M_{mb} T_{na} g\}_{m,n \in \mathbb{Z}}$$

is a frame for $L^2(\mathbb{R})$ we call this a **Gabor frame**. Gabor frames are used in signal processing. It is a very deep question which values of a, b, g give Gabor frames. There are some necessary requirements however.

THEOREM 10.24. *If the family given by (g, a, b) yields a Gabor frame then:*
(1) $ab \le 1$.
(2) If $ab = 1$ then this family is a frame if and only if it is a Riesz basis.

Also, the Balian-Low Theorem puts some restrictions on the function $g \in L^2(\mathbb{R})$ for the case $ab = 1$.

THEOREM 10.25 (Balian-Low Theorem). *If $g \in L^2(\mathbb{R})$, $ab = 1$ and (g, a, b) generates a Gabor frame, then either $xg(x) \notin L^2(\mathbb{R})$ or $g' \notin L^2(\mathbb{R})$.*

The Balian-Low Theorem implies that Gaussian functions e^{-ax^2} cannot yield Gabor frames for $ab = 1$.

The main problem here is:

PROBLEM 10.26. Find all functions g and positive constants a, b so that (g, a, b) forms a Gabor frame for $L^2(\mathbb{R})$.

Recently, a significant advance was made on this problem by Dai/Sun [**44**] when they solved the old and famous **abc-problem**. We refer to [**44**] for the history of the problem. In particular, Dai/Sun classified all triples (a, b, c) so that

$$\mathcal{G}(\chi_I, a, b) \text{ is a Gabor frame when } |I| = c.$$

10.8. Phase Retrieval. Phase retrieval is one of the largest areas of engineering with applications to **x-ray crystallography, Electron Microscopy, Coherence Theory, Diffractive Imaging, Astronomical Imaging, x-ray tomography, Optics, Digital Holography, Speech Recognition** and more [**7, 12, 51, 52, 70–72**]. For an introduction to this subject see [**17**].

Phase retrieval is the problem of recovering a signal from the absolute values of linear measurement coefficients called **intensity measurements**. Note multiplying a signal by a global phase factor does not affect these coefficients, so we seek signal recovery mod a global phase factor.

There are two main approaches to this problem of phase retrieval. One is to restrict the problem to a subclass of signals on which the intensity measurements become injective. The other is to use a larger family of measurements so that the intensity measurements map any signal injectively. The latter approach in phase retrieval first appears in [**3**] where the authors examine injectivity of intensity measurements for finite Hilbert spaces. The authors completely characterize measurement vectors in the real case which yield such injectivity, and they provide a surprisingly small upper bound on the minimal number of measurements required for the complex case. This sparked an incredible volume of current phase retrieval research [**1, 2, 4, 20–22, 46, 49**] focused on algorithms and conditions guaranteeing injective and stable intensity measurements.

DEFINITION 10.27. A family of vectors $\Phi = \{\varphi_i\}_{i=1}^M$ does **phase retrieval** on $\mathbb{H}^N$ if whenever $x, y \in \mathbb{H}^N$ satisfy

$$|\langle x, \varphi_i\rangle| = |\langle y, \varphi_i\rangle|, \text{ for all } i = 1, 2 \cdots, M,$$

then $x = cy$ where $|c| = 1$.

A fundamental result in phase retrieval involves the **complement property**.

DEFINITION 10.28. A family of vectors $\{\varphi_i\}_{i=1}^M$ in $\mathbb{H}^N$ has the **complement property** if whenever we choose $I \subset \{1, 2, \cdots, M\}$, at least one of the sets $\{\varphi_i\}_{i\in I}$ or $\{\varphi_i\}_{i\in I^c}$ spans $\mathbb{H}^N$.

Note that the complement property implies $M \geq 2N - 1$. For if $M \leq 2N - 2$ then we can choose $I = 1, 2, \cdots, N - 1$ and since the two induced subsets of our vectors each has only $N - 1$ vectors, neither can span $\mathbb{H}^N$.

The fundamental result here is due to Balan/Casazza/Edidin [**3**]:

THEOREM 10.29. *In $\mathbb{R}^N$, a family of vectors $\{\varphi_i\}_{i=1}^M$ does phase retrieval if and only if it has the complement property. Moreover, there is a dense set of families of vectors $\{\varphi_i\}_{i=1}^{2N-1}$ which do phase retrieval.*

In the complex case, [**3**] showed that a dense sent of families of $(4N-2)$-vectors does phase retrieval. Later, Bodmann [**9**] showed that phase retrieval can be done in the complex case with $(4N-4)$ vectors. This was then improved by Conca/Edidin/Hering/Vinzant [**42**].

THEOREM 10.30. *In $\mathbb{C}^N$, there are families (in fact a dense set of families) of vectors $\{\varphi_i\}_{i=1}^{4N-4}$ which do phase retrieval.*

Again, phase retrieval cannot be done with fewer vectors than $4N-4$.

Given a signal x in a Hilbert space, intensity measurements may also be thought of as norms of x under rank one projections. Here the spans of measurement vectors serve as the one dimensional range of the projections. In some applications, however, a signal must be reconstructed from the norms of higher dimensional components. In X-ray crystallography for example, such a problem arises with crystal twinning [**47**]. In this scenario, there exists a similar phase retrieval problem: given subspaces $\{W_n\}_{n=1}^M$ of an N-dimensional Hilbert space $\mathbb{H}^N$ and orthogonal projections $P_n : \mathbb{H}^N \to W_n$, can we recover any $x \in \mathbb{H}^N$ (up to a global phase factor) from the measurements $\{\|P_n x\|\}_{n=1}^M$? This problem was recently studied in [**6**] where the authors use semidefinite programming to develop a reconstruction algorithm for when the $\{W_n\}_{n=1}^M$ are equidimensional random subspaces. Most results using random intensity measurements require the cardinality of measurements to scale linearly with the dimension of the signal space along with an additional logarithmic factor [**22**], but this logarithmic factor was recently removed in [**21**]. Similarly, signal reconstruction from the norms of equidimensional random subspace components are possible with the cardinality of measurements scaling linearly with the dimension [**6**].

In [**17**] it was shown:

THEOREM 10.31. *Phase retrieval can be done on $\mathbb{R}^N$ with $2N-1$ orthogonal projections of arbitrary rank.*

This theorem raises an important question:

PROBLEM 10.32. Can phase retrieval be done on $\mathbb{R}^N$ (respectively, $\mathbb{C}^N$) with fewer than $2N-1$ (respectively, $4N-4$) projections? If so, what are the fewest number of projections needed in both the real and complex case?

References

[1] B. Alexeev, A. S. Bandeira, M. Fickus, D. G. Mixon, *Phase retrieval with polarization*, Available online: arXiv:1210.7752

[2] R. Balan, B. G. Bodmann, P. G. Casazza, and D. Edidin, *Painless reconstruction from magnitudes of frame coefficients*, J. Fourier Anal. Appl. **15** (2009), no. 4, 488–501, DOI 10.1007/s00041-009-9065-1. MR2549940 (2010m:42066)

[3] R. Balan, P. Casazza, and D. Edidin, *On signal reconstruction without phase*, Appl. Comput. Harmon. Anal. **20** (2006), no. 3, 345–356, DOI 10.1016/j.acha.2005.07.001. MR2224902 (2007b:94054)

[4] A. S. Bandeira, J. Cahill, D. G. Mixon, and A. A. Nelson, *Saving phase: injectivity and stability for phase retrieval*, Appl. Comput. Harmon. Anal. **37** (2014), no. 1, 106–125, DOI 10.1016/j.acha.2013.10.002. MR3202304

[5] A. S. Bandeira, M. Fickus, D. G. Mixon, and P. Wong, *The road to deterministic matrices with the restricted isometry property*, J. Fourier Anal. Appl. **19** (2013), no. 6, 1123–1149, DOI 10.1007/s00041-013-9293-2. MR3132908

[6] C. Bachoc and M. Ehler, *Signal reconstruction from the magnitude of subspace components*, IEEE Trans. Inform. Theory **61** (2015), no. 7, 4015–4027, DOI 10.1109/TIT.2015.2429634. MR3367817

[7] C. Becchetti and L. P. Ricotti. *Speech recognition theory and C++ implementation.* Wiley (1999).

[8] J. J. Benedetto and M. Fickus, *Finite normalized tight frames*, Adv. Comput. Math. **18** (2003), no. 2-4, 357–385, DOI 10.1023/A:1021323312367. Frames. MR1968126 (2004c:42059)

[9] B. Bodmann, *Stable phase retrieval with low-redundancy frames*, Prepriint.

[10] B. G. Bodmann and P. G. Casazza, *The road to equal-norm Parseval frames*, J. Funct. Anal. **258** (2010), no. 2, 397–420, DOI 10.1016/j.jfa.2009.08.015. MR2557942 (2010j:42060)

[11] B. Bodmann, P. G. Casazza, D. Edidin and R. Balan, *Frames for linear reconstruction without phase*, Preprint.

[12] R. H. Bates and D. Mnyama. *The status of practical Fourier phase retrieval*, in W. H. Hawkes, ed., Advances in Electronics and Electron Physics, 67:1-64, 1986.

[13] B. G. Bodmann and V. I. Paulsen, *Frames, graphs and erasures*, Linear Algebra Appl. **404** (2005), 118–146, DOI 10.1016/j.laa.2005.02.016. MR2149656 (2006a:42047)

[14] J. Bourgain, S. Dilworth, K. Ford, S. Konyagin, and D. Kutzarova, *Explicit constructions of RIP matrices and related problems*, Duke Math. J. **159** (2011), no. 1, 145–185, DOI 10.1215/00127094-1384809. MR2817651 (2012k:11010)

[15] J. Cahill and X. Chen, *A note on scalable frames*, Proceedings of the 10th International Conference on Sampling Theory and Applications, 93 - 96.

[16] J. Cahill and P. G. Casazza, *The Paulsen problem in operator theory*, Oper. Matrices **7** (2013), no. 1, 117–130, DOI 10.7153/oam-07-06. MR3076462

[17] J. Cahill, P. G. Casazza, J. Peterson and L. Woodland, **Phase retrieval by projections**, prepring.

[18] M. Fickus, D. G. Mixon, M. J. Poteet, and N. Strawn, *Constructing all self-adjoint matrices with prescribed spectrum and diagonal*, Adv. Comput. Math. **39** (2013), no. 3-4, 585–609, DOI 10.1007/s10444-013-9298-z. MR3116042

[19] J. Cahill, M. Fickus, D. G. Mixon, M. J. Poteet, and N. Strawn, *Constructing finite frames of a given spectrum and set of lengths*, Appl. Comput. Harmon. Anal. **35** (2013), no. 1, 52–73, DOI 10.1016/j.acha.2012.08.001. MR3053746

[20] E. J. Candès, Y. C. Eldar, T. Strohmer, and V. Voroninski, *Phase retrieval via matrix completion*, SIAM J. Imaging Sci. **6** (2013), no. 1, 199–225, DOI 10.1137/110848074. MR3032952

[21] E. J. Candès and X. Li, *Solving quadratic equations via PhaseLift when there are about as many equations as unknowns*, Found. Comput. Math. **14** (2014), no. 5, 1017–1026, DOI 10.1007/s10208-013-9162-z. MR3260258

[22] E. J. Candès, T. Strohmer, and V. Voroninski, *PhaseLift: exact and stable signal recovery from magnitude measurements via convex programming*, Comm. Pure Appl. Math. **66** (2013), no. 8, 1241–1274, DOI 10.1002/cpa.21432. MR3069958

[23] P. G. Casazza, *The art of frame theory*, Taiwanese J. Math. **4** (2000), no. 2, 129–201. MR1757401 (2001f:42046)

[24] P. G. Casazza, *Modern tools for Weyl-Heisenberg (Gabor) frame theory*, Advances in Imaging and Electron Physics **115** (2000) 1-127.

[25] P. G. Casazza, *Custom building finite frames*, Wavelets, frames and operator theory, Contemp. Math., vol. 345, Amer. Math. Soc., Providence, RI, 2004, pp. 61–86, DOI 10.1090/conm/345/06241. MR2066822 (2005f:42078)

[26] P. G. Casazza, *Consequences of the Marcus/Spielman/Srivastava solution to the Kadison-Singer Problem*, Preprint.

[27] P. G. Casazza, M. Fickus, D. G. Mixon, Y. Wang, and Z. Zhou, *Constructing tight fusion frames*, Appl. Comput. Harmon. Anal. **30** (2011), no. 2, 175–187, DOI 10.1016/j.acha.2010.05.002. MR2754774 (2012c:42069)

[28] P. G. Casazza, M. Fickus, J. C. Tremain, and E. Weber, *The Kadison-Singer problem in mathematics and engineering: a detailed account*, Operator theory, operator algebras, and applications, Contemp. Math., vol. 414, Amer. Math. Soc., Providence, RI, 2006, pp. 299–355, DOI 10.1090/conm/414/07820. MR2277219 (2007j:42016)

[29] P. G. Casazza, M. Fickus, J. Kovačević, M. T. Leon, and J. C. Tremain, *A physical interpretation of tight frames*, Harmonic analysis and applications, Appl. Numer. Harmon. Anal., Birkhäuser Boston, Boston, MA, 2006, pp. 51–76, DOI 10.1007/0-8176-4504-7_4. MR2249305 (2007d:42053)

[30] P. G. Casazza and J. Kovačević, *Equal-norm tight frames with erasures*, Adv. Comput. Math. **18** (2003), no. 2-4, 387–430, DOI 10.1023/A:1021349819855. Frames. MR1968127 (2004e:42046)

[31] P. G. Casazza and G. Kutyniok, *Frames of subspaces*, Wavelets, frames and operator theory, Contemp. Math., vol. 345, Amer. Math. Soc., Providence, RI, 2004, pp. 87–113, DOI 10.1090/conm/345/06242. MR2066823 (2005e:42090)
[32] *Finite frames*, Applied and Numerical Harmonic Analysis, Birkhäuser/Springer, New York, 2013. Theory and applications; Edited by Peter G. Casazza and Gitta Kutyniok. MR2964005
[33] P. G. Casazza, G. Kutyniok, and S. Li, *Fusion frames and distributed processing*, Appl. Comput. Harmon. Anal. **25** (2008), no. 1, 114–132, DOI 10.1016/j.acha.2007.10.001. MR2419707 (2009d:42094)
[34] P. G. Casazza and G. Kutyniok, *A generalization of Gram-Schmidt orthogonalization generating all Parseval frames*, Adv. Comput. Math. **27** (2007), no. 1, 65–78, DOI 10.1007/s10444-005-7478-1. MR2317921 (2008f:42031)
[35] P. G. Casazza and M. T. Leon, *Existence and construction of finite tight frames*, J. Concr. Appl. Math. **4** (2006), no. 3, 277–289. MR2224599 (2006k:42062)
[36] P. G. Casazza and M. T. Leon, *Existence and construction of finite frames with a given frame operator*, Int. J. Pure Appl. Math. **63** (2010), no. 2, 149–157. MR2683591 (2011h:42042)
[37] P.G. Casazza, D. Redmond and J.C. Tremain, *Real equiangular frames*, Preprint.
[38] P. G. Casazza and J. C. Tremain, *The Kadison-Singer problem in mathematics and engineering*, Proc. Natl. Acad. Sci. USA **103** (2006), no. 7, 2032–2039 (electronic), DOI 10.1073/pnas.0507888103. MR2204073 (2006j:46074)
[39] P. G. Casazza and L. Woodland, *The fundamentals of spectral tetris frame constructions*, Preprint.
[40] X. Chen, G. Kutyniok, K. A. Okoudjou, F. Philipp, and R. Wang, *Measures of scalability*, IEEE Trans. Inform. Theory **61** (2015), no. 8, 4410–4423, DOI 10.1109/TIT.2015.2441071. MR3372361
[41] O. Christensen, *An introduction to frames and Riesz bases*, Applied and Numerical Harmonic Analysis, Birkhäuser Boston, Inc., Boston, MA, 2003. MR1946982 (2003k:42001)
[42] A. Conco, D. Edidin, M. Hering, and C. Vinzant, *An algebraic characterization of injectivity in phase retrieval*, arXiv:1312:0158v1.
[43] M. S. Craig and R. L. Genter, *Geophone array formation and semblance evaluation*, Geophysics **71** (2006), 1–8.
[44] X. R. Dai and Q. Sun, *The abc-problem for Gabor Systems*, arXiv:1304.7750v1.
[45] R. A. DeVore, *Deterministic constructions of compressed sensing matrices*, J. Complexity **23** (2007), no. 4-6, 918–925, DOI 10.1016/j.jco.2007.04.002. MR2371999 (2008k:94014)
[46] L. Demanet and P. Hand, *Stable optimizationless recovery from phaseless linear measurements*, J. Fourier Anal. Appl. **20** (2014), no. 1, 199–221, DOI 10.1007/s00041-013-9305-2. MR3180894
[47] J. Drenth, *Principles of protein x-ray crystallography*, Springer, 2010.
[48] R. J. Duffin and A. C. Schaeffer, *A class of nonharmonic Fourier series*, Trans. Amer. Math. Soc. **72** (1952), 341–366. MR0047179 (13,839a)
[49] Y. C. Eldar and S. Mendelson, *Phase retrieval: stability and recovery guarantees*, Appl. Comput. Harmon. Anal. **36** (2014), no. 3, 473–494, DOI 10.1016/j.acha.2013.08.003. MR3175089
[50] J. Jasper, D. G. Mixon, and M. Fickus, *Kirkman equiangular tight frames and codes*, IEEE Trans. Inform. Theory **60** (2014), no. 1, 170–181, DOI 10.1109/TIT.2013.2285565. MR3150919
[51] J. R. Fienup. *Reconstruction of an object from the modulus of its fourier transform*, Optics Letters, **3** (1978), 27-29.
[52] J. R. Fienup. *Phase retrieval algorithms: A comparison*, Applied Optics, **21** (15) (1982), 2758-2768.
[53] *Frame Research Center*, http://www.framerc.org/
[54] D. Gabor, *Theory of Communications*, J. Inst. Elec. Engrg. **93** (1946) 429-457.
[55] K. H. Gröchenig, *Foundations of time-frequency analysis*, Birkhäuser, Boston, 2000.
[56] S. Foucart and H. Rauhut, *A mathematical introduction to compressive sensing*, Applied and Numerical Harmonic Analysis, Birkhäuser/Springer, New York, 2013. MR3100033
[57] R. W. Heath Jr., T. Strohmer, and A. J. Paulraj, *On quasi-orthogonal signatures for CDMA systems*, IEEE Trans. Inform. Theory **52** (2006), no. 3, 1217–1226, DOI 10.1109/TIT.2005.864469. MR2238086 (2007a:94232)
[58] J. Haantjes, *Equilateral point-sets in elliptic two- and three-dimensional spaces*, Nieuw Arch. Wiskunde (2) **22** (1948), 355–362. MR0023530 (9,369c)

[59] D. Han, K. Kornelson, D. Larson, and E. Weber, *Frames for undergraduates*, Student Mathematical Library, vol. 40, American Mathematical Society, Providence, RI, 2007. MR2367342 (2010e:42044)

[60] D. Han and D. R. Larson, *Frames, bases and group representations*, Mem. Amer. Math. Soc. **147** (2000), no. 697, x+94, DOI 10.1090/memo/0697. MR1686653 (2001a:47013)

[61] R. B. Holmes and V. I. Paulsen, *Optimal frames for erasures*, Linear Algebra Appl. **377** (2004), 31–51, DOI 10.1016/j.laa.2003.07.012. MR2021601 (2004j:42028)

[62] S. S. Iyengar and R. R. Brooks, eds., *Distributed Sensor Networks*, Chapman & Hall/CRC, Baton Rouge, 2005.

[63] R. V. Kadison and I. M. Singer, *Extensions of pure states*, Amer. J. Math. **81** (1959), 383–400. MR0123922 (23 #A1243)

[64] J. Kovačevićand A. Chebira, *An introduction to frames*, in *Foundations and Trends in Signal Processing* (2008) NOW publishers.

[65] G. Kutyniok, K. A. Okoudjou, and F. Philipp, *Scalable frames and convex geometry*, Operator methods in wavelets, tilings, and frames, Contemp. Math., vol. 626, Amer. Math. Soc., Providence, RI, 2014, pp. 19–32, DOI 10.1090/conm/626/12507. MR3329091

[66] P. W. H. Lemmens and J. J. Seidel, *Equiangular lines*, J. Algebra **24** (1973), 494–512. MR0307969 (46 #7084)

[67] S. Li, *On general frame decompositions*, Numer. Funct. Anal. Optim. **16** (1995), no. 9-10, 1181–1191, DOI 10.1080/01630569508816668. MR1374971 (97b:42055)

[68] J. H. van Lint and J. J. Seidel, *Equiangular point sets in elliptic geometry*, Proc. Nederl. Akad. Wetensch. Series A **69** (1966) 335-348.

[69] A. W. Marcus, D. A. Spielman, and N. Srivastava, *Interlacing families II: Mixed characteristic polynomials and the Kadison-Singer problem*, Ann. of Math. (2) **182** (2015), no. 1, 327–350, DOI 10.4007/annals.2015.182.1.8. MR3374963

[70] J. G. Proakis, J. R. Deller and J. H. L. Hansen. *Discrete-Time processing of speech signals.* IEEE Press (2000).

[71] L. Rabiner and B. H. Juang. *Fundamentals of speech recognition.* Prentice Hall Signal Processing Series (1993).

[72] J. M. Renes, R. Blume-Kohout, A. J. Scott, and C. M. Caves, *Symmetric informationally complete quantum measurements*, J. Math. Phys. **45** (2004), no. 6, 2171–2180, DOI 10.1063/1.1737053. MR2059685 (2004m:81043)

[73] C. J. Rozell and D. H. Johnson, *Analyzing the robustness of redundant population codes in sonsory and feature extraction systems*, Neurocomputing **69** (2006), 1215–1218.

[74] M. Rudelson and R. Vershynin, *On sparse reconstruction from Fourier and Gaussian measurements*, Comm. Pure Appl. Math. **61** (2008), no. 8, 1025–1045, DOI 10.1002/cpa.20227. MR2417886 (2009e:94034)

[75] W. Rudin, *Functional analysis*, 2nd ed., International Series in Pure and Applied Mathematics, McGraw-Hill, Inc., New York, 1991. MR1157815 (92k:46001)

[76] T. Strohmer and R. W. Heath Jr., *Grassmannian frames with applications to coding and communication*, Appl. Comput. Harmon. Anal. **14** (2003), no. 3, 257–275, DOI 10.1016/S1063-5203(03)00023-X. MR1984549 (2004d:42053)

[77] M. A. Sustik, J. A. Tropp, I. S. Dhillon, and R. W. Heath Jr., *On the existence of equiangular tight frames*, Linear Algebra Appl. **426** (2007), no. 2-3, 619–635, DOI 10.1016/j.laa.2007.05.043. MR2350682 (2008f:15066)

[78] J. C. Tremain, *Concrete Constructions of Equiangular Line Sets*, in preparation.

[79] L. R. Welch, *Lower bounds on the maximum cross-correlation of signals*, IEEE Trans. Inform. Theory **20** (1974) 397-399.

DEPARTMENT OF MATHEMATICS, UNIVERSITY OF MISSOURI, COLUMBIA, MISSOURI 65211-4100
E-mail address: `casazzap@missouri.edu`

DEPARTMENT OF MATHEMATICS, UNIVERSITY OF MISSOURI, COLUMBIA, MISSOURI 65211-4100
E-mail address: `rglz82@mail.missouri.edu`

Proceedings of Symposia in Applied Mathematics
Volume **73**, 2016
http://dx.doi.org/10.1090/psapm/073/00628

Unit norm tight frames in finite-dimensional spaces

Dustin G. Mixon

Abstract. There are several problems in signal processing which motivate the study of unit norm tight frames in finite-dimensional spaces. We discuss some of these problems and then observe various constructions and properties that have been identified in the last decade. Special attention is given to the frame potential, eigensteps, and equiangular tight frames.

1. Introduction

A **frame** is a countable collection of vectors $\Phi = \{\varphi_i\}_{i\in I}$ in a separable Hilbert space $\mathcal{H}$ with the property that there exist **frame bounds** $0 < A \leq B < \infty$ such that

$$A\|x\|^2 \leq \sum_{i\in I} |\langle x, \varphi_i\rangle|^2 \leq B\|x\|^2 \qquad \forall x \in \mathcal{H}.$$

Here, $\|\cdot\|$ denotes the norm induced by the inner product $\langle\cdot,\cdot\rangle$ corresponding to $\mathcal{H}$. Intuitively, a frame allows the mapping $x \mapsto \{\langle x, \varphi_i\rangle\}_{i\in I}$ to capture the energy of any $x \in \mathcal{H}$, and this property enables any such x to be reconstructed with the help of some **dual frame**. In particular, for every frame $\Phi = \{\varphi_i\}_{i\in I}$ in $\mathcal{H}$, there exists a dual frame $\Psi = \{\psi_i\}_{i\in I}$ in $\mathcal{H}$ such that

$$x = \sum_{i\in I} \langle x, \varphi_i\rangle \psi_i \qquad \forall x \in \mathcal{H}.$$

For example, any orthonormal basis Φ of $\mathcal{H}$ is a frame with frame bounds $A = B = 1$, and a corresponding dual frame is simply $\Psi = \Phi$. In general, we say a frame is **tight** if $A = B$. In this case, choosing the dual frame $\Psi = (1/A)\Phi$ allows for a particularly painless reconstruction formula:

$$x = \frac{1}{A}\sum_{i\in I} \langle x, \varphi_i\rangle \varphi_i \qquad \forall x \in \mathcal{H}.$$

2010 *Mathematics Subject Classification.* Primary 42C15.

Key words and phrases. Unit norm tight frames, frame potential, eigensteps, equiangular tight frames.

The author thanks Matt Fickus and Pete Casazza for introducing him to frames with contagious enthusiasm. The author also thanks the anonymous referee for thorough and thoughtful suggestions. This work was supported by an AFOSR Young Investigator Research Program award, NSF Grant No. DMS-1321779, and AFOSR Grant No. F4FGA05076J002. The views expressed in this chapter are those of the author and do not reflect the official policy or position of the United States Air Force, Department of Defense, or the U.S. Government.

Finally, we say a frame Φ is **unit norm** if each of the frame elements satisfies $\|\varphi_i\| = 1$. In this way, a unit norm tight frame (UNTF) is a natural generalization of an orthonormal basis, and as we will see, there is an assortment of UNTFs which are not orthonormal bases.

This chapter considers the case where $\mathcal{H}$ has finite dimension, meaning $\mathcal{H}$ is isomorphic to either $\mathbb{R}^M$ or $\mathbb{C}^M$ for some positive integer M. In this setting, the finitude of the upper frame bound together with the unit-norm property forces the number of frame elements to be finite, say N, and it is convenient to express the frame as an $M \times N$ matrix whose columns are the frame elements. By abuse of notation, we will denote this matrix by $\Phi = [\varphi_1 \cdots \varphi_N]$. The frame bounds are then lower and upper bounds on the quantity

$$\frac{\|\Phi^* x\|^2}{\|x\|^2} = \frac{\langle \Phi^* x, \Phi^* x\rangle}{\|x\|^2} = \frac{\langle x, \Phi\Phi^* x\rangle}{\|x\|^2}, \qquad x \neq 0,$$

which is minimized and maximized at the smallest and largest eigenvalues of $\Phi\Phi^*$, respectively. As such, the optimal frame bounds are precisely these extreme eigenvalues. At this point, we note that $A > 0$ precisely when the frame elements span, and so any frame of N elements in M dimensions must satisfy $N \geq M$. Also, it is clear that Φ is tight precisely when $\Phi\Phi^*$ is a multiple A of the $M \times M$ identity matrix I. Equivalently, the rows of Φ are orthogonal, each with norm $\sqrt{A}$. In addition, we may easily determine the necessary value of A by appealing to the cyclic property of the trace:

$$MA = \mathrm{Tr}[AI] = \mathrm{Tr}[\Phi\Phi^*] = \mathrm{Tr}[\Phi^*\Phi] = N,$$

where the last step follows from the fact that each frame element has norm 1. Overall, a UNTF in finite-dimensional space is an $M \times N$ matrix such that

- the rows are orthogonal,
- each row has norm $\sqrt{N/M}$, and
- each column has norm 1.

As one might expect, the simultaneous conditions on the rows and columns make the construction of UNTFs somewhat challenging.

In the next section, we will provide a series of applications that motivate the study of UNTFs in finite-dimensional spaces. In Section 3, we will then provide a list of examples to build up an intuition for what UNTFs look like. This will prompt two main questions. First, how do UNTFs correspond to the notion of equidistribution on the unit sphere? This is answered in Section 4, in which we discuss the frame potential. Second, how might one construct all possible UNTFs? This is answered in Section 5, where we discuss the theory of eigensteps. We conclude in Section 6 by briefly reviewing a particular class of UNTFs called equiangular tight frames.

2. Motivating applications

In this section, we discuss several applications of UNTFs to problems in signal processing. This will motivate the rest of this chapter, which considers various constructions and properties of UNTFs.

2.1. Robustness to noise or erasures. Suppose Alice has a message that she wishes to transmit to Bob, but the channel through which she can send her message will invariably corrupt the transmission. The goal is to develop an encoding–decoding scheme that will ensure Bob will receive the message.

To model this scenario, suppose the channel acts by adding random noise to each entry of the transmitted signal. For example, perhaps a sequence of i.i.d. $N(0, \sigma^2)$ random variables is added. Then Alice might redundantly encode her signal $x \in \mathbb{R}^M$ using a frame, meaning $\Phi^* x \in \mathbb{R}^N$ is sent over the channel. What Bob then receives has the form $y = \Phi^* x + e$, where e denotes the channel's random noise vector. Considering the isotropy of the Gaussian distribution, Bob can easily determine the maximum likelihood estimate $\hat{x}$ of x by solving a least squares problem:

$$\hat{x} := \arg\min_z \|\Phi^* z - y\|.$$

If we take Ψ to be the pseudoinverse of Φ, i.e., $\Psi = (\Phi\Phi^*)^{-1}\Phi$, then $\hat{x} = \Psi y$. Notice that Ψ is a dual frame of Φ since $\Psi\Phi^* = (\Phi\Phi^*)^{-1}\Phi\Phi^* = I$; in fact, this Ψ is the **canonical dual frame** of Φ.

Now that an optimal dual frame (decoder) has been identified for each frame (encoder), we will optimize the encoder frame. For this, we measure performance by mean-squared error:

$$\begin{aligned}\text{MSE} &:= \mathbb{E}\|\hat{x} - x\|^2 \\ &= \mathbb{E}\|\Psi(\Phi^* x + e) - x\|^2 \\ &= \mathbb{E}\|\Psi e\|^2.\end{aligned}$$

Intuitively, this is a measure of average distortion, and we wish to minimize this quantity. To this end, it has been shown that of all unit norm frames Φ, the minimizers of MSE are UNTFs; see Theorem 3.1 in [**26**].

For another model of the channel between Alice and Bob, suppose that after noise is added to the signal, some arbitrary entry of the signal is set to zero. This is called an erasure channel, as depicted in the following:

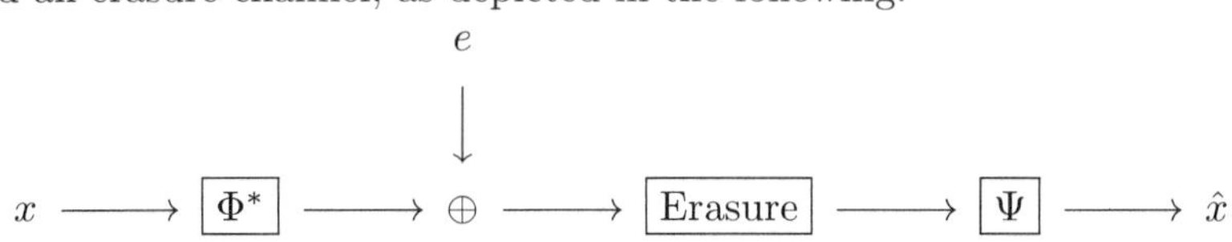

Notice that in this setup, Bob intends to blindly decode by applying the canonical dual frame Ψ of Φ regardless of what erasure he may perceive. Here, if we let D_n denote the diagonal matrix of all 1s, but with a 0 at the nth diagonal entry, then the following quantity measures performance in the event that the nth entry is erased:

$$\text{MSE}_n := \mathbb{E}\|\hat{x} - x\|^2 = \mathbb{E}\|\Psi D_n(\Phi^* x + e) - x\|^2.$$

At this point, we can consider two extreme models for how an entry is selected for erasing. If an entry is selected at random for erasure, then we would consider the average MSE over all n. On the other hand, if an adversary decides which entry to erase after observing x, then we would maximize MSE over all n. Interestingly, both average- and worst-case MSE are minimized when Φ is a UNTF; see Theorem 4.4 in [**26**].

2.2. Fingerprinting to defeat piracy. Suppose you are a famous recording artist, and you are about to release a new single. Since you are aware that media piracy is a cause for concern, you decide to send a slightly different copy of your song file to each recipient—that way, if a particular version of your song becomes popular on the internet, you will know which recipient of your song was the original culprit. If $x \in \mathbb{R}^M$ denotes your song file, then you make different copies of the file by adding a personalized fingerprint to the file, namely $x+\varphi_n$, where $\{\varphi_n\}_{n=1}^N$ is the total collection of fingerprints you will use.

Unfortunately, some of the recipients might conspire to produce a forgery of your song file. Suppose $\mathcal{K}\subseteq\{1,\ldots,N\}$ denotes the indices of these conspiring culprits. Then they might attempt to forge your file by making a noisy linear combination of their personalized fingerprinted versions:

$$\hat{x}=\sum_{n\in\mathcal{K}}c_n(x+\varphi_n)+e,\qquad \sum_{n\in\mathcal{K}}c_n=1,\qquad e\sim N(0,I).$$

Here, the purpose of the noise is to "cover their tracks" so as to keep you from determining $\mathcal{K}$.

Suppose a forgery $\hat{x}$ surfaces on the internet. Then its difference from the true file has the form

$$\hat{x}-x=\sum_{n\in\mathcal{K}}c_n\varphi_n+e.$$

To test whether the nth recipient is a possible culprit, you can take the inner product $\langle\hat{x}-x,\varphi_n\rangle$. Then the largest such inner product determines the most guilty-looking recipient. If this recipient is a member of $\mathcal{K}$, then you can interrogate him to determine the remainder of the culprits. Knowing this, the culprits might select the scalars $\{c_n\}_{n\in\mathcal{K}}$ in their linear combination so as to frame an innocent recipient. Theorem 7 in **[33]** shows that this is actually impossible provided the fingerprints form a certain type of UNTF, called an **equiangular** UNTF; this is a UNTF with the additional property that there is some $c\geq 0$ such that $|\langle\varphi_i,\varphi_j\rangle|=c$ for every $i,j\in\{1,\ldots,N\}$ with $i\neq j$. In particular, the probability of accidentally accusing the most guilty-looking innocent recipient is always small, regardless of how the culprits choose $\{c_n\}_{n\in\mathcal{K}}$.

2.3. Sparse decomposition. Consider a radar antenna that transmits a ping $\varphi\in\mathbb{C}^M$, which is then re-radiated from various aircraft in a region of interest. Then the radar antenna can listen for the superposition of these different echoes, each being a different translation and modulation of the original ping (corresponding to a time delay due to distance and a Doppler shift due to velocity). As such, one can consider a frame whose frame elements are the original ping φ along with all translates and modulates of the ping: $\Phi=\{T^aE^b\varphi\}_{a,b\in\mathbb{Z}/M\mathbb{Z}}$; here, we use cyclic translates and modulates for simplicity:

$$\begin{aligned}(Tx)(m)&:=x(m-1),\\(Ex)(m)&:=e^{2\pi im/M}x(m),\end{aligned}$$

which is reasonable if we zero-pad the original ping appropriately. Such a frame Φ is called a **Gabor frame**, and provided the ping φ has unit norm, this frame will necessarily be a UNTF. With this notation, each target corresponds to a different pair (a,b) according to its distance and velocity, and we receive the superposition

$$y=\sum_{(a,b)\in\mathcal{K}}x_{(a,b)}T^aE^b\varphi=\Phi x,$$

where $x\in\mathbb{C}^{M^2}$ has $|\mathcal{K}|$ nonzero entries. The purpose of radar is to determine information about the various objects in the region of interest, so we would like to determine x from y, since x encodes the distances and velocities of the $|\mathcal{K}|$ objects that echoed the ping. However, $y = \Phi x$ is a severely underdetermined linear system, and so to do this, we would need to exploit the fact that x has few nonzero entries.

This is an instance of a more general problem called **sparse decomposition**, where we seek to express y as a sparse combination of columns of Φ. This problem of solving a linear system with a sparsity prior has received a lot of attention recently under the name **compressed sensing**, and today, we have several algorithms of reconstructing x from y provided x is sufficiently sparse and the columns of Φ are sufficiently incoherent, meaning no two columns of Φ look alike. More explicitly, the various algorithms tend to perform better when the **worst-case coherence**

$$\mu(\Phi) := \max_{\substack{i,j\in\{1,\ldots,N\}\\ i\neq j}} |\langle\varphi_i, \varphi_j\rangle|$$

is small. For the application of radar, this suggests the use of a ping that looks very little like its various translates and modulates, as in [**28**]. As established in [**43**], the general problem of sparse decomposition is particularly solvable when Φ is a UNTF with small worst-case coherence.

3. Examples

The previous section illustrated how UNTFs can be used in a variety of signal processing applications. Recognizing the importance of UNTFs, this leaves one longing for a nontrivial example. From the introduction, the reader is already aware that every orthonormal basis is an example of a UNTF. This section offers an extensive list of additional examples for the reader's satisfaction. We start with UNTFs in two-dimensional real space:

EXAMPLE 3.1. Pick $N \geq 3$ and consider the Nth roots of unity. These are N complex numbers, each of unit modulus, and we can view each as a vector in $\mathbb{R}^2$ whose coordinates are the real and imaginary parts. Then the resulting N vectors form a unit norm tight frame in $\mathbb{R}^2$.

This example is intuitively pleasing because it suggests that UNTFs satisfy some notion of equidistribution on the unit sphere in $\mathbb{R}^M$, which is discussed at length in the next section. It also leads one to wonder whether there are any other UNTFs in $\mathbb{R}^2$. This is answered in the following theorem:

THEOREM 3.2 (Theorem 2.7 in [**26**]). *The vectors $\{\varphi_n\}_{n=1}^N$ in $\mathbb{R}^2$ form a unit norm tight frame if and only if the corresponding complex numbers $\{z_n\}_{n=1}^N$ satisfy*

$$\sum_{n=1}^N z_n^2 = 0.$$

The proof of this theorem amounts to a straightforward exercise in trigonometry. Not only can we use this result to verify the previous example, it also gives us an intuition for the entire set of $2 \times N$ UNTFs for any fixed $N \geq 2$. Indeed, we can add $\{z_n^2\}_{n=1}^N$ head-to-tail as if they are unit vectors, and then the above characterization gives that the sum goes back to the origin, meaning the unit vectors form a closed chain. One can imagine deforming this chain as if it were a necklace made of sticks, and by the above theorem, the set of all possible configurations corresponds to the set of all possible UNTFs in $\mathbb{R}^2$. The fact that two unit vectors sum to zero precisely when they are negatives of each other corresponds to the fact that orthonormal bases are the only UNTFs in $\mathbb{R}^2$ with $N = 2$. Also, three unit

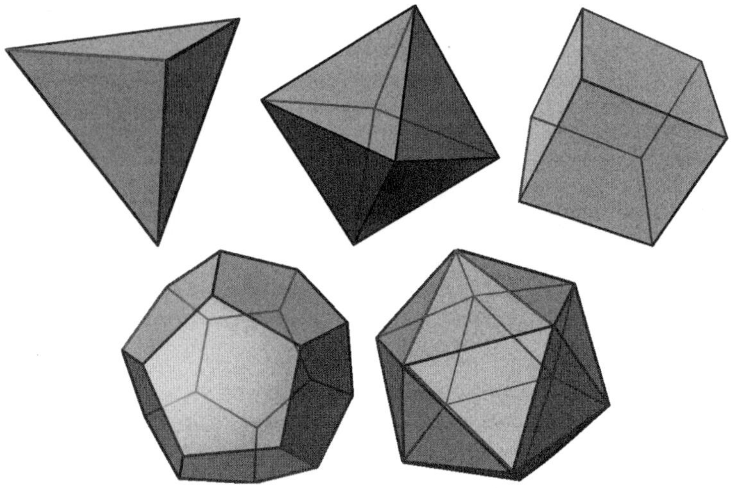

FIGURE 1. Platonic solids (from [**35**]). When circumscribed by the unit sphere, each vertex is a vector of unit norm, and the collection of vertices forms a unit norm tight frame in $\mathbb{R}^3$. The fact that these vertices are well-distributed on the sphere suggests that UNTFs might satisfy some notion of equidistribution, which is investigated further in Section 4.

vectors sum to zero precisely when they form the legs of an equilateral triangle, indicating that the cube roots of unity form the only UNTF of 3 vectors in $\mathbb{R}^2$ up to rotation and negation; for obvious reasons, this is called the **Mercedes-Benz frame**. When $N \geq 4$, the necklace of sticks has a lot more freedom. For example, when $N = 4$, unit vectors sum to zero precisely when they form the legs of a rhombus, and intuitively, we can push or pull at opposite corners to deform the rhombus into a continuous ensemble of rhombi, which correspond to a continuous ensemble of distinct UNTFs. Intuitively, increasing N produces additional degrees of freedom, leading to even more UNTFs.

Now that we have a complete characterization of UNTFs in $\mathbb{R}^2$, we continue by considering examples in $\mathbb{R}^3$:

EXAMPLE 3.3. Consider a Platonic solid which is circumscribed by the unit sphere in $\mathbb{R}^3$ (for examples, see Figure 1). Then the vertices of this solid form a unit norm tight frame.

This example was verified in [**19**], and it can be viewed as a 3-dimensional analog of the roots of unity, again exhibiting an intuitive notion of equidistribution on the sphere. In general, if vectors exhibit enough symmetry, you can expect them to form a UNTF. The following result makes this more precise:

THEOREM 3.4 (Theorem 6.3 in [**44**]). *Let $\mathcal{H}$ denote either $\mathbb{R}^M$ or $\mathbb{C}^M$, and let G be a finite group of norm-preserving linear operators over $\mathcal{H}$. Suppose G is irreducible, that is, $\{U\varphi\}_{U\in G}$ spans $\mathcal{H}$ for every $\varphi \in \mathcal{H}$ of unit norm. Then every such spanning set is a unit norm tight frame.*

The symmetry group of each platonic solid is irreducible, as is the (dihedral) symmetry group of the Nth roots of unity, and so the fact that these correspond to UNTFs is a consequence of this result. In general, a finite group need not be irreducible for there to exist a vector whose orbit under the group forms a UNTF:

EXAMPLE 3.5. Consider the group generated by a diagonal matrix whose diagonal entries are distinct Nth roots of unity. Then the orbit of the first (or any) identity basis fails to span, whereas the orbit of the normalized all-ones vector forms a unit norm tight frame.

For instance, denoting $\omega = e^{2\pi i/7}$, we might have

$$D = \begin{bmatrix} \omega^3 & 0 & 0 \\ 0 & \omega & 0 \\ 0 & 0 & \omega^6 \end{bmatrix}, \qquad \Phi = \frac{1}{\sqrt{3}} \begin{bmatrix} 1 & \omega^3 & \omega^6 & \omega^2 & \omega^5 & \omega & \omega^4 \\ 1 & \omega & \omega^2 & \omega^2 & \omega^4 & \omega^5 & \omega^6 \\ 1 & \omega^6 & \omega^5 & \omega^4 & \omega^3 & \omega^2 & \omega \end{bmatrix}.$$

Notice that the orbit of the normalized all-ones vector under such a group will always produce a matrix whose rows come from the discrete Fourier transform matrix (suitably scaled). With this identification, it is easy to verify that such matrices, called **harmonic frames**, are necessarily UNTFs. The construction of harmonic frames establishes that a complex UNTF exists for every pair (M, N) such that $M \leq N$. For the real case, one can leverage the real and imaginary parts of a harmonic frame [**49**]. The following example gives another general construction of real UNTFs:

EXAMPLE 3.6. Given (M, N) such that $N \geq 2M$, construct each row of an $M \times N$ UNTF using a greedy process known as **Spectral Tetris**[1]. In particular, design the matrix one row at a time, each time putting as much energy as possible on the left. For example, $(M, N) = (3, 7)$ corresponds to the following UNTF:

$$\begin{bmatrix} 1 & 1 & \frac{1}{\sqrt{6}} & \frac{1}{\sqrt{6}} & 0 & 0 & 0 \\ 0 & 0 & \frac{\sqrt{5}}{\sqrt{6}} & -\frac{\sqrt{5}}{\sqrt{6}} & \frac{\sqrt{2}}{\sqrt{6}} & \frac{\sqrt{2}}{\sqrt{6}} & 0 \\ 0 & 0 & 0 & 0 & \frac{\sqrt{2}}{\sqrt{3}} & -\frac{\sqrt{2}}{\sqrt{3}} & 1 \end{bmatrix}.$$

To get this, we first note that each entry must have size at most 1 (by the unit-norm constraint), and so we populate the first two entries of the first row with 1s accordingly. Since each row must have norm $\sqrt{7/3}$, we may not put a third 1 in the first row. We also need to allow for the second row to be orthogonal to the first, and so we evenly distribute the remaining energy across the next two entries. For the second row, the unit-norm constraint forces the first two entries to be 0, and then the orthogonality constraint (along with the greedy objective of putting as much energy on the left as possible) forces the next two entries to be $\pm\sqrt{5/6}$. The energy remaining for the second row is less than 1, and so we evenly distribute it over the next two entries. Finally, the last row is forced to have 0s in the first four entries by the unit-norm constraint, and then the next two entries become $\pm\sqrt{2/3}$, leaving a unit of energy that we put in the last entry.

Spectral Tetris is guaranteed to produce a UNTF provided $N \geq 2M$ [**13**], and the generated UNTFs necessarily enjoy two important properties. First, the greedy method ensures that each row has at most $N/M + 2$ nonzero entries, meaning the entire matrix is particularly sparse, having $O(N)$ nonzero entries, thereby allowing

[1]Spectral Tetris was named after the tile-matching puzzle video game *Tetris* because each frame element can be viewed as a localized contribution to the spectrum of $\Phi\Phi^*$, which we want to be flat so that Φ is tight.

for efficient matrix-vector multiplication. In fact, the UNTFs which come from Spectral Tetris are known to be the sparsest possible [**14**]. The second important property to note is that columns which are far apart tend to have disjoint support, and so they are orthogonal to each other; this is useful to the construction of a generalization of UNTFs called **tight fusion frames** or tight "frames of subspaces" [**13**].

Having observed several examples of UNTFs, we are led to two natural questions. First, is there a notion of equidistribution on the unit sphere that UNTFs necessarily satisfy? We have seen several examples of UNTFs for which this seems plausible, but then the fact that orthonormal bases are also examples of UNTFs seems to suggest otherwise. The next section will illustrate that, indeed, there is a natural notion of equidistribution which completely characterizes UNTFs, even despite orthonormal bases. Second, Theorem 3.2 corresponds to a method of constructing all possible UNTFs in $\mathbb{R}^2$, which leads one to wonder: Is there a more general construction available for $\mathbb{R}^M$ and $\mathbb{C}^M$? Indeed there is, as we detail in Section 5.

4. The frame potential

Imagine a collection of N electrons which are free to move within a spherical shell. By Coulomb's law, each electron imposes a force on each of the other $N-1$ electrons, and the magnitude of this force is inversely proportional to the square of the distance between the two electrons. As such, if all of the electrons are initialized in a small region of the sphere, they will push each other away and eventually settle into some arrangement of positions that attempts to minimize potential energy. Of course, the electrons may fail to converge to an arrangement of minimal potential because their pursuit is greedy, thereby leading to a local minimizer, which we will say exhibits **equilibrium** under Coulomb's force. Still, there exists a global minimizer by the extreme value theorem, which we will say exhibits **equidistribution** under Coulomb's force. Finding equidistributed arrangements under Coulomb's force is known as the Thomson problem [**42**], and is a special case of Smale's 7th problem [**38**].

With sufficient imagination, one may conjure any number of force laws and observe which arrangements on the sphere minimize potential energy. For our purposes, it is convenient to restrict one's imagination so that the force that $y \in \mathbb{R}^M$ applies to $x \in \mathbb{R}^M$ has direction $(x-y)/\|x-y\|$ and magnitude determined by $\|x-y\|$:

$$F(x,y) := f(\|x-y\|)\frac{x-y}{\|x-y\|}.$$

(We allow $f(\cdot)$ to take negative outputs, as is the case for gravitation.) The potential energy corresponding to points at x and y is then given by

$$\begin{aligned} P(x,y) &= -\int_0^1 F(tx+(1-t)y,y)\cdot(x-y)dt \\ &= -\int_0^{\|x-y\|} f(u)du, \end{aligned}$$

meaning $P(x,y) = p(\|x-y\|)$, where $p(\cdot)$ satisfies $p(0)=0$ and $p'(r) = -f(r)$. Given a force, which in turn determines a potential, we seek arrangements of points

$\{\varphi_n\}_{n=1}^N$ on the sphere which minimize the total potential energy:

$$\sum_{n=1}^{N}\sum_{\substack{n'=1\\n'\neq n}}^{N} P(\varphi_n,\varphi_{n'}). \tag{4.1}$$

In pursuit of global minimizers, we first consider the arrangements which satisfy the Lagrange equations (the "critical points"). For the class of forces we are considering, Theorem 4.6 in [**4**] gives that these arrangements necessarily have the property that each φ_n is a scalar multiple of

$$\sum_{\substack{n'=1\\n'\neq n}}^{N} F(\varphi_n,\varphi_{n'}). \tag{4.2}$$

We note that due to their symmetries, the vertices of each Platonic solid satisfies this condition, regardless of f, and so one is compelled to find a particular f for which every UNTF is equidistributed. If there were such an f, it must also allow orthonormal bases to be equidistributed, which is certainly not the case for Coulomb's force. However, we do have $\Phi\Phi^* = (N/M)I$ for any UNTF Φ, which implies that

$$\left(\frac{N}{M}-1\right)\varphi_n = \sum_{\substack{n'=1\\n'\neq n}}^{N} \langle\varphi_n,\varphi_{n'}\rangle\varphi_{n'}. \tag{4.3}$$

As such, we seek a force F that allows us to use (4.3) to show that (4.2) is a constant multiple of φ_n. To this end, we define the **frame force** to be

$$\mathrm{FF}(x,y) := \langle x,y\rangle(x-y).$$

Notice that this choice of force takes $f(r) = r - r^3/2$. Then we can use (4.3) to get

$$\begin{aligned}\sum_{\substack{n'=1\\n'\neq n}}^{N} \mathrm{FF}(\varphi_n,\varphi_{n'}) &= \sum_{\substack{n'=1\\n'\neq n}}^{N} \langle\varphi_n,\varphi_{n'}\rangle(\varphi_n-\varphi_{n'})\\ &= \sum_{\substack{n'=1\\n'\neq n}}^{N} \langle\varphi_n,\varphi_{n'}\rangle\varphi_n - \sum_{\substack{n'=1\\n'\neq n}}^{N} \langle\varphi_n,\varphi_{n'}\rangle\varphi_{n'}\\ &= \left(\sum_{\substack{n'=1\\n'\neq n}}^{N} \langle\varphi_n,\varphi_{n'}\rangle - \frac{N}{M}+1\right)\varphi_n,\end{aligned}$$

which is a scalar multiple of φ_n. As such, UNTFs satisfy the Lagrange equations for the frame force, as is necessary for equidistribution.

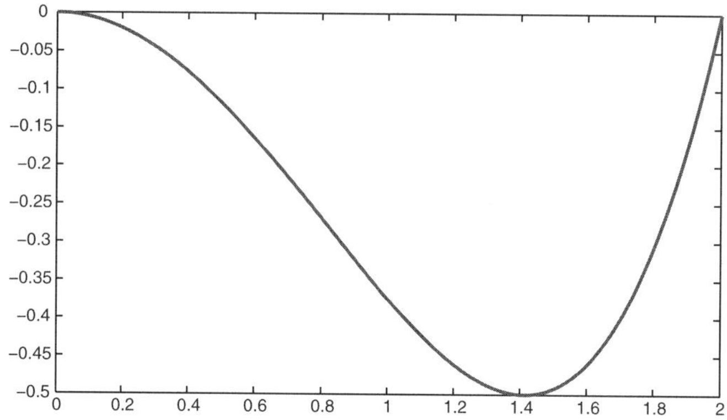

FIGURE 2. Potential function $p(r) = -r^2/2+r^4/8$. If the distance between a pair of points is r, then this potential encourages the distance to approach $\sqrt{2}$, i.e., the potential's minimizer. If the points are confined to the unit sphere, then this corresponds to the points being orthogonal. Indeed, the frame potential, which is the total potential of a collection of points on the unit sphere (up to an additive constant) simultaneously encourages all pairs of points to be orthogonal. Interestingly, unit norm tight frames are precisely the global minimizers of the frame potential, and furthermore, every local minimizer is necessarily global.

To test equidistribution, we consider the corresponding potential: Integrating and negating $f(r) = r - r^3/2$ gives $p(r) = -r^2/2 + r^4/8$ (see Figure 2), and so

$$\begin{aligned} P(x,y) &= p(\|x-y\|) \\ &= -\frac{\|x-y\|^2}{2}\bigg(1-\frac{\|x-y\|^2}{4}\bigg) \\ &= \frac{1}{2}\Big(|\langle x,y\rangle|^2-1\Big). \end{aligned}$$

Ignoring constant terms, this means that minimizing total potential energy (4.1) is equivalent to minimizing the **frame potential**:

$$\mathrm{FP}(\Phi) := \sum_{n=1}^{N}\sum_{n'=1}^{N}|\langle\varphi_n,\varphi_{n'}\rangle|^2 = \|\Phi^*\Phi\|_{\mathrm{HS}}^2.$$

With this, the following establishes the requirements for equidistribution:

$$\begin{aligned} 0 &\leq \bigg\|\Phi\Phi^* - \frac{N}{M}I\bigg\|_{\mathrm{HS}}^2 \\ &= \mathrm{Tr}[(\Phi\Phi^*)^2] - 2\frac{N}{M}\,\mathrm{Tr}[\Phi\Phi^*] + \frac{N^2}{M^2}\,\mathrm{Tr}[I] \\ &= \mathrm{FP}(\Phi) - \frac{N^2}{M}. \end{aligned}$$

Indeed, rearranging gives that the frame potential satisfies $\mathrm{FP}(\Phi) \geq N^2/M$ with equality precisely when $\Phi\Phi^* = (N/M)I$, i.e., Φ is a UNTF. This inequality is sometimes called the zeroth-order Welch bound [**47**]. By construction, we know

UNTFs exist whenever (M, N) satisfies $M \leq N$, and so UNTFs are precisely the global minimizers of the frame potential, i.e., UNTFs are the arrangements of points on the sphere which exhibit equidistribution under the frame force. Next, we turn to the more difficult problem of equilibrium. Indeed, for general forces (such as the Coulomb force), points often settle for a local minimizer of total potential which is not global. Perhaps surprisingly, the following result establishes that this is not an issue when applying the frame force:

THEOREM 4.1 (Theorem 7.1 in [**4**]). *Consider the frame potential of all arrangements of N points in the unit sphere in $\mathbb{R}^M$ or $\mathbb{C}^M$. Then every local minimizer is also a global minimizer.*

Note that while the motivation for the frame potential came from a frame force defined between points in real space, the above theorem is valid for both real and complex spaces. The proof of this theorem uses the contrapositive: Given any arrangement which is not tight, find a family of arbitrarily close arrangements, each having strictly smaller frame potential; then the original arrangement is not a local minimizer. The utility of this result is demonstrated in part by its application to the so-called Paulsen problem.

4.1. Application: The Paulsen problem. Given a frame which is nearly unit norm and nearly tight, how far is the closest UNTF? This question was first posed by Vern Paulsen [**6**], and has since been dubbed **the Paulsen problem**. To evaluate what it means to be nearly unit norm and nearly tight, one must first choose a notion of "distance" from unit-norm tightness. Consider any continuous gauge function $\rho\colon \mathbb{C}^{M\times N} \to [0,\infty)$ such that $\rho(\Phi) = 0$ precisely when Φ is a UNTF and furthermore $\rho(\Phi)$ is bounded away from zero whenever $\|\Phi\|_{\mathrm{HS}}$ is sufficiently large. For example, one may take

$$\rho(\Phi) = \|\operatorname{diag}(\Phi^*\Phi) - \mathbf{1}\|_2 + \left\|\Phi\Phi^* - \frac{N}{M}I\right\|_{\mathrm{HS}}.$$

Then the Paulsen problem asks for a function $\delta = \delta(\epsilon)$ such that $\rho(\Phi) < \delta$ implies the existence of a UNTF that is within ϵ of Φ.

There is a nice argument due to Don Hadwin [**6**] which guarantees the existence of such a function: Suppose otherwise—that there is a sequence $\{\Phi_i\}_{i=1}^{\infty}$ for which $\rho(\Phi_i) \to 0$ but the distance between each Φ_i and each UNTF is bounded away from zero. Then a tail of this sequence lies in a compact set of the form $\{\Phi : \rho(\Phi) \leq \eta\}$, and so Bolzano–Weierstrass guarantees a convergent subsequence. Let Φ_∞ denote the limit point of this subsequence. By the continuity of ρ, we then have $\rho(\Phi_\infty) = 0$, meaning Φ_∞ is a UNTF, thereby contradicting the assumption that the distance between each Φ_i and each UNTF is bounded away from zero.

(We can establish more about the function $\delta(\epsilon)$ by observing that set of UNTFs forms an algebraic variety. Indeed, one may leverage a powerful result from algebraic geometry called Lojesiewicz's inequality [**18**] to conclude that for certain choices of ρ, the function $\delta(\epsilon)$ should have the form $\delta(\epsilon) = C\epsilon^\alpha$, where C and α possibly depend on M and N.)

Note that it is easy to find a nearby unit norm frame by normalizing the frame elements, and in fact, one can control how much tightness is lost in this process [**6**]. As such, one may seek a solution to the Paulsen problem by assuming without loss of generality that the original frame is a unit norm frame. This leads to the following problem:

PROBLEM 4.2. For every (M, N) such that $M \leq N$, find δ, C and α such that for every unit norm frame Φ satisfying $\|\Phi\Phi^* - (N/M)I\|_{\mathrm{HS}} \leq \delta$, there exists a UNTF $\Phi^\star$ such that

$$\|\Phi^\star - \Phi\|_{\mathrm{HS}} \leq C\bigg\|\Phi\Phi^* - \frac{N}{M}I\bigg\|_{\mathrm{HS}}^{\alpha}.$$

Having distilled the problem of interest, we now identify the applicability of the frame potential: Since local minimizers of the frame potential are UNTFs, it makes sense to attempt gradient descent to approach a nearby UNTF from an initial unit norm frame.[2] Unfortunately, gradient descent is not guaranteed to converge from any arbitrary initial arrangement; indeed, the gradient vanishes at any arrangement which satisfies the Lagrange equations (the "critical points"), which occurs precisely when the arrangement can be partitioned into mutually orthogonal sub-arrangements which form UNTFs for their spans [**4**].

Call an arrangement **orthogonally partitionable (OP)** if there is a nontrivial partition of the arrangement into mutually orthogonal sub-arrangements (note that the term *orthodecomposable* is used elsewhere in the literature), and ϵ-OP if one can partition into sub-arrangements such that inner products between members of different sub-arrangements are less than ϵ in absolute value. Intuitively, gradient descent produces a sequence of unit norm frames which are not ϵ-OP, then the gradient at these arrangements will never vanish, and so we should expect good performance; indeed, in this case, we get convergence at a linear rate, and furthermore, the limiting UNTF is within a factor of $\|\Phi\Phi^* - (N/M)I\|_{\mathrm{HS}}$ away from the initial arrangement Φ (i.e., $\alpha = 1$ in Problem 4.2, which is optimal) [**12**]. Additionally, if $\|\Phi\Phi^* - (N/M)I\|_{\mathrm{HS}}$ is sufficiently small, then Φ is necessarily not ϵ-OP, provided M and N are relatively prime; this makes intuitive sense since a UNTF cannot be OP when M and N are relatively prime (why not?). As such, one may perform gradient descent of the frame potential to solve the Paulsen problem when M and N are relatively prime.

In the remaining case where M and N are not relatively prime, there is no ϵ for which we can guarantee that the gradient descent iterations never become ϵ-OP (see Example 9 in [**12**]), and so we need a different approach. In this case, if ever the arrangement becomes ϵ-OP, then we can "jump" to a nearby arrangement which is exactly OP, and if the original was sufficiently tight, we can ensure that the new arrangement's partitions have equal redundancies (that is, for each sub-arrangement, the ratio of number of points to the dimension of their span is constant). We can then run gradient descent on each sub-arrangement individually, keeping its span fixed, and jumping again when necessary. Each time we need to jump, we incur a penalty in our estimate of the distance to the limiting UNTF. In particular, $\alpha = 1/7^J$ in Problem 4.2 if J jumps are used, and since $J \leq M$, we can at least guarantee $\alpha = 1/7^M$ in this non-relatively-prime case [**12**]. Improving on this value of α remains an open problem.

[2]The astute physicist will notice that in a physical system, the gradient of the potential informs the acceleration of a trajectory, rather than its velocity. In this section, the goal is to use the frame potential (and Theorem 4.1 in particular) to solve the Paulsen problem, and so we ignore whether the steepest-descent trajectory follows the physics-based intuition behind the frame potential. The reader is invited to investigate further with papers by Strawn [**39**] and Bodmann and Haas [**7**].

5. Eigensteps

The previous section completely characterized the set of UNTFs as local minimizers of the frame potential. We also saw that one can perform gradient descent on the frame potential (with intermittent "jumps" as necessary) to find a UNTF close to a given unit norm frame which is nearly tight. While this characterization is useful (both for intuition and for solving the Paulsen problem), it is not explicit. This section provides an explicit construction of every UNTF of N vectors in M dimensions, real or complex.

Before explaining the theory behind this general construction, it is instructive to make an attempt without the theory. The following is the beginning of an arbitrary real construction with $(M, N) = (3, 5)$:

$$\begin{bmatrix} 1 & \cos\theta & ? & ? & ? \\ 0 & \sin\theta & ? & ? & ? \\ 0 & 0 & ? & ? & ? \end{bmatrix}.$$

Indeed, the first two frame elements are completely determined (up to orthogonal rotation) by the angle θ between them. We should keep track of the spectrum of $\Phi\Phi^*$ since the addition of columns will only increase the eigenvalues, and in the end, we need all of the eigenvalues to be $5/3$. At the moment, the spectrum is completely determined by θ:

$$\lambda(\Phi_2\Phi_2^*) = (1 + \cos\theta, 1 - \cos\theta, 0).$$

(We are denoting the first n columns of Φ by Φ_n.) Also, the eigenspaces of $\Phi_2\Phi_2^*$ are the bisector of the frame elements, the line in the xy-plane which is orthogonal to this bisector, and the z-axis. The corresponding eigenvalues indicate how much of the frame elements' energy lie in these subspaces. In particular, since the eigenvalues only increase with the addition of more frame elements, we know that $1 + \cos\theta \leq 5/3$, and so θ cannot be too small. We now need to carefully pick a third frame element so as to not have too much energy pointing in the direction of the leading eigenvector. But what choices are available? And are the choices parameterized by some parameter like θ?

At this point, we recall that it is difficult to find analytic expressions for eigenvalues of $M \times M$ matrices when $M \geq 5$. As such, one might suspect that keeping track of the spectrum of $\Phi\Phi^*$ is infeasible. This suspicion happens to be wrong; while it may be difficult to find a closed-form expression for eigenvalues in terms of matrix entries, it is easy to analytically express the entries of a diagonalizable matrix in terms of its eigenvalues and eigenvectors. In similar spirit, if we know how we want the spectrum of $\Phi_n\Phi_n^*$ to evolve as n ranges from 1 to N, then we can use this to produce analytic expressions for the frame elements. In particular, if we want the spectrum to change from $(\alpha_1, \ldots, \alpha_M)$ to $(\beta_1, \ldots, \beta_M)$ by adding a frame element φ_{n+1}, then by Theorem 2 in **[10]**, we can determine the necessary size of the projection of φ_{n+1} onto each of the eigenspaces of $\Phi_n\Phi_n^*$:

$$\|P_{n;\lambda}\varphi_{n+1}\|^2 = -\lim_{x\to\lambda}(x - \lambda)\frac{(x - \beta_1)\cdots(x - \beta_M)}{(x - \alpha_1)\cdots(x - \alpha_M)},$$

where $P_{n;\lambda}$ denotes the projection onto the eigenspace of $\Phi_n\Phi_n^*$ corresponding to eigenvalue λ. Given these sizes of projections, we have the freedom to choose the orientations of the projections, and any such choice will uniquely determine the new frame element φ_{n+1} and update the spectrum as desired. Once this choice has been made, there is a corresponding analytic expression for how the eigenspaces of $\Phi_{n+1}\Phi_{n+1}^*$ are updated from those of $\Phi_n\Phi_n^*$; this expression is a bit more complicated, as it depends on the dimensions of the eigenspaces, and so the interested reader should consult Theorem 7 in [**10**].

Overall, we have a method of constructing all UNTFs with a prescribed sequence of spectra $\{\lambda(\Phi_n\Phi_n^*)\}_{n=1}^N$, and so it remains to determine all possible sequences of spectra. Since $\Phi_1\Phi_1^*$ is a rank-1 projection, we know that the first spectrum is simply $(1,0,\ldots,0)$. Also, since $\Phi_N\Phi_N^* = \Phi\Phi^* = (N/M)I$, the last spectrum must be $(N/M,\ldots,N/M)$. Every time we add a frame element, the energy of that frame element contributes to the spectrum:

$$\begin{aligned}\mathrm{Tr}[\Phi_{n+1}\Phi_{n+1}^*] &= \mathrm{Tr}[\Phi_n\Phi_n^* + \varphi_{n+1}\varphi_{n+1}^*]\\ &= \mathrm{Tr}[\Phi_n\Phi_n^*] + \mathrm{Tr}[\varphi_{n+1}\varphi_{n+1}^*]\\ &= \mathrm{Tr}[\Phi_n\Phi_n^*] + 1.\end{aligned}$$

In words, the sum of the eigenvalues increases by 1 with each frame element. The least obvious observation to make is a classical result that the addition of a positive semidefinite rank-1 matrix to a matrix with spectrum $\alpha = (\alpha_1,\ldots,\alpha_M)$ produces a new matrix whose spectrum $\beta = (\beta_1,\ldots,\beta_M)$ **interlaces** the original:

$$\alpha_N \le \beta_N \le \alpha_{N-1} \le \beta_{N-1} \le \cdots \le \alpha_2 \le \beta_2 \le \alpha_1 \le \beta_1.$$

As shorthand, we write $\alpha \sqsubseteq \beta$. With these four observations, we define **eigensteps**[3] to be any sequence of sequences $\{\{\lambda_{m;n}\}_{m=1}^M\}_{n=1}^N$ such that

- the first sequence is $(\lambda_{1;1},\lambda_{2;1},\ldots,\lambda_{M;1}) = (1,0,\ldots,0)$,
- the last sequence is $(\lambda_{1;N},\ldots,\lambda_{M;N}) = (N/M,\ldots,N/M)$,
- the trace is consistent with unit-norm frame elements, i.e., $\sum_{m=1}^M \lambda_{m;n} = n$ for each n, and
- each sequence interlaces the previous: $\{\lambda_{m;n}\}_{m=1}^M \sqsubseteq \{\lambda_{m;n+1}\}_{m=1}^M$.

By design, we know that for every UNTF Φ, the sequence of spectra $\{\lambda(\Phi_n\Phi_n^*)\}_{n=1}^N$ is necessarily a sequence of eigensteps. It turns out that the converse also holds: For every sequence of eigensteps, there exists a UNTF with a matching sequence of spectra (this is the main result in [**10**]). In summary, every UNTF can be constructed by first selecting eigensteps, and then iteratively selecting frame elements in terms of these eigensteps using the process described earlier. For the sake of example, we return to the case of 5 frame elements in $\mathbb{R}^3$:

EXAMPLE 5.1. Instead of attempting to write down a general family of UNTFs directly, we start by finding eigensteps and then use these eigensteps as instructions for producing all UNTFs. In the following, we express eigensteps in an $M \times N$ matrix Λ whose (m,n)th entry is $\lambda_{m;n}$. The first and second properties of eigensteps determine the first and last columns:

$$\Lambda = \begin{bmatrix} 1 & ? & ? & ? & \frac{5}{3}\\ 0 & ? & ? & ? & \frac{5}{3}\\ 0 & ? & ? & ? & \frac{5}{3}\end{bmatrix}.$$

[3] There are two meanings behind the name *eigensteps*. First, each spectrum in the sequence is another step toward the desired UNTF. Second, when working with the interlacing constraint, it is helpful to visualize each spectrum in the sequence in terms of a bar graph, which resembles a staircase.

Next by interlacing, we know that equal eigenvalues in one sequence determines an eigenvalue in the next/previous:

$$\Lambda = \begin{bmatrix} 1 & ? & \frac{5}{3} & \frac{5}{3} & \frac{5}{3} \\ 0 & ? & ? & \frac{5}{3} & \frac{5}{3} \\ 0 & 0 & ? & ? & \frac{5}{3} \end{bmatrix}.$$

The trace condition then forces the entries in each column to sum to 1:

$$\Lambda = \begin{bmatrix} 1 & 2-y & \frac{5}{3} & \frac{5}{3} & \frac{5}{3} \\ 0 & y & \frac{4}{3}-x & \frac{5}{3} & \frac{5}{3} \\ 0 & 0 & x & \frac{2}{3} & \frac{5}{3} \end{bmatrix}. \tag{5.1}$$

The variables x and y are not completely free—one must ensure that the eigensteps they generate satisfy interlacing (the other three properties of eigensteps are already satisfied at this point). In this case, the interlacing inequalities involving x and y are

$$\begin{aligned} &x \le \frac{2}{3} \le \frac{4}{3} - x \le \frac{5}{3}, \\ &0 \le x \le y \le \frac{4}{3} - x \le 2 - y \le \frac{5}{3}, \\ &0 \le y \le 1 \le 2 - y. \end{aligned} \tag{5.2}$$

Overall, the (x, y) which produce eigensteps form a convex polygon, as depicted in Figure 3. In general, the set of eigensteps for a given (M, N) form a convex polytope, and this convex polytope is parameterized in general in [**23**]. Now that we have identified every possible choice of eigensteps, we can use them as instructions to construct all possible UNTFs. Intuitively, if we were to standardize the directions along eigenspaces that we select in each step, then varying x and y would produce a 2-dimensional submanifold of the set of UNTFs; indeed, for this example, one such standardized selection produces the following continuum of 3×5 UNTFs, all the entries of which are relatively simple functions of x and y:

$$\begin{aligned} \varphi_1(1) &= 1 \\ \varphi_1(2) &= 0 \\ \varphi_1(3) &= 0 \end{aligned}$$

$$\begin{aligned} \varphi_2(1) &= 1 - y \\ \varphi_2(2) &= \sqrt{y(2-y)} \\ \varphi_2(3) &= 0 \end{aligned}$$

$$\begin{aligned} \varphi_3(1) &= \frac{\sqrt{(3y-1)(2+3x-3y)(2-x-y)}}{6\sqrt{1-y}} - \frac{\sqrt{(5-3y)(4-3x-3y)(y-x)}}{6\sqrt{1-y}} \\ \varphi_3(2) &= \frac{\sqrt{y(3y-1)(2+3x-3y)(2-x-y)}}{6\sqrt{(1-y)(2-y)}} + \frac{\sqrt{(5-3y)(2-y)(4-3x-3y)(y-x)}}{6\sqrt{y(1-y)}} \\ \varphi_3(3) &= \frac{\sqrt{5x(4-3x)}}{3\sqrt{y(2-y)}} \end{aligned}$$

$$\begin{aligned}
\varphi_4(1) &= -\frac{\sqrt{(4-3x)(3y-1)(2-x-y)(4-3x-3y)}}{12\sqrt{(2-3x)(1-y)}} - \frac{\sqrt{(4-3x)(5-3y)(y-x)(2+3x-3y)}}{12\sqrt{(2-3x)(1-y)}} \\
&\quad - \frac{\sqrt{x(3y-1)(y-x)(2+3x-3y)}}{4\sqrt{3(2-3x)(1-y)}} + \frac{\sqrt{x(5-3y)(2-x-y)(4-3x-3y)}}{4\sqrt{3(2-3x)(1-y)}} \\
\varphi_4(2) &= -\frac{\sqrt{(4-3x)y(3y-1)(2-x-y)(4-3x-3y)}}{12\sqrt{(2-3x)(1-y)(2-y)}} \\
&\quad + \frac{\sqrt{(4-3x)(2-y)(5-3y)(y-x)(2+3x-3y)}}{12\sqrt{(2-3x)y(1-y)}} \\
&\quad - \frac{\sqrt{xy(3y-1)(y-x)(2+3x-3y)}}{4\sqrt{3(2-3x)(1-y)(2-y)}} - \frac{\sqrt{x(2-y)(5-3y)(2-x-y)(4-3x-3y)}}{4\sqrt{3(2-3x)y(1-y)}} \\
\varphi_4(3) &= \frac{\sqrt{5x(2+3x-3y)(4-3x-3y)}}{6\sqrt{(2-3x)y(2-y)}} + \frac{\sqrt{5(4-3x)(y-x)(2-x-y)}}{2\sqrt{3(2-3x)y(2-y)}}
\end{aligned}$$

$$\begin{aligned}
\varphi_5(1) &= \frac{\sqrt{(4-3x)(3y-1)(2-x-y)(4-3x-3y)}}{12\sqrt{(2-3x)(1-y)}} + \frac{\sqrt{(4-3x)(5-3y)(y-x)(2+3x-3y)}}{12\sqrt{(2-3x)(1-y)}} \\
&\quad - \frac{\sqrt{x(3y-1)(y-x)(2+3x-3y)}}{4\sqrt{3(2-3x)(1-y)}} + \frac{\sqrt{x(5-3y)(2-x-y)(4-3x-3y)}}{4\sqrt{3(2-3x)(1-y)}} \\
\varphi_5(2) &= \frac{\sqrt{(4-3x)y(3y-1)(2-x-y)(4-3x-3y)}}{12\sqrt{(2-3x)(1-y)(2-y)}} \\
&\quad - \frac{\sqrt{(4-3x)(2-y)(5-3y)(y-x)(2+3x-3y)}}{12\sqrt{(2-3x)y(1-y)}} \\
&\quad - \frac{\sqrt{xy(3y-1)(y-x)(2+3x-3y)}}{4\sqrt{3(2-3x)(1-y)(2-y)}} - \frac{\sqrt{x(2-y)(5-3y)(2-x-y)(4-3x-3y)}}{4\sqrt{3(2-3x)y(1-y)}} \\
\varphi_5(3) &= -\frac{\sqrt{5x(2+3x-3y)(4-3x-3y)}}{6\sqrt{(2-3x)y(2-y)}} + \frac{\sqrt{5(4-3x)(y-x)(2-x-y)}}{2\sqrt{3(2-3x)y(2-y)}}
\end{aligned}$$

5.1. Application: Schur–Horn theorem.

Eigensteps can be used to provide a constructive proof of the Schur–Horn theorem. To state this theorem, we require a definition: $\{\lambda_n\}_{n=1}^N$ is said to **majorize** $\{\mu_n\}_{n=1}^N$ if

$$\begin{aligned}
\sum_{n'=1}^{n} \lambda_{n'} &\geq \sum_{n'=1}^{n} \mu_{n'} \qquad \forall n \in \{1,\ldots,N-1\}, \\
\sum_{n'=1}^{N} \lambda_{n'} &= \sum_{n'=1}^{N} \mu_{n'}.
\end{aligned}$$

As shorthand, we write $\{\lambda_n\}_{n=1}^N \succeq \{\mu_n\}_{n=1}^N$.

THEOREM 5.2 (Schur–Horn theorem, [**30**, **37**]). *There exists a self-adjoint matrix with spectrum $\{\lambda_n\}_{n=1}^N$ and diagonal entries $\{\mu_n\}_{n=1}^N$ if and only if $\{\lambda_n\}_{n=1}^N \succeq \{\mu_n\}_{n=1}^N$.*

To prove this theorem, first note that it suffices to prove it in the special case where the spectrum is nonnegative. Indeed, for any α, we obviously have that the self-adjoint matrix G exists if and only if $G+\alpha I$ exists. Also, $\{\lambda_n\}_{n=1}^N \succeq \{\mu_n\}_{n=1}^N$ is almost immediately equivalent to $\{\lambda_n+\alpha\}_{n=1}^N \succeq \{\mu_n+\alpha\}_{n=1}^N$. As such, we may pick α so that the spectrum is nonnegative without loss of generality. When it exists, the resulting positive semidefinite matrix will enjoy a Cholesky factorization

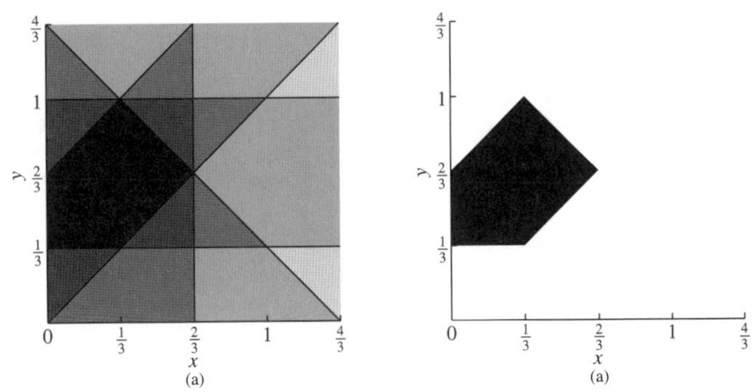

FIGURE 3. Pairs of parameters (x, y) for which the eigensteps (5.1) satisfy the interlacing constraints (5.2). Specifically, (a) illustrates the various half-spaces implicated by the interlacing inequalities, whereas (b) provides their intersection. Picking any (x, y) in this pentagon and plugging into (5.1) will produce a valid sequence of eigensteps, which can then be used as instructions for constructing a 3×5 unit norm tight frame. Also, for every such unit norm tight frame, there is a corresponding point in this pentagon from which it can be constructed. In this way, eigensteps provide an explicit construction of every unit norm tight frame.

$\Phi^*\Phi$, in which case μ_n corresponds to the norm squared of the nth column of Φ. As such, the Schur–Horn theorem reduces to a generalization of the UNTF existence problem: When does there exist a matrix Φ with prescribed norms such that $\Phi^*\Phi$ has a prescribed spectrum? (Note that in the case where Φ is a UNTF, the spectrum of $\Phi^*\Phi$ is simply a zero-padded version of that of $\Phi\Phi^* = (M/N)I$.) This can be determined by generalizing the theory of eigensteps (i.e., change the final spectrum and the trace condition appropriately); this more general form is provided in [**10**]. From this perspective, majorization essentially follows from a repeated application of interlacing [**31**]. The proof then follows from the fact that a particular greedy choice of eigensteps, called **Top Kill**[4], will always exist when the spectrum majorizes the squared lengths, and that no eigensteps exist otherwise.

5.2. Application: Optimal frame completion. Consider a scenario in which you currently have a certain collection of measurement vectors, and you are given a budget to construct additional measurement vectors. How should you select the additional measurement vectors? In order to enjoy robust reconstruction, you are inclined to use the additional measurement vectors to complete your current system to a tight frame; this idea was first studied in [**17**]. Sometimes, it is impossible to complete to a tight frame. For example, if your current frame is

$$\Phi = \begin{bmatrix} 1 & 0 & 0 & 1 & 0 & 0 & 1 \\ 0 & 1 & 0 & 0 & 1 & 0 & 0 \\ 0 & 0 & 1 & 0 & 0 & 1 & 0 \end{bmatrix},$$

[4]Top Kill was named after the procedure used in an attempt to seal an oil well in response to the 2010 Deepwater Horizon oil spill, which was prominently featured in the news at the time of its development.

then the spectrum of $\Phi\Phi^*$ is $(3, 2, 2)$, and so a budget of one additional vector is insufficient thanks to interlacing: the largest eigenvalue after completion must be ≥ 3, whereas the smallest eigenvalue must be 2. In such situations, one is inclined to find a frame completion which is the "best possible." This problem of **optimal frame completion** was first posed in [**22**].

Recently, this problem was completely solved using eigensteps [**21**]. The solution has two parts. First, [**21**] characterizes every possible completion of a given frame with a budget of additional frame elements of prescribed lengths; in the special case where the additional frame elements each have length zero, this result recovers the Schur–Horn theorem, and so this characterization is a generalization of sorts. Second, [**21**] explicitly identifies the completion whose spectrum is majorized by the spectrum of every other completion; the classical theory of Schur convexity then guarantees that, for any function of the spectrum which is both convex and symmetric (e.g., the frame potential, the sum of the reciprocals of the eigenvalues, etc.), this majorization-minimal completion also minimizes this function of the spectrum. As such, the identified completion is optimal, essentially independent of the notion of optimality.

5.3. Application: UNTF connectivity. For a fixed (M, N), is the set of real (complex) UNTFs path-connected? This question was first posed by David Larson in a Research Experiences for Undergraduates summer program in 2002, motivated by the desire for interpolations between frames with desirable properties. UNTFs are natural generalizations of orthonormal bases, and so this question is a natural analog to the facts that the orthogonal group is not connected, while the unitary group is. However, only special cases of this **frame homotopy problem** were solved until recently, when [**11**] provided a complete solution using eigensteps.

To understand how eigensteps play a role in the solution of this problem, it is instructive to recall the proof that the unitary group is connected. Every unitary matrix U is normal (since $UU^* = I = U^*U$), and so it is diagonalized by some unitary matrix S. Since U acts as an isometry, its eigenvalues necessarily have unit modulus, and so we have a factorization $U = SDS^*$, where the diagonal entries of D have unit modulus. At this point, we observe that D can be continuously deformed to the identity matrix by letting the diagonal entries traverse the complex unit circle at appropriate rates; denote this matrix path by $D(t)$ for $0 \leq t \leq 1$. Then $SD(t)S^*$ is unitary for every t since it is a product of unitary matrices, and $SD(1)S^* = SS^* = I$. Since every unitary matrix is path-connected to the identity matrix, every pair of unitary matrices then enjoys a connecting path by transitivity.

Now consider what the theory of eigensteps provides: a two-step procedure to construct every possible UNTF. The first step is to choose eigensteps, and the second step is to choose directions along each iteration's eigenspaces. One may think of this as a sort of polar decomposition of UNTFs. Just like the unitary group's connectivity is proved by fixing certain factors in the requisite unitary diagonalization, the collection of complex UNTFs is proved to be connected by fixing directionalities and then fixing eigensteps in the polar decomposition. Indeed, given two complex UNTFs Φ and Ψ, one may continuously deform the eigensteps of Φ to match those of Ψ (this is possible since the set of all eigensteps is a convex polytope), and then continuously deform the directionalities by leveraging the unitary group's connectivity. Actually, the argument is a little more subtle if either frame's eigensteps lies on the boundary of the polytope, and the argument is a bit more complicated in the real case since the orthogonal group is not connected; the interested reader is invited to find more details in [**11**].

6. Equiangular tight frames

Recall the motivating applications of unit norm tight frames in Section 2, and notice that most of the applications we discussed require a unit norm tight frame which satisfies additional properties. For example, fingerprinting makes use of an equiangular UNTF (commonly called an **equiangular tight frame** or ETF). Additionally, ETFs are more robust to erasures than UNTFs in general [**29**]. In the context of sparse decomposition, one seeks a unit norm tight frame with small worst-case coherence

$$\mu(\Phi) := \max_{\substack{i,j\in\{1,\dots,N\}\\ i\neq j}} |\langle\varphi_i,\varphi_j\rangle|,$$

and as we will soon see, ETFs have minimial worst-case coherence whenever they exist. Due to their widespread applicability, we conclude this chapter with a brief review of ETFs.

First, consider the worst-case coherence $\mu(\Phi)$ over all unit norm frames of N vectors in M-dimensional space. Since this set is compact, there necessarily exist unit norm frames which minimize worst-case coherence, and these are called **Grassmannian frames** [**41**]. One method of identifying Grassmannian frames uses the **Welch bound** [**41,47**], which is a lower bound on the worst-case coherence:

$$\mu(\Phi) \geq \sqrt{\frac{N-M}{M(N-1)}}.$$

The proof of the Welch bound is instructive: Identifying Φ with the $M\times N$ matrix $[\varphi_1\cdots\varphi_N]$ gives

$$N+N(N-1)\mu(\Phi)^2 \geq \sum_{i=1}^{N}\sum_{j=1}^{N}|\langle\varphi_i,\varphi_j\rangle|^2 = \|\Phi^*\Phi\|_{\mathrm{HS}}^2 \geq \frac{N^2}{M},$$

where the last inequality follows from rearranging the following inequality:

$$0 \leq \|\Phi\Phi^* - \frac{N}{M}I\|_{\mathrm{HS}}^2 = \|\Phi^*\Phi\|_{\mathrm{HS}}^2 - \frac{N^2}{M}.$$

As such, equality in the Welch bound occurs precisely when

- the cosines $|\langle\varphi_i,\varphi_j\rangle|$ are the same for every i and $j\neq i$, and
- the frame Φ is unit norm and tight.

Again, we call such ensembles equiangular tight frames, and since they achieve equality in a lower bound of $\mu(\Phi)$, we conclude that they are necessarily Grassmannian. This result is significant because Grassmannian frames are very difficult to construct in general (see [**5**], for example), and the additional structural information afforded by ETFs make them more accessible; indeed, as we will see, there are currently several infinite families of known ETFs.

Examples of ETFs include the cube roots of unity (viewed as vectors in $\mathbb{R}^2$) and the vertices of the origin-centered tetrahedron (viewed as vectors in $\mathbb{R}^3$). The apparent beauty of ETFs coupled with their importance as Grassmannian frames has made them the subject of active research recently.

What follows is a list of trivial ETFs:

- **Orthonormal bases.** This case takes $N = M$, and it is easy to verify the ETF conditions.
- **Regular simplices.** In this case, $N = M + 1$. For a simple construction of this example, take $N-1$ rows from an $N \times N$ discrete Fourier transform matrix. Then the resulting columns, after being scaled to have unit norm, form an ETF.
- **Frames in one dimension.** When $M = 1$, any unit norm frame amounts to a list of scalars of unit modulus, and such frames are necessarily ETFs.

To date, there are only a few known infinite families of nontrivial complex ETFs. Interestingly, despite ETFs being characterized in terms of functional analysis (namely, equality in the Welch bound), each of the known infinite families is based on some sort of combinatorial design:

- **Strongly regular graphs.** A (v, k, λ, μ)-strongly regular graph is a k-regular graph with v vertices such that every pair of adjacent vertices has λ common neighbors, whereas every pair of non-adjacent vertices has μ common neighbors. One may manipulate the adjacency matrix of a strongly regular graph with appropriate parameters to find an embedding of some M-dimensional real ETF of $N = v+1$ vectors in $\mathbb{R}^N$ (here, M is a complicated function of the graph parameters; see [**46**] for details). In fact, real ETFs are in one-to-one correspondence with a subclass of strongly regular graphs in this way. As an example, the case where $N = 2M$ corresponds to the so-called conference graphs.
- **Difference sets.** Let G be a finite abelian group. Then $D \subseteq G$ is said to be a (G, k, λ)-difference set if $|D| = k$ and for every nonzero $g \in G$, there are exactly λ different pairs $(d_1, d_2) \in D \times D$ such that $g = d_1 - d_2$. One may use any difference set to construct an ETF with $M = |D|$ and $N = |G|$ (see [**16, 48**]). In particular, each vector in the ETF is obtained by taking a character of G and restricting its domain to D (before scaling to have unit norm).
- **Steiner systems.** A (t, k, v)-Steiner system is a v-element set S of points together with a collection $\mathcal{B}$ of k-element subsets of S called blocks with the property that each t-element subset of S is contained in exactly one block. It is not difficult to show that each point is necessarily contained in exactly $r = (v-1)/(k-1)$ blocks. One may use any $(2, k, v)$-Steiner system to construct an ETF in $\mathbb{C}^{\mathcal{B}}$ (see [**20**]). Specifically, for each point $p \in S$, they embed an r-dimensional regular simplex into $\mathbb{C}^{\mathcal{B}}$ so as to be supported on the blocks that contain p. The union of these embedded simplices then forms an ETF of $v(r + 1)$ vectors in $\mathbb{C}^{\mathcal{B}}$. Every such construction necessarily has $N > 2M$.

In the remainder of this section, we describe what is known about real equiangular tight frames. Throughout, we use $\exists\,\mathrm{RETF}(M, N)$ to denote the statement "there exists a real equiangular tight frame with parameters (M, N)." We start with some basic properties:

THEOREM 6.1 (see [**40**]). *$\exists\,\mathrm{RETF}(M, N)$ implies each of the following:*

(a) $N \leq M(M + 1)/2$.
(b) $\exists\,\mathrm{RETF}(N - M, N)$.

Part (a) above can be seen by observing that the rank-1 matrices $\{\varphi_n\varphi_n^*\}_{n=1}^N$ are necessarily linearly independent in the $M(M+1)/2$-dimensional space of $M\times M$ symmetric matrices (this follows from computing the spectrum of their Gram matrix). Part (b) uses a concept known as the **Naimark complement**. In particular, since any ETF Φ satisfies $\Phi\Phi^* = \frac{N}{M}I$, the rows of Φ can be viewed as orthonormal vectors (suitably scaled). As such, one may complete the orthonormal basis with $N-M$ other row vectors. Collecting these rows into an $(N-M)\times N$ matrix and normalizing the columns results in another ETF (called the Naimark complement of the original).

Real ETFs are intimately related to graphs. Given any real ETF, negate some of the vectors so that each one has positive inner product with the last vector (this process produces another ETF Φ). Next, remove the last vector to get a subcollection of vectors Ψ (this is no longer an ETF). Use $\Psi = \{\psi_n\}_{n=1}^{N-1}$ to build a graph in the following way: Take $v = N-1$ vertices and say vertex i is adjacent to vertex j if $\langle\psi_i,\psi_j\rangle < 0$. It turns out that this graph is necessarily strongly regular with parameters determined by M and N:

THEOREM 6.2 (Corollary 5.6 in [**46**]). $\exists\,\mathrm{RETF}(M,N)$ *if and only if* $\exists\,\mathrm{srg}(N-1,k,\frac{3k-N}{2},\frac{k}{2})$ *with*

$$k = \frac{N}{2} - 1 + \Big(1-\frac{N}{2M}\Big)\sqrt{\frac{M(N-1)}{N-M}}.$$

The spectrum of a strongly regular graph can be expressed in terms of its graph parameters. In fact, it turns out that the eigenvalues must be integer, which in turn implies the following integrality conditions:

THEOREM 6.3 (Theorem A in [**40**]). *Suppose* $N\neq 2M$. *Then* $\exists\,\mathrm{RETF}(M,N)$ *implies that*

$$\sqrt{\frac{M(N-1)}{N-M}},\qquad \sqrt{\frac{(N-M)(N-1)}{M}}$$

are both odd integers.

Since we identify real ETFs with certain strongly regular graphs, we can leverage necessary conditions for existence of the latter to inform existence of the former:

THEOREM 6.4 (see [**8, 9**]). *Given* v, k, λ *and* μ, *let* $r\geq 0$ *and* $s\leq -1$ *denote the solutions to*

$$x^2 + (\mu-\lambda)x + (\mu-k) = 0,$$

and take

$$f := \frac{s(v-1)+k}{s-r},\qquad q_{11}^1 := \frac{f^2}{v}\bigg(1+\frac{r^3}{k^2}-\frac{(r+1)^3}{(v-k-1)^2}\bigg).$$

Then $\exists\,\mathrm{srg}(v,k,\lambda,\mu)$ *implies each of the following:*

(a) *The* ***Krein conditions*** *are satisfied:*

$$(r+1)(k+r+2rs) \leq (k+r)(s+1)^2,$$
$$(s+1)(k+s+2rs) \leq (k+s)(r+1)^2.$$

(b) $v \leq \begin{cases} \frac{1}{2}f(f+3) & \text{if } q_{11}^1 = 0 \\ \frac{1}{2}f(f+1) & \text{if } q_{11}^1 \neq 0 \end{cases}.$

(c) *If* $\mu = 1$, *then* $\frac{vk}{(\lambda+1)(\lambda+2)}$ *is integer.*

In the case of real ETFs, $\mu = k/2$, and so $\mu = 1$ implies $k = 2$, thereby implying $3 - N/2 = \lambda \geq 0$, i.e., $N \leq 6$. However, an exhaustive search shows that $\mu > 1$ for every (M, N) satisfying the conditions in Theorems 6.1 and 6.3 with $N \leq 6$, and so part (c) above is moot. It is unclear whether part (a) or part (b) is also covered by the previous conditions. Interestingly, for part (b), the case $q_{11}^1 = 0$ is important when discerning the existence of real ETFs; for example, when $(M, N) = (21, 28)$ and $(253, 276)$, v lies between $\frac{1}{2}f(f+1)$ and $\frac{1}{2}f(f+3)$, but $q_{11}^1 = 0$ in both instances (also, real ETFs are known to exist in both instances).

6.1. Redundancy 2. When $N = 2M$, we have $k = \frac{N}{2} - 1$, and the above strongly regular graph is called a **conference graph**. A bit more is known about this special type of strongly regular graph:

THEOREM 6.5 (see [**3**]). *$\exists\,\mathrm{RETF}(M, 2M)$ implies that M is odd and $2M - 1$ is a sum of two squares.*

Known constructions are summarized in the following theorem:

THEOREM 6.6 (see [**1**]). *Each of the following implies $\exists\,\mathrm{RETF}(M, N)$ with $N = 2M$:*

(a) *$N = q + 1$ for some prime power $q \equiv 1 \bmod 4$.*
(b) *$N = q^2(q + 2) + 1$ for some prime power $q \equiv 3 \bmod 4$ and prime power $q + 2$.*
(c) *$N = 5 \cdot 9^{2t+1} + 1$ for some integer $t \geq 0$.*
(d) *$N = (h - 1)^{2s} + 1$ for some integer $s \geq 1$, where h is the order of a skew-Hadamard matrix.*
(e) *$N = (n - 1)^s + 1$ for some integer $s \geq 2$, where n is the order of a conference matrix.*

6.2. Maximal real ETFs. Let S^{M-1} denote the unit sphere in $\mathbb{R}^M$. A **spherical t-design** in S^{M-1} is a finite collection $X \subseteq S^{M-1}$ satisfying

$$\frac{1}{|X|}\sum_{x \in X} f(x) = \frac{1}{\omega_{M-1}}\int_{S^{M-1}} f(x)dS$$

for every polynomial $f(x)$ of degree at most t; on the right-hand side, the integral is taken with respect to the Haar measure of S^{M-1}, and ω_{M-1} denotes the measure of S^{M-1}. For each t, there is a Fisher-type inequality that provides a lower bound on $|X|$. For example, a spherical 5-design X necessarily satisfies

$$|X| \geq M(M + 1),$$

see [**24**]. We say a spherical t-design is **tight** if it satisfies equality in the Fisher-type inequality.

THEOREM 6.7 (stated without proof in [**36**]). *Every tight spherical 5-design X is of the form $X = \Phi \cup (-\Phi)$ for some ETF Φ of $M(M - 1)/2$ elements in $\mathbb{R}^M$. Conversely, for every such ETF Φ, the collection $\Phi \cup (-\Phi)$ forms a tight spherical 5-design.*

PROOF. The first direction follows from discussions in [**27**]. In particular, Theorem 5.12 in [**15**] gives that X necessarily enjoys a partition into pairs of antipodal vectors. Furthermore, it is a consequence of Theorem 5.11 in [**24**] that every $x, y \in X$ with $x \neq \pm y$ satisfies $\langle x, y\rangle = \pm 1/\sqrt{M + 2}$, which is the Welch bound when $N = M(M + 1)/2$.

For the other direction, we follow the theory developed in [**45**] (much like [**34**]). In particular, it is known that $\Phi \cup (-\Phi)$ is a spherical 5-design if and only if

$$\sum_{\varphi\in\Phi}\langle x,\varphi\rangle^2=\frac{N}{M}\|x\|^2,\qquad \sum_{\varphi\in\Phi}\langle x,\varphi\rangle^4=\frac{3N}{M(M+2)}\|x\|^4\qquad \forall x\in\mathbb{R}^M. \tag{6.1}$$

The first identity follows immediately from the fact that Φ is a unit norm tight frame. The second identity follows from the fact that the outer products $\{\varphi\varphi^*\}_{\varphi\in\Phi}$ form a regular simplex once they are projected onto the orthogonal complement of the identity matrix. We flesh out this argument below.

Consider the $M(M+1)/2$-dimensional vector space of $M\times M$ real symmetric matrices, take P_I to be the orthogonal projection operator onto the span of the $M\times M$ identity matrix I, and let $P_{I^\perp}$ denote the projection onto the orthogonal complement. Then

$$P_I X=\frac{\langle X,I\rangle}{\|I\|_{\mathrm{HS}}^2}I=\frac{\mathrm{Tr}[X]}{M}I$$

for every $M\times M$ real symmetric matrix X. We use this along with the Pythagorean theorem to decompose the second sum in (6.1):

$$\begin{aligned}\sum_{\varphi\in\Phi}\langle x,\varphi\rangle^4&=\sum_{\varphi\in\Phi}\langle xx^*,\varphi\varphi^*\rangle_{\mathrm{HS}}^2\\&=\sum_{\varphi\in\Phi}\Big(\langle xx^*,P_I(\varphi\varphi^*)\rangle_{\mathrm{HS}}^2+\langle xx^*,P_{I^\perp}(\varphi\varphi^*)\rangle_{\mathrm{HS}}^2\Big)\\&=\frac{M+1}{2M}\|x\|^4+\sum_{\varphi\in\Phi}\langle P_{I^\perp}(xx^*),P_{I^\perp}(\varphi\varphi^*)\rangle_{\mathrm{HS}}^2.\end{aligned}\tag{6.2}$$

At this point, we recall that $\langle\varphi\varphi^*,\psi\psi^*\rangle_{\mathrm{HS}}=\langle\varphi,\psi\rangle^2=1/(M+2)$ whenever $\varphi,\psi\in\Phi$ with $\varphi\neq\psi$ (when they are equal, we get 1 since they lie in the unit sphere). Also,

$$\begin{aligned}\langle\varphi\varphi^*,\psi\psi^*\rangle_{\mathrm{HS}}&=\langle P_I(\varphi\varphi^*),P_I(\psi\psi^*)\rangle_{\mathrm{HS}}+\langle P_{I^\perp}(\varphi\varphi^*),P_{I^\perp}(\psi\psi^*)\rangle_{\mathrm{HS}}\\&=\frac{1}{M}+\langle P_{I^\perp}(\varphi\varphi^*),P_{I^\perp}(\psi\psi^*)\rangle_{\mathrm{HS}}.\end{aligned}$$

This implies that $\{\sqrt{M/(M-1)}P_{I^\perp}(\varphi\varphi^*)\}_{\varphi\in\Phi}$ is a collection of matrices of unit norm, and the inner product between any two of these matrices is

$$\frac{-2}{(M-1)(M+2)}.$$

Considering this matches the Welch bound for $M(M+1)/2$ unit vectors in a $(M(M+1)/2-1)$-dimensional space (e.g., the orthogonal complement of I), we may conclude that these matrices form a regular simplex, which is an example of a unit norm tight frame. As such, we may simplify the second term in (6.2):

$$\begin{aligned}\sum_{\varphi\in\Phi}\langle P_{I^\perp}(xx^*),P_{I^\perp}(\varphi\varphi^*)\rangle_{\mathrm{HS}}^2&=\frac{M-1}{M}\sum_{\varphi\in\Phi}\bigg\langle P_{I^\perp}(xx^*),\sqrt{\frac{M}{M-1}}P_{I^\perp}(\varphi\varphi^*)\bigg\rangle_{\mathrm{HS}}^2\\&=\frac{M-1}{M}\cdot\frac{M(M+1)/2}{M(M+1)/2-1}\|P_{I^\perp}(xx^*)\|_{\mathrm{HS}}^2.\end{aligned}\tag{6.3}$$

At this point, we apply the Pythagorean theorem to get

$$\|P_{I^\perp}(xx^*)\|_{\mathrm{HS}}^2=\|xx^*\|_{\mathrm{HS}}^2-\|P_I(xx^*)\|_{\mathrm{HS}}^2=\Big(1-\frac{1}{M}\Big)\|x\|^4,$$

where the last step follows from the definition of P_I. Substituting this into (6.3) and (6.2) and then simplifying gives the desired identity. $\square$

In this special case where $N = M(M+1)/2$, it is straightforward to verify that Theorem 6.3 implies something special about the form of M. In particular, provided $N \neq 2M$ (i.e., $M \neq 3$), $\exists\,\mathrm{RETF}(M, M(M+1)/2)$ requires an integer $m \geq 1$ such that $M = (2m+1)^2 - 2$. Overall, $\exists\,\mathrm{RETF}(M, M(M+1)/2)$ implies $M \in \{3, 7, 23, 47, 79, 119, 167, \ldots\}$. The following theorem summarizes what is known about the existence of these ETFs:

THEOREM 6.8 (see [**2**, **25**, **32**]).
(a) $M \in \{3, 7, 23\}$ *implies* $\exists\,\mathrm{RETF}(M, M(M+1)/2)$.
(b) $M = 47$ *implies* $\nexists\,\mathrm{RETF}(M, M(M+1)/2)$.
(c) *Suppose* $k \equiv 2 \bmod 3$, k *and* $2k+1$ *are both square-free, and take* $m = 2k$. *Then* $M = (2m+1)^2 - 2$ *implies* $\nexists\,\mathrm{RETF}(M, M(M+1)/2)$.

Part (b) above was originally proved by Makhnev [**32**] in terms of strongly regular graphs, and soon thereafter, Bannai, Munemasa and Venkov [**2**] found an alternative proof in terms of spherical 5-designs (along with a proof of part (c) above). Other than the dimension bound (Theorem 6.1) and the integrality conditions (Theorem 6.3), this is the only known nonexistence result for real ETFs with $N > 2M$. In fact, this disproves a conjecture that was posed in [**40**] and reiterated in [**46**].

References

[1] N. A. Balonin, J. Seberry, A review and new symmetric conference matrices, Available online: `http://ro.uow.edu.au/cgi/viewcontent.cgi?article=3757&context=eispapers`

[2] E. Bannai, A. Munemasa, and B. Venkov, *The nonexistence of certain tight spherical designs*, Algebra i Analiz **16** (2004), no. 4, 1–23, DOI 10.1090/S1061-0022-05-00868-X; English transl., St. Petersburg Math. J. **16** (2005), no. 4, 609–625. MR2090848 (2005e:05022)

[3] V. Belevitch, Theorem of $2n$-terminal networks with application to conference telephony, Electr. Commun. 26 (1950) 231–244.

[4] J. J. Benedetto and M. Fickus, *Finite normalized tight frames*, Adv. Comput. Math. **18** (2003), no. 2-4, 357–385, DOI 10.1023/A:1021323312367. Frames. MR1968126 (2004c:42059)

[5] J. J. Benedetto, J. D. Kolesar, Geometric properties of Grassmannian frames for $\mathbb{R}^2$ and $\mathbb{R}^3$, EURASIP J. Appl. Signal Process. 2006 (2006) 1–17.

[6] B. G. Bodmann and P. G. Casazza, *The road to equal-norm Parseval frames*, J. Funct. Anal. **258** (2010), no. 2, 397–420, DOI 10.1016/j.jfa.2009.08.015. MR2557942 (2010j:42060)

[7] B. G. Bodmann and J. Haas, *Frame Potentials and the Geometry of Frames*, J. Fourier Anal. Appl. **21** (2015), no. 6, 1344–1383. MR3421919

[8] A. E. Brouwer, Strongly regular graphs, In: Handbook of Combinatorial Designs, 2nd ed., 2007, 852–868.

[9] A. E. Brouwer and W. H. Haemers, *Spectra of graphs*, Universitext, Springer, New York, 2012. MR2882891

[10] J. Cahill, M. Fickus, D. G. Mixon, M. J. Poteet, and N. Strawn, *Constructing finite frames of a given spectrum and set of lengths*, Appl. Comput. Harmon. Anal. **35** (2013), no. 1, 52–73, DOI 10.1016/j.acha.2012.08.001. MR3053746

[11] J. Cahill, D. G. Mixon, N. Strawn, Connectivity and irreducibility of algebraic varieties of finite unit norm tight frames, Available online: arXiv:1311.4748

[12] P. G. Casazza, M. Fickus, and D. G. Mixon, *Auto-tuning unit norm frames*, Appl. Comput. Harmon. Anal. **32** (2012), no. 1, 1–15, DOI 10.1016/j.acha.2011.02.005. MR2854158

[13] P. G. Casazza, M. Fickus, D. G. Mixon, Y. Wang, and Z. Zhou, *Constructing tight fusion frames*, Appl. Comput. Harmon. Anal. **30** (2011), no. 2, 175–187, DOI 10.1016/j.acha.2010.05.002. MR2754774 (2012c:42069)

[14] P. G. Casazza, A. Heinecke, F. Krahmer, and G. Kutyniok, *Optimally sparse frames*, IEEE Trans. Inform. Theory **57** (2011), no. 11, 7279–7287, DOI 10.1109/TIT.2011.2160521. MR2883655 (2012h:94043)

[15] P. Delsarte, J. M. Goethals, and J. J. Seidel, *Spherical codes and designs*, Geometriae Dedicata **6** (1977), no. 3, 363–388. MR0485471 (58 #5302)

[16] C. Ding and T. Feng, *A generic construction of complex codebooks meeting the Welch bound*, IEEE Trans. Inform. Theory **53** (2007), no. 11, 4245–4250, DOI 10.1109/TIT.2007.907343. MR2446568 (2010c:94023)

[17] D.-J. Feng, L. Wang, and Y. Wang, *Generation of finite tight frames by Householder transformations*, Adv. Comput. Math. **24** (2006), no. 1-4, 297–309, DOI 10.1007/s10444-004-7637-9. MR2222273 (2006m:42057)

[18] J. F. Fernando, J. M. Gamboa, On Lojasiewicz's inequality and the nullstellenstatz for rings of semialgebraic functions, Math. Res. Lett. 16 (2010) 10001–10015.

[19] M. Fickus, Constructions of normalized tight frames, preprint.

[20] M. Fickus, D. G. Mixon, and J. C. Tremain, *Steiner equiangular tight frames*, Linear Algebra Appl. **436** (2012), no. 5, 1014–1027, DOI 10.1016/j.laa.2011.06.027. MR2890902

[21] M. Fickus, J. Marks, M. J. Poteet, A generalized Schur-Horn theorem and optimal frame completions, Available online: arXiv:1408.2882

[22] M. Fickus, D. G. Mixon, M. J. Poteet, Frame completions for optimally robust reconstruction, Proc. SPIE 8138 (2011) 81380Q/1–8.

[23] M. Fickus, D. G. Mixon, M. J. Poteet, and N. Strawn, *Constructing all self-adjoint matrices with prescribed spectrum and diagonal*, Adv. Comput. Math. **39** (2013), no. 3-4, 585–609, DOI 10.1007/s10444-013-9298-z. MR3116042

[24] J.-M. Goethals and J. J. Seidel, *Spherical designs*, Relations between combinatorics and other parts of mathematics (Proc. Sympos. Pure Math., Ohio State Univ., Columbus, Ohio, 1978), Proc. Sympos. Pure Math., XXXIV, Amer. Math. Soc., Providence, R.I., 1979, pp. 255–272. MR525330 (82h:05014)

[25] J.-M. Goethals and J. J. Seidel, *The regular two-graph on* 276 *vertices*, Discrete Math. **12** (1975), 143–158. MR0384597 (52 #5471)

[26] V. K. Goyal, J. Kovačević, and J. A. Kelner, *Quantized frame expansions with erasures*, Appl. Comput. Harmon. Anal. **10** (2001), no. 3, 203–233, DOI 10.1006/acha.2000.0340. MR1829801 (2002h:94012)

[27] P. de la Harpe and C. Pache, *Cubature formulas, geometrical designs, reproducing kernels, and Markov operators*, Infinite groups: geometric, combinatorial and dynamical aspects, Progr. Math., vol. 248, Birkhäuser, Basel, 2005, pp. 219–267, DOI 10.1007/3-7643-7447-0_6. MR2195455 (2007i:05037)

[28] M. A. Herman and T. Strohmer, *High-resolution radar via compressed sensing*, IEEE Trans. Signal Process. **57** (2009), no. 6, 2275–2284, DOI 10.1109/TSP.2009.2014277. MR2641823 (2011a:94028)

[29] R. B. Holmes and V. I. Paulsen, *Optimal frames for erasures*, Linear Algebra Appl. **377** (2004), 31–51, DOI 10.1016/j.laa.2003.07.012. MR2021601 (2004j:42028)

[30] A. Horn, *Doubly stochastic matrices and the diagonal of a rotation matrix*, Amer. J. Math. **76** (1954), 620–630. MR0063336 (16,105c)

[31] R. A. Horn and C. R. Johnson, *Matrix analysis*, Cambridge University Press, Cambridge, 1985. MR832183 (87e:15001)

[32] A. A. Makhnev, *On the nonexistence of strongly regular graphs with the parameters* $(486, 165, 36, 66)$ (Russian, with English and Ukrainian summaries), Ukraïn. Mat. Zh. **54** (2002), no. 7, 941–949, DOI 10.1023/A:1022066425998; English transl., Ukrainian Math. J. **54** (2002), no. 7, 1137–1146. MR2015515 (2004i:05161)

[33] D. G. Mixon, C. J. Quinn, N. Kiyavash, and M. Fickus, *Fingerprinting with equiangular tight frames*, IEEE Trans. Inform. Theory **59** (2013), no. 3, 1855–1865, DOI 10.1109/TIT.2012.2229781. MR3030758

[34] G. Nebe and B. Venkov, *On tight spherical designs*, Algebra i Analiz **24** (2012), no. 3, 163–171, DOI 10.1090/S1061-0022-2013-01249-0; English transl., St. Petersburg Math. J. **24** (2013), no. 3, 485–491. MR3014131

[35] Platonic solid, Wikipedia, Available online: `http://en.wikipedia.org/wiki/Platonic_solid`

[36] D. J. Redmond, *Existence and construction of real-valued equiangular tight frames*, ProQuest LLC, Ann Arbor, MI, 2009. Thesis (Ph.D.)–University of Missouri - Columbia. MR2890163

[37] I. Schur, Über eine Klasse von Mittelbildungen mit Anwendungen auf die Determinationen-Theorie, Sitzungsber. Berl. Math. Ges. 22 (1923) 9–20.

[38] S. Smale, *Mathematical problems for the next century*, Math. Intelligencer **20** (1998), no. 2, 7–15, DOI 10.1007/BF03025291. MR1631413 (99h:01033)

[39] N. Strawn, *Optimization over finite frame varieties and structured dictionary design*, Appl. Comput. Harmon. Anal. **32** (2012), no. 3, 413–434, DOI 10.1016/j.acha.2011.09.001. MR2892742

[40] M. A. Sustik, J. A. Tropp, I. S. Dhillon, and R. W. Heath Jr., *On the existence of equiangular tight frames*, Linear Algebra Appl. **426** (2007), no. 2-3, 619–635, DOI 10.1016/j.laa.2007.05.043. MR2350682 (2008f:15066)

[41] T. Strohmer and R. W. Heath Jr., *Grassmannian frames with applications to coding and communication*, Appl. Comput. Harmon. Anal. **14** (2003), no. 3, 257–275, DOI 10.1016/S1063-5203(03)00023-X. MR1984549 (2004d:42053)

[42] J. J. Thomson, On the structure of the atom: An investigation of the stability and periods of oscillation of a number of corpuscles arranged at equal intervals around the circumference of a circle; With application of the results to the theory of atomic structure, Philosophical Mag. 6, vol. 7, (1904) 237–265.

[43] J. A. Tropp, *On the conditioning of random subdictionaries*, Appl. Comput. Harmon. Anal. **25** (2008), no. 1, 1–24, DOI 10.1016/j.acha.2007.09.001. MR2419702 (2009e:60014)

[44] R. Vale and S. Waldron, *Tight frames and their symmetries*, Constr. Approx. **21** (2005), no. 1, 83–112, DOI 10.1007/s00365-004-0560-y. MR2105392 (2005h:42063)

[45] B. Venkov, *Réseaux et designs sphériques* (French, with English and French summaries), Réseaux euclidiens, designs sphériques et formes modulaires, Monogr. Enseign. Math., vol. 37, Enseignement Math., Geneva, 2001, pp. 10–86. MR1878745 (2002m:11061)

[46] S. Waldron, *On the construction of equiangular frames from graphs*, Linear Algebra Appl. **431** (2009), no. 11, 2228–2242, DOI 10.1016/j.laa.2009.07.016. MR2567829 (2010k:42064)

[47] L. R. Welch, Lower bounds on the maximum cross correlation of signals, IEEE Trans. Inform. Theory 20 (1974) 397–399.

[48] P. Xia, S. Zhou, and G. B. Giannakis, *Achieving the Welch bound with difference sets*, IEEE Trans. Inform. Theory **51** (2005), no. 5, 1900–1907, DOI 10.1109/TIT.2005.846411. MR2235693 (2007b:94148a)

[49] G. Zimmermann, *Normalized tight frames in finite dimensions*, Recent progress in multivariate approximation (Witten-Bommerholz, 2000), Internat. Ser. Numer. Math., vol. 137, Birkhäuser, Basel, 2001, pp. 249–252. MR1877512

Department of Mathematics and Statistics, Air Force Institute of Technology, Wright-Patterson AFB, Ohio

Proceedings of Symposia in Applied Mathematics
Volume **73**, 2016
http://dx.doi.org/10.1090/psapm/073/00629

Algebro-Geometric Techniques and Geometric Insights for Finite Frames

Nate Strawn

ABSTRACT. Through a series of examples, we explore algebraic varieties of finite unit norm tight frames (FUNTFs) using techniques from Geometry and Algebraic Geometry. First, we identify the FUNTF varieties as almost-everywhere transversal intersections of tori and Stiefel manifolds. This allows us to characterize the singular points on these varieties, compute the tangent spaces at regular points, and determine when the FUNTF variety is a manifold. Next, using elimination theory, we explicitly solve the system of equations defining the FUNTF varieties. We show an alternative parameterization using eigensteps (also known as Gelfand-Tsetlin patterns), and we use eigensteps to examine the path connectivity of FUNTF varieties and their non-singular points. We conclude by using these facts to show that the FUNTF varieties are irreducible.

CONTENTS

1. Background
2. Notation
3. Intersecting tori and Stiefel manifolds in the Hilbert-Schmidt sphere
4. Explicit, locally-defined, analytic coordinate functions on $\mathcal{F}_{d,N}$
5. Connectivity and irreducibility of $\mathcal{F}_{d,N}$
6. A final challenge
References

1. Background

Some recent breakthroughs in Finite Frame Theory have invoked Algebro-Geometric and Geometric techniques. The line of reasoning traversed in these notes first began with the work of Dykema and Strawn [**E**]. In this paper, spaces of FUNTFs were first examined as algebraic varieties and their geometric structure was examined closely. In that same paper, path-connectivity was demonstrated for the spaces of FUNTFs in two dimensions with at least four vectors, and path-connectivity was conjectured to hold for spaces of FUNTFs in arbitrary dimension

2010 *Mathematics Subject Classification.* Primary 42C15, 47B99; Secondary 14M99.
Key words and phrases. Frame theory, real algebraic geometry, frame homotopy problem, eigensteps.

as long as the number of frame vectors exceeded the dimension by at least two. Some partial results were obtained on this homotopy problem by Giol et al. [**I**], and the conjecture was completely verified in [**D**].

The pivotal development that enabled the verification of this conjecture has been explicit parameterizations of spaces of FUNTFs. In [**L**], explicit parameterizations of FUNTF spaces were constructed for the first time. In particular, that work used techniques from elimination theory to solve the system of polynomials encoding the constraints the FUNTFs. While solving this system of equations could be carried out, the solution involved solutions of quartic equations, and writing out the equations required dozens of pages. As such, the utility of the parameterization remains somewhat limited.

In [**C**], a powerful parameterization of finite frames was uncovered using the theory of eigensteps. If one considers a rank-one update of a symmetric positive-definite matrix

$$A + xx^*,$$

there is no known general closed-form solution for the updated eigenvalues and eigenvectors. On the other hand, if one specifies the desired eigenvalues for the rank-one update, then there are convenient equations for x and the eigenvectors of the updated system. Eigensteps are a specification of eigenvalue sequences for partial sums of the form

$$x_1x_1^*,\ x_1x_1^* + x_2x_2^*,\ x_1x_1^* + x_2x_2^* + x_3x_3^*, \dots$$

Parameterizations of FUNTF varieties using eigensteps are relatively simple, and having control over the eigenvalues of a frame allows one to control the structure of the frame inside of a trajectory. Because of this, eigenstep parameterizations allow for the construction of paths connecting arbitrary frames to special "hubs" inside of the FUNTF varieties, and this observation leads to the solution of the frame homotopy problem .

FUNTFs are generally used in applications because they satisfy certain optimality properties (see [**H**, **K**, **N**]). Since there are potentially many degrees of freedom on a FUNTF variety, it is desirable to perform optimization on these spaces for various applications. A general procedure for such optimization utilizes the geometry of FUNTF varieties to perform geometric gradient descent directly on spaces of FUNTFs [**M**]. The fact that these spaces are connected means that such a procedure is not doomed to fail because it is initialized on a component that does not contain the optimizer. While [**M**] develops a convergence theory for this procedure, it is only valid in the case when the spaces of FUNTFs are manifolds. In [**E**] it was shown that FUNTF varieties of N vectors in d dimensional space are manifolds if and only if N and d are relatively prime. Because of the abundance of singularities, developing a proper theory for optimization over these spaces shall first require a deeper understanding of the singular points on these varieties.

In this chapter, we will survey the various parameterizations of the FUNTF varieties, describe the singular points of these varieties, and show how these parameterizations lead to complete solutions to the frame homotopy problem. We also describe how similar techniques may be used to show that generic FUNTFs are full-spark , which has consequences for compressed sensing applications of FUNTFs.

2. Notation

In these notes, we consider only real-valued frames, and we identify finite frames of N members in $\mathbb{R}^d$ with their realization as d by N matrices (the synthesis operator). We denote the set of all d by N matrices with real entries by $M_{d,N}$, we let $\mathcal{F}_{d,N} \subset M_{d,N}$ denote the space of N-member FUNTFs in $\mathbb{R}^d$. That is, $\mathcal{F}_{d,N}$ consists of all d by N matrices whose columns $\{f_i\}_{i=1}^N$ satisfy the unit norm condition ($\|f_i\| = 1$ for all $i = 1, \dots, N$) and tightness condition:

$$\sum_{i=1}^{N} |\langle x, f_i \rangle|^2 = A\|x\|^2$$

for all $x \in \mathbb{R}^d$, where $A > 0$ is a fixed constant.

We let I_d denote the d by d identity matrix, $\mathbf{0}_d \in \mathbb{R}^d$ denotes the zero vector, and $\mathbf{1}_d \in \mathbb{R}^d$ denotes the vector with all entries equal to 1. For a d by d matrix A, we let $\operatorname{tr}(A)$ denote the trace of the matrix A and $\operatorname{diag}(A) \in \mathbb{R}^d$ denotes the vector consisting of all diagonal entries of A.

For $F = (f_1\, f_2\, \cdots\, f_N) \in \mathcal{F}_{d,N}$ we have that FF^* is the frame operator of F and F^*F is the Grammian, where F^* is the transpose of F (the analysis operator). Also, note that

$$FF^* = \sum_{i=1}^{N} f_i f_i^*.$$

Throughout these notes, we shall use the frames

$$\mathcal{E} = (\phi_1\, \phi_2\, \phi_3\, \phi_4\, \phi_5\, \phi_6) = \begin{pmatrix} 1 & \frac{\sqrt{3}}{3} & 0 & \frac{\sqrt{2}}{2} & 0 & \frac{\sqrt{6}}{6} \\ 0 & \frac{\sqrt{3}}{3} & 1 & 0 & 0 & -\frac{\sqrt{6}}{3} \\ 0 & \frac{\sqrt{3}}{3} & 0 & -\frac{\sqrt{2}}{2} & 1 & \frac{\sqrt{6}}{6} \end{pmatrix}$$

and

$$\Xi = \begin{pmatrix} 1 & 0 & 0 & 1 & 0 & 0 \\ 0 & 1 & 0 & 0 & 1 & 0 \\ 0 & 0 & 1 & 0 & 0 & 1 \end{pmatrix}$$

as our central examples. It is not difficult to check that $\mathcal{E}, \Xi \in \mathcal{F}_{3,6}$. The Gram matrix of $\mathcal{E}$ is

$$\begin{pmatrix} 1 & \frac{\sqrt{3}}{3} & 0 & \frac{\sqrt{2}}{2} & 0 & \frac{\sqrt{6}}{6} \\ \frac{\sqrt{3}}{3} & 1 & \frac{\sqrt{3}}{3} & 0 & \frac{\sqrt{3}}{3} & 0 \\ 0 & \frac{\sqrt{3}}{3} & 1 & 0 & 0 & -\frac{\sqrt{6}}{3} \\ \frac{\sqrt{2}}{2} & 0 & 0 & 1 & -\frac{\sqrt{2}}{2} & 0 \\ 0 & \frac{\sqrt{3}}{3} & 0 & -\frac{\sqrt{2}}{2} & 1 & \frac{\sqrt{6}}{6} \\ \frac{\sqrt{6}}{6} & 0 & -\frac{\sqrt{6}}{3} & 0 & \frac{\sqrt{6}}{6} & 1 \end{pmatrix}.$$

Setting $\rho_{d,N} = \frac{N}{d}$, we let

$$\mathrm{St}_{d,N} = \{F \in M_{d,N} : FF^* = \rho_{d,N} I_d\}$$

denote the scaled Stiefel manifold , which is in correspondence with the tight frames of N members in $\mathbb{R}^d$ with frame bound $\rho_{d,N}$. We let

$$\mathcal{S}_d = \{u \in \mathbb{R}^d : \|u\| = 1\}$$

and

$$\mathcal{S}_{d,N} = \{V \in M_{d,N} : \operatorname{tr}(VV^*) = N\}$$

respectively denote the unit spheres in $\mathbb{R}^d$ under the Euclidean norm, and the radius $\sqrt{N}$ sphere in $M_{d,N}$ under the Hilbert-Schmidt norm . Finally, define the torus

$$\mathbb{T}_{d,N} = \{V \in M_{d,N} : \operatorname{diag}(V^*V) = \mathbf{1}_N\}.$$

A real algebraic variety is any subset of $\mathbb{R}^d$ which is the zero set of a system of polynomials, and an irreducible algebraic variety is a variety that is not a non-trivial union of two varieties. For a reference on algebraic geometry, refer to [**G**], and for a reference on manifolds, diffeomorphisms, and transversal intersections of manifolds, see [**F**].

3. Intersecting tori and Stiefel manifolds in the Hilbert-Schmidt sphere

By definition, $F \in \mathbb{T}_{d,N}$ if and only if the diagonal of F^*F consists entirely of 1's. Thus,

$$\operatorname{tr}(FF^*) = \operatorname{tr}(F^*F) = N.$$

Similarly, for $F \in \mathrm{St}_{d,N}$, we have that

$$\operatorname{tr}(FF^*) = \operatorname{tr}(\rho_{d,N} I_d) = d\rho_{d,N} = N.$$

Aggregating these accumulated facts, we obtain Proposition 3.1.

PROPOSITION 3.1. *For $N \geq d$, we have that $\mathbb{T}_{d,N} \subset \mathcal{S}_{d,N}$, $St_{d,N} \subset \mathcal{S}_{d,N}$, and $\mathcal{F}_{d,N} = \mathbb{T}_{d,N} \cap St_{d,N}$. This also verifies that $\mathcal{F}_{d,N}$ is a variety defined by quadratic polynomials.*

Because both $\mathbb{T}_{d,N}$ and $\mathrm{St}_{d,N}$ lie in the $\mathcal{S}_{d,N}$, we may examine their intersection and the transversality of that intersection relative to $\mathcal{S}_{d,N}$. It is now well-known that $\mathbb{T}_{d,N} \cap \mathrm{St}_{d,N}$ is not empty for all $N \geq d \geq 1$.

3.1. Local transversality of the intersection of $\mathbb{T}_{d,N}$ and $\mathrm{St}_{d,N}$ in $\mathcal{S}_{d,N}$.

We now show that the intersection of $\mathbb{T}_{3,6}$ and $\mathrm{St}_{3,6}$ is transversal at $\mathcal{E}$, but not at Ξ. First, let $\mathcal{O}(k)$ denote the group of orthogonal k by k matrices and note that the Lie algebra (tangent space at the identity element) of this group is the set of k by k skew-symmetric matrices. By observing the respective transitive actions of

$$\mathcal{O}(3) \times \mathcal{O}(3) \times \mathcal{O}(3) \times \mathcal{O}(3) \times \mathcal{O}(3) \times \mathcal{O}(3) \text{ and } \mathcal{O}(6)$$

on $\mathbb{T}_{3,6}$ and $\mathrm{St}_{3,6}$ given by

$$(f_1\, f_2\, f_3\, f_4\, f_5\, f_6) \mapsto (U_1 f_1\, U_2 f_2\, U_3 f_3\, U_4 f_4\, U_5 f_5\, U_6 f_6) \text{ and } F \mapsto FU$$

respectively, we see that the tangent space of $\mathbb{T}_{3,6}$ at $\mathcal{E}$ is equal to

$$T_{\mathcal{E}}\mathbb{T}_{3,6} = \{Y \in M_{3,6} : \operatorname{diag}(Y^*\mathcal{E}) = \mathbf{0}_N\},$$

and also that the tangent space of $\mathrm{St}_{3,6}$ at $\mathcal{E}$ equals

$$T_{\mathcal{E}}\mathrm{St}_{3,6} = \{\mathcal{E}Z \in M_{3,6} : Z \in M_{6,6},\ Z = -Z^*\}.$$

To show local transversality of the intersection of $\mathbb{T}_{3,6}$ and $\mathrm{St}_{3,6}$ at $\mathcal{E}$, we must show that, for each $X = (x_1\, x_2\, x_3\, x_4\, x_5\, x_6) \in T_{\mathcal{E}}\mathcal{S}_{3,6} = \{X \in M_{3,6} : \operatorname{tr}(X^*\mathcal{E}) = 0\}$, there is a skew-symmetric $Z \in M_{6,6}$ and a $Y = (y_1\ y_2\ y_3\ y_4\ y_5\ y_6) \in T_{\mathcal{E}}\mathbb{T}_{3,6}$ such that $X = \mathcal{E}Z + Y$. In columns, this becomes the system of equations

$$\begin{array}{ccccccccccccccc}
x_1 & = & 0 & + & z_{21}\phi_2 & + & z_{31}\phi_3 & + & z_{41}\phi_4 & + & z_{51}\phi_5 & + & z_{61}\phi_6 & + & y_1 \\
x_2 & = & -z_{21}\phi_1 & + & 0 & + & z_{32}\phi_3 & + & z_{42}\phi_4 & + & z_{52}\phi_5 & + & z_{62}\phi_6 & + & y_2 \\
x_3 & = & -z_{31}\phi_1 & - & z_{32}\phi_2 & + & 0 & + & z_{43}\phi_4 & + & z_{53}\phi_5 & + & z_{63}\phi_6 & + & y_3 \\
x_4 & = & -z_{41}\phi_1 & - & z_{42}\phi_2 & - & z_{43}\phi_3 & + & 0 & + & z_{54}\phi_5 & + & z_{64}\phi_6 & + & y_4 \\
x_5 & = & -z_{51}\phi_1 & - & z_{52}\phi_2 & - & z_{53}\phi_3 & - & z_{54}\phi_4 & + & 0 & - & z_{65}\phi_6 & + & y_5 \\
x_6 & = & -z_{61}\phi_1 & - & z_{62}\phi_2 & - & z_{63}\phi_3 & - & z_{64}\phi_4 & - & z_{65}\phi_5 & + & 0 & + & y_6
\end{array}.$$

Noting that $\langle y_i, \phi_i \rangle = 0$, these equations imply the system of equations

(3.2)

$$\begin{array}{ccccccccccccc}
\langle x_1, \phi_1\rangle & = & 0 & + & \frac{\sqrt{3}}{3}z_{21} & + & 0 & + & \frac{\sqrt{2}}{2}z_{41} & + & 0 & + & \frac{\sqrt{6}}{6}z_{61} \\
\langle x_2, \phi_2\rangle & = & -\frac{\sqrt{3}}{3}z_{21} & + & 0 & + & \frac{\sqrt{3}}{3}z_{32} & + & 0 & + & \frac{\sqrt{3}}{3}z_{52} & + & 0 \\
\langle x_3, \phi_3\rangle & = & 0 & - & \frac{\sqrt{3}}{3}z_{32} & + & 0 & + & 0 & + & 0 & - & \frac{\sqrt{6}}{3}z_{63} \\
\langle x_4, \phi_4\rangle & = & -\frac{\sqrt{2}}{2}z_{41} & + & 0 & + & 0 & + & 0 & - & \frac{\sqrt{2}}{2}z_{54} & + & 0 \\
\langle x_5, \phi_5\rangle & = & 0 & - & \frac{\sqrt{3}}{3}z_{52} & + & 0 & + & \frac{\sqrt{2}}{2}z_{54} & + & 0 & + & \frac{\sqrt{6}}{6}z_{65} \\
\langle x_6, \phi_6\rangle & = & -\frac{\sqrt{6}}{6}z_{61} & + & 0 & + & \frac{\sqrt{6}}{3}z_{63} & + & 0 & - & \frac{\sqrt{6}}{6}z_{65} & + & 0
\end{array}.$$

We can express this system more compactly in the form

$$\begin{pmatrix} \langle x_1, \phi_1\rangle \\ \langle x_2, \phi_2\rangle \\ \langle x_3, \phi_3\rangle \\ \langle x_4, \phi_4\rangle \\ \langle x_5, \phi_5\rangle \\ \langle x_6, \phi_6\rangle \end{pmatrix} = (\Gamma_{\mathcal{E}} \odot Z)\mathbf{1}_6 = \begin{pmatrix} 0 & \frac{\sqrt{3}}{3}z_{21} & 0 & \frac{\sqrt{2}}{2}z_{41} & 0 & \frac{\sqrt{6}}{6}z_{61} \\ -\frac{\sqrt{3}}{3}z_{21} & 0 & \frac{\sqrt{3}}{3}z_{32} & 0 & \frac{\sqrt{3}}{3}z_{52} & 0 \\ 0 & -\frac{\sqrt{3}}{3}z_{32} & 0 & 0 & 0 & -\frac{\sqrt{6}}{3}z_{63} \\ -\frac{\sqrt{2}}{2}z_{41} & 0 & 0 & 0 & -\frac{\sqrt{2}}{2}z_{54} & 0 \\ 0 & -\frac{\sqrt{3}}{3}z_{32} & 0 & \frac{\sqrt{2}}{2}z_{54} & 0 & \frac{\sqrt{6}}{6}z_{65} \\ -\frac{\sqrt{6}}{6}z_{61} & 0 & \frac{\sqrt{6}}{3}z_{63} & 0 & -\frac{\sqrt{6}}{6}z_{65} & 0 \end{pmatrix} \begin{pmatrix} 1 \\ 1 \\ 1 \\ 1 \\ 1 \\ 1 \end{pmatrix},$$

where $\odot$ denotes the Hadamard, or entry-wise product and $\Gamma_{\mathcal{E}}$ is the Gram matrix of $\mathcal{E}$. Note that $\Gamma_{\mathcal{E}} \odot Z$ is skew-symmetric. Because of the skew-symmetry of the terms arising in these equations, there is a very intuitive analogy for the system. Consider the banks B_1, B_2, B_3, B_4, B_5, and B_6. Suppose that we want to ship money in between the banks so that by the end of the month the net changes in the total cash deposits (excluding all other transactions) at the banks are $\langle x_1, \phi_1\rangle$, $\langle x_2, \phi_2\rangle$, $\langle x_3, \phi_3\rangle$, $\langle x_4, \phi_4\rangle$, $\langle x_5, \phi_5\rangle$, and $\langle x_6, \phi_6\rangle$ respectively. Skew-symmetry in the above equations enforces the condition that shipping money from bank B_i to bank B_j decreases the deposits at B_i by the specified amount, and increases the deposits at B_j by the specified amount. Every 0 in the above equations represents a shipping route that is prohibited (it takes too long to ship between those routes, or maybe it is not safe to ship money along those routes). Finally, the net transactions over the entire network must be zero.

Now, if we can find z_{ij} values that solve this system, then we immediately obtain $X - \mathcal{E}Z \in T_{\mathcal{E}}\mathbb{T}_{3,6}$. Thus, our problem has been reduced to determining whether this particular set of equations has a solution. We shall now illustrate a strategy that ultimately always produces the solution, provided that one exists. Moreover, the success of this strategy may be immediately related to the structure of the Gram matrix.

Noting that $X \in T_{\mathcal{E}}\mathcal{S}_{3,6}$ if and only if $\sum_{i=1}^{6}\langle x_i, \phi_i\rangle = 0$, we naively begin constructing Z column by column to satisfy all the equations. For the first column, we set $z_{61} = \sqrt{6}\langle x_1, \phi_1\rangle$ and all the remaining entries of the column are set to zero. Similarly, we can set $z_{52} = \sqrt{3}\langle x_2, \phi_2\rangle$, $z_{63} = -\frac{\sqrt{6}}{2}\langle x_3, \phi_3\rangle$, $z_{54} = -\sqrt{2}\langle x_4, \phi_4\rangle$, and all the other entries of the first four columns equal to zero. This ensures that the first four equations are all satisfied, and by imposing skew-symmetry we have the following partial definition of Z:

$$Z = \begin{pmatrix} 0 & 0 & 0 & 0 & 0 & -\sqrt{6}\langle x_1,\phi_1\rangle \\ 0 & 0 & 0 & 0 & -\sqrt{3}\langle x_2,\phi_2\rangle & 0 \\ 0 & 0 & 0 & 0 & 0 & \frac{\sqrt{6}}{2}\langle x_3,\phi_3\rangle \\ 0 & 0 & 0 & 0 & \sqrt{2}\langle x_4,\phi_4\rangle & 0 \\ 0 & \sqrt{3}\langle x_2,\phi_2\rangle & 0 & -\sqrt{2}\langle x_4,\phi_4\rangle & 0 & -z_{65} \\ \sqrt{6}\langle x_1,\phi_1\rangle & 0 & -\frac{\sqrt{6}}{2}\langle x_3,\phi_3\rangle & 0 & z_{65} & 0 \end{pmatrix}$$

Substituting in these values for z_{ij}, the fifth and sixth equations become

$$\begin{aligned} \langle x_5,\phi_5\rangle &= -\frac{\sqrt{3}}{3}\left(\sqrt{3}\langle x_2,\phi_2\rangle\right) + \frac{\sqrt{2}}{2}\left(-\sqrt{2}\langle x_4,\phi_4\rangle\right) + \frac{\sqrt{6}}{6}z_{65} \\ \langle x_6,\phi_6\rangle &= -\frac{\sqrt{6}}{6}\left(\sqrt{6}\langle x_1,\phi_1\rangle\right) + \frac{\sqrt{6}}{3}\left(-\frac{\sqrt{6}}{2}\langle x_3,\phi_3\rangle\right) - \frac{\sqrt{6}}{6}z_{65}. \end{aligned}$$

Solving both equations for z_{65} yields

$$\begin{aligned} z_{65} &= \sqrt{6}\left(\langle x_2,\phi_2\rangle + \langle x_4,\phi_4\rangle + \langle x_5,\phi_5\rangle\right) \\ z_{65} &= -\sqrt{6}\left(\langle x_1,\phi_1\rangle + \langle x_3,\phi_3\rangle + \langle x_6,\phi_6\rangle\right). \end{aligned}$$

By virtue of the fact that $\sum_{i=1}^{6}\langle x_i,\phi_i\rangle = 0$, we have that

$$\langle x_2,\phi_2\rangle + \langle x_4,\phi_4\rangle + \langle x_5,\phi_5\rangle = -\left(\langle x_1,\phi_1\rangle + \langle x_3,\phi_3\rangle + \langle x_6,\phi_6\rangle\right)$$

and we see that these expressions for z_{65} are consistent. Consequently, we have a general formula for Z solving the desired equations, and we conclude that $\mathbb{T}_{3,6}$ and $\mathrm{St}_{3,6}$ intersect transversally at $\mathcal{E}$ in $\mathcal{S}_{3,6}$.

Now, let us consider the same scenario for Ξ. By considering the Gram matrix of Ξ, the equations are

$$\begin{aligned} \langle x_1,e_1\rangle &= z_{41} \\ \langle x_2,e_2\rangle &= z_{52} \\ \langle x_3,e_3\rangle &= z_{63} \\ \langle x_4,e_1\rangle &= -z_{41} \\ \langle x_5,e_2\rangle &= -z_{52} \\ \langle x_6,e_3\rangle &= -z_{63}. \end{aligned}$$

By simply choosing $x_1 = e_1$, $x_2 = -2e_2$, $x_3 = e_1$, $x_4 = 0$, $x_5 = 0$, and $x_6 = 0$, we clearly have that

$$\langle x_1,e_1\rangle + \langle x_2,e_2\rangle + \langle x_3,e_3\rangle + \langle x_4,e_1\rangle + \langle x_5,e_2\rangle + \langle x_6,e_3\rangle = 0$$

but the above system of equations is inconsistent because $1 = z_{41}$ and $1 = -z_{41}$. Thus, we see that the transversality of the intersection fails for Ξ. Returning to the bank analogy from before, it becomes intuitively clear why the system of equations does not have a general solution. The banks B_1 and B_4 are only connected to each other, so it is not possible to ship money to either of them from bank B_2. Since the zeros in the system of equations are governed precisely by the zeros of the Gram matrix, we are naturally led to the following definition.

DEFINITION 3.3. For a frame $F = (f_1\, f_2\, \cdots f_N) \in M_{d,N}$, the **correlation network** or **frame graph** is the undirected graph $\mathcal{G}_F = (V,E)$ with vertices $V = \{1,2,\ldots,N\}$ and edge set

$$E = \{(i,j) : \langle f_i, f_j\rangle \neq 0\}.$$

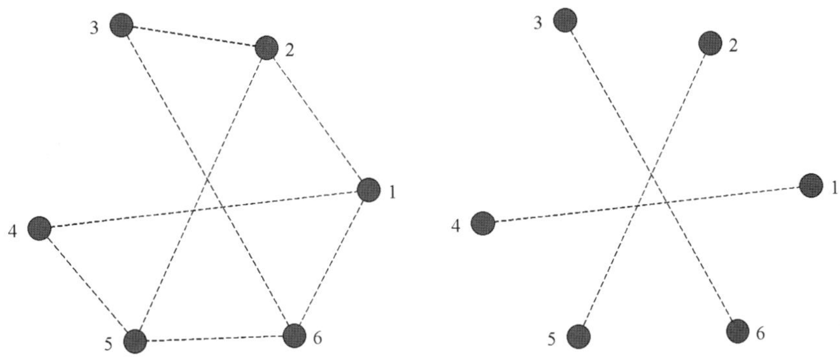

(A) $\mathcal{G}_{\mathcal{E}}$ (B) $\mathcal{G}_{\Xi}$

FIGURE 1. Correlation networks of our examples.

Figure 1 depicts the correlation networks of $\mathcal{E}$ and Ξ, and Figure 2 depicts the reduction of $\mathcal{G}_{\mathcal{E}}$ used to solve the linear system for our demonstration of local transversality. It is also clear from that this system cannot be solved for Ξ because $\mathcal{G}_{\Xi}$ is not connected. With the definition of the correlation network in hand, we may prove the following theorem that connects the correlation network and transversal intersection at F, and as a corollary we obtain the local dimension of the smooth structure which was first shown in [**L**].

THEOREM 3.4 (S., 2011). *Suppose $N \geq d \geq 2$. The manifolds $\mathbb{T}_{d,N}$ and $St_{d,N}$ intersect transversally in $\mathcal{S}_{d,N}$ at $F \in \mathcal{F}_{d,N}$ if and only if $\mathcal{G}_F$ is connected. Moreover, the local dimension of $\mathcal{F}_{d,N}$ around such an F is given by*

$$(d-1)N + \left(dN - \binom{d+1}{2}\right) - (dN-1) = (d-1)N - \binom{d+1}{2} + 1.$$

Connectivity of $\mathcal{G}_F$ thus provides us with a combinatorial condition for local transversal intersection. On the other hand, connectivity of $\mathcal{G}_F$ depends on the structure of the orthogonal pairs inside of F. It is therefore straightforward to obtain the following, equivalent geometric condition for local transversal intersection.

PROPOSITION 3.5. *The correlation network $\mathcal{G}_F$ is connected if and only if F cannot be partitioned into two non-trivial subsets of matrices with orthogonal column spaces. If F does admit such a partition, we say that F is* **orthodecomposable**.

Very little is known about the local geometry of $\mathcal{F}_{d,N}$ around orthodecomposable frames, and the existence of these points introduces an additional level of complexity when one would like to carry out optimization programs on $\mathcal{F}_{d,N}$.

PROBLEM 3.6. *If $\mathcal{F}_{d,N}$ is not a manifold, describe the local geometry around the orthodecomposable frames in $\mathcal{F}_{d,N}$.*

3.2. When is $\mathcal{F}_{d,N}$ a manifold? Now that we have a characterization of the singular points on $\mathcal{F}_{d,N}$, we can determine exactly when $\mathcal{F}_{d,N}$ is a manifold.

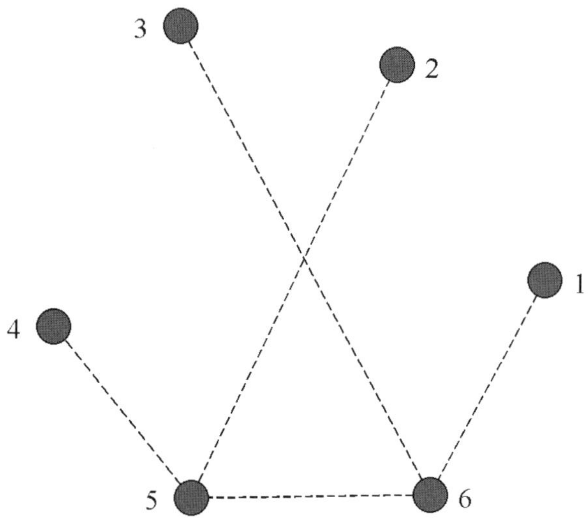

FIGURE 2. To solve the system (3.2), we ignore all the edges from $\mathcal{G}_{\mathcal{E}}$ that do not connect to the vertices labeled 5 and 6. We then fix the transfers to/from $1, 2, 3, 4$ from/to 5 and 6 that satisfy the first four equations, and then the required transfer between 5 and 6 is shown to be consistent by invoking the fact that the net change over the entire graph is zero.

Suppose $F \in \mathcal{F}_{d,N}$ is orthodecomposable, so there is a partition of F into $F_1 \in M_{d,N_1}$ and $F_2 \in M_{d,N_2}$, where $N_1, N_2 > 0$. Moreover, $\mathrm{col}(F_1) \perp \mathrm{col}(F_2)$ and hence

$$FF^*x_1 = F_1F_1^*x_1 = \frac{N}{d}x_1 \text{ and } FF^*x_2 = F_2F_2^*x_2 = \frac{N}{d}x_2$$

for $x_1 \in \mathrm{col}(F_1)$ and $x_2 \in \mathrm{col}(F_2)$. Consequently, we see that F_1 and F_2 are FUNTFs for subspaces of dimension $d_1 > 0$ and $d_2 > 0$, respectively. Moreover, the above can occur if and only if the frame bounds satisfy

$$\frac{N_1}{d_1} = \frac{N_2}{d_2} = \frac{N}{d} = \frac{N_1 + N_2}{d_1 + d_2}.$$

Since $N_1 = N - N_2 < N$, these equations imply the existence of an integer c so that $cN_1 = N$ and $cd_1 = d$, and we conclude that N and d are not relatively prime.

On the other hand, if N and d are not relatively prime with $N = cN_1$ and $d = cd_1$ for some $c > 1$, we set $N_2 = (c-1)N_1$ and $d_2 = (c-1)d_1$, choose $F_1 \in \mathcal{F}_{d_1,N_1}$ and $F_2 \in \mathcal{F}_{d_2,N_2}$, and observe that

$$\begin{pmatrix} F_1 & 0 \\ 0 & F_2 \end{pmatrix} \in \mathcal{F}_{d,N}$$

is orthodecomposable. Gathering together all of our observations, we obtain the following proposition.

PROPOSITION 3.7. *The variety $\mathcal{F}_{d,N}$ is a manifold if and only if N and d are relatively prime.*

4. Explicit, locally-defined, analytic coordinate functions on $\mathcal{F}_{d,N}$

In this section we show how to construct explicit local coordinate systems around all non-orthodecomposable $F \in \mathcal{F}_{d,N}$. While different optimization techniques may be carried out by simply understanding the tangent spaces of the FUNTF varieties, explicit coordinate systems were originally explored to demonstrate that the FUNTF varieties are indeed connected.

4.1. Local coordinates via elimination theory. First, let us build up our intuition by exploring perturbations of $\mathcal{E}$. Since $\mathcal{E}$ is a non-singular point of $\mathcal{F}_{d,N}$, Theorem 3.4 gives us that the local dimension of the manifold structure is

$$(3-1)6 - \binom{4}{2} + 1 = 7.$$

Now, perturbations of $\mathcal{E}$ that remain in $\mathbb{T}_{3,6}$ consist of angular perturbations (i.e. perturbations that keep norms fixed) of each column in $\mathcal{E}$. Because of the dimension constraint, we see that we can perform simultaneous angular perturbations of at most three vectors in $\mathcal{E}$. Let

$$\Theta = \begin{pmatrix} \theta_{11} & \theta_{12} & \theta_{13} \\ \theta_{21} & \theta_{22} & \theta_{23} \end{pmatrix}$$

denote the parameters associated to the angular perturbation of the first three columns of $\mathcal{E}$, and set

$$\phi_i(\Theta) = \phi_i(\theta_{1i}, \theta_{2i}) \text{ and } \phi_i(0,0) = \phi_i$$

for $i = 1, 2, 3$. This only accounts for six of the seven parameters, so we let τ denote the additional parameter and assume that the first three columns are independent of τ. Our goal is to find functions $\phi_i(\Theta, \tau)$ for $i = 4, 5, 6$ such that

$$(\Theta, \tau) \longrightarrow \Phi(\Theta, \tau) = (\phi_1(\Theta) \quad \phi_2(\Theta) \quad \phi_3(\Theta) \quad \phi_4(\Theta,\tau) \quad \phi_5(\Theta,\tau) \quad \phi_6(\Theta,\tau)) \in \mathcal{F}_{3,6}$$

is an embedding at $(\Theta_0, \tau_0) = (0, 0)$ and $\Phi(\Theta_0, \tau_0) = \mathcal{E}$. Now, a necessary condition is that

$$\Phi(\Theta,\tau)\Phi(\Theta,\tau)^* = \sum_{i=1}^{3} \phi_i(\Theta)\phi_i(\Theta)^* + \sum_{i=4}^{6} \phi_i(\Theta,\tau)\phi_i(\Theta,\tau)^* = 2I_3$$

which is equivalent to

$$\begin{aligned}
\widetilde{\Phi}(\Theta,\tau)\widetilde{\Phi}(\Theta,\tau)^* &= \left(\phi_4(\Theta,\tau) \quad \phi_5(\Theta,\tau) \quad \phi_6(\Theta,\tau)\right)\left(\phi_4(\Theta,\tau) \quad \phi_5(\Theta,\tau) \quad \phi_6(\Theta,\tau)\right)^* \\
&= \sum_{i=4}^{6} \phi_i(\Theta,\tau)\phi_i(\Theta,\tau)^* \\
&= 2I_3 - \sum_{i=1}^{3} \phi_i(\Theta)\phi_i(\Theta)^* \\
&= \widetilde{S}(\Theta)
\end{aligned}$$

Since $\widetilde{S}(0) = 2I_3 - \sum_{i=1}^{3} \phi_i\phi_i^*$ is positive definite and invertible, in a small enough neighborhood around Θ_0 we shall also have that $\widetilde{S}(\Theta)$ is positive definite and invertible. It then follows that any $\widetilde{\Phi}(\Theta,\tau)$ satisfying this condition is also invertible.

Thus, for Θ in a small neighborhood around Θ_0, we may perform a few algebraic manipulations to obtain the equivalent condition

$$\widetilde{\Phi}(\Theta,\tau)^*\widetilde{S}(\Theta)^{-1}\widetilde{\Phi}(\Theta,\tau) = I_3$$

or

$$\langle \widetilde{S}(\Theta)^{-1}\phi_i(\Theta,\tau), \phi_j(\Theta,\tau)\rangle = \begin{cases} 1 & i=j \\ 0 & i \neq j \end{cases}$$

for all $i,j = 4,5,6$. Consider the conditions that only involve $\phi_4(\Theta,\tau)$:

$$\begin{aligned} \phi_4(\Theta,\tau)^*\widetilde{S}(\Theta)^{-1}\phi_4(\Theta,\tau) &= 1 \\ \phi_4(\Theta,\tau)^*\phi_4(\Theta,\tau) &= 1. \end{aligned}$$

Thus, the possible choices for $\phi_4(\Theta,\tau)$ come from the intersection of an ellipsoid and a sphere, but where the ellipsoid is a function of Θ. We see that the parameter τ comes from the single degree of freedom obtained in this intersection, and this motivates the set of conditions

$$\begin{aligned} \phi_4(\Theta,\tau)^*\widetilde{S}(\Theta)^{-1}\phi_4(\Theta,\tau) &= 1 \\ \phi_4(\Theta,\tau)^*\phi_4(\Theta,\tau) &= 1 \\ \phi_4(\Theta,\tau)^*\eta &= \tau \end{aligned}$$

where η is a unit vector satisfying $\phi_4^*\widetilde{S}(0)^{-1}\eta = 0$ and $\phi_4^*\eta = 0$, or equivalently η is tangent to the intersection of the ellipsoid and the sphere at (Θ_0, τ_0). For each fixed τ, we can use the third equation to eliminate one of the variables in the first two equations, which ultimately reduces the first two equations to a system of the form

$$\begin{aligned} \alpha_2(x)y^2 + \alpha_1(x)y + \alpha_0(x) &= 0 \\ \beta_2(x)y^2 + \beta_1(x)y + \beta_0(x) &= 0 \end{aligned}$$

where the polynomials α_i, β_i have degree at most $2-i$ for $i = 0,1,2$ and the coefficients of the polynomials α_i, β_i are functions of (Θ,τ). Here, x and y are variables that effectively specify a consistent system of linear equations on $\phi_4(\Theta,\tau)$:

$$\phi_4(\Theta,\tau)^*\eta_x = x,\ \phi_4(\Theta,\tau)^*\eta_y = y,\ \text{and}\ \phi_4(\Theta,\tau)^*\eta = \tau.$$

We state the following standard result without proof:

PROPOSITION 4.1. *The above system admits a solution y if and only if there is an x such that the Bézout determinant*

$$(\alpha_2(x)\beta_1(x) - \alpha_1(x)\beta_2(x))(\alpha_1(x)\beta_0(x) - \alpha_0(x)\beta_1(x)) - (\alpha_2(x)\beta_0(x) - \alpha_0(x)\beta_2(x))^2$$

vanishes.

By considering the degrees of α_i and β_i, we see that $\mathrm{Bz}(x)$ is a polynomial of degree at most four. Consequently, we may use the quartic formula to analytically solve for $x = x(\Theta,\tau)$, and then we can back solve for y and the variable eliminated by the linear constraint. In this manner, we see that it is formally possible to construct an analytic solution to $\phi_4(\Theta,\tau)$. It then remains to solve the systems

$$\begin{aligned} \phi_i(\Theta,\tau)^*\widetilde{S}(\Theta)^{-1}\phi_i(\Theta,\tau) &= 1 \\ \phi_i(\Theta,\tau)^*\phi_i(\Theta,\tau) &= 1 \\ \phi_i(\Theta,\tau)^*\widetilde{S}(\Theta)^{-1}\phi_4(\Theta,\tau) &= 0 \end{aligned}$$

for $i = 5, 6$. Since the final equation is a linear constraint, an analytic solution is obtained using the same approach we used to construct $\phi_4(\Theta, \tau)$.

Thus, we can formally construct analytic coordinates on $\mathcal{F}_{3,6}$ in a neighborhood around $\mathcal{E}$. In [**L**], it was shown that these formal coordinates provide a local parameterization of $\mathcal{F}_{3,6}$ around $\mathcal{E}$. While the coordinate functions themselves can be computed explicitly, the equations use the quartic formula, which is a very involved expression. Consequently, these explicit coordinate functions can fill many pages.

4.2. Local coordinates from eigensteps. In contrast to the above coordinate systems, liftings of eigensteps provides explicit parameterizations of $\mathcal{F}_{d,N}$ that are far more concise. On the other hand, it is not clear that there is such a lifting exists for every non-singular point in $\mathcal{F}_{d,N}$, which leads to an interesting open problem.

Let $\lambda' : M_{d,d} \to \mathbb{R}^d$ denote the map taking a matrix to its eigenvalues listed in non-increasing order (note that the list is not strictly decreasing because the map includes multiplicities). Now, we define the eigensteps map $\lambda : \mathcal{F}_{d,N} \to M_{d,N}$ by setting

$$\lambda(F) = \begin{pmatrix} \lambda'(F_1 F_1^*) & \lambda'(F_2 F_2^*) & \cdots & \lambda'(FF^*) \end{pmatrix}$$

where $F_k = (f_1\ f_2\ \cdots\ f_k)$ are the first k columns of F. For example, a somewhat laborious calculation give us

$$\lambda(\mathcal{E}) = \begin{pmatrix} 1 & (3+\sqrt{3})/3 & (3+\sqrt{6})/3 & 2 & 2 & 2 \\ 0 & (3-\sqrt{3})/3 & 1 & (6+\sqrt{6})/6 & 2 & 2 \\ 0 & 0 & (3-\sqrt{6})/3 & (6-\sqrt{6})/6 & 1 & 2 \end{pmatrix}.$$

The following lemma (proven in [**C**]) characterizes the image of $\mathcal{F}_{d,N}$ under this map.

LEMMA 4.2 (Cahill, Fickus, Mixon, Poteet, S., 2013). *Let $\Delta_{d,N}$ denote the set of matrices $\Lambda = (\lambda_1\ \cdots\ \lambda_N) = (\lambda_{ij})$ such that the column interlacing conditions*

$$\lambda_{1,j+1} \geq \lambda_{1,j} \geq \lambda_{2,j+1} \geq \lambda_{2,j} \geq \cdots \geq \lambda_{d,j+1} \geq \lambda_{d,j} \geq 0$$

hold for $j = 1, \ldots, N-1$, the trace conditions

$$\sum_{i=1}^{d} \lambda_{i,j} = j$$

hold for $j = 1, \ldots, N$, and $\lambda_{11} = 1$ and $\lambda_{1N} = \frac{N}{d}$. Then $\lambda(\mathcal{F}_{d,N}) = \Delta_{d,N}$.

Letting $\mathcal{F}_{d,N}^B$ denote the set of all frames whose first d columns form a basis of $\mathbb{R}^d$, we note that the Q matrix in the QR decomposition of the first d columns of $F \in \mathcal{F}_{d,N}^B$ is unique under the assumption that the diagonal of R is strictly positive. Moreover, this identification is a smooth function of F and we obtain the continuous map $\mathcal{Q} : \mathcal{F}_{d,N}^B \to \mathcal{O}(d)$ by setting $\mathcal{Q}(F)$ equal to the Q in this QR decomposition.

Now, define $\Psi : \mathcal{F}_{d,N}^B \to \mathcal{O}(d) \times \Delta_{d,N}$ by setting $\Psi(F) = (\mathcal{Q}(F), \lambda(F))$. Note that if $F \in \mathcal{F}_{d,N}^B$ is such that $\lambda(F) \notin \partial\Delta_{d,N}$, then Ψ is locally analytic at F.

We shall now construct $\Phi : \mathcal{F}_{d,N}^B \to \mathcal{O}(3) \times \Delta_{3,6}$ that is the local inverse of Ψ at $\mathcal{E}$. The following theorem establishes the existence and continuity of this map.

THEOREM 4.3 (c.f. Cahill, Mixon, S. 2014). *Let $\lambda(F) \in \Delta_{d,N}^\circ$, where $F \in \mathcal{F}_{d,N}$ and $\Delta_{d,N}^\circ$ denotes the interior of $\Delta_{d,N}$. Then there are sequences of vector-valued functions $v_k : \Delta_{d,N}^\circ \to \mathbb{R}^d$ and $w_k : \Delta_{d,N}^\circ \to \mathbb{R}^d$, a sequence of matrix-valued* *functions* $W_k : \Delta^\circ_{d,N} \to M_{d,d}$, *and sequences of orthogonal matrices* V_k, P_k, *and* Q_k *so that when we define the sequences*

$$\begin{aligned} U_1(U,\mu) &= U \\ \phi_1(U,\mu) &= U_1(U,\mu)e_1 \\ \phi_{k+1}(U,\mu) &= U_k(U,\mu)V_kP_k^*v_k(\mu) \\ U_{k+1}(U,\mu) &= U_k(U,\mu)V_kP_k^*W_k(\mu)Q_k \end{aligned}$$

for $k = 1, \dots, N-1$ *and all* $(U,\mu) \in \mathcal{O}(d) \times \Delta^\circ_{d,N}$, *then the continuous function*

$$\Phi(U,\mu) = (\phi_1(U,\mu) \ \cdots \ \phi_N(U,\mu)) \in \mathcal{F}_{d,N}$$

satisfies $\lambda(\Phi(U,\mu)) = \mu$ *and* $\Phi(\mathcal{Q}(F),\lambda(F)) = F$ *where* $\mathcal{Q}(F)$ *is* Q *from the QR decomposition of the first* d *columns of* F.

The actual V_k, P_k, Q_k, v_k, w_k, and W_k all have formulas described by the main result in [**C**]. We shall now exhibit these different values and functions in our specific example. First, we fix the matrices V_k, P_k, and Q_k according to the table in Figure 3.

k	V_k	P_k	Q_k
1	$\begin{pmatrix}1&0&0\\0&1&0\\0&0&1\end{pmatrix}$	$\begin{pmatrix}1&0&0\\0&1&0\\0&0&1\end{pmatrix}$	$\begin{pmatrix}1&0&0\\0&1&0\\0&0&1\end{pmatrix}$
2	$\begin{pmatrix}1&0&0\\0&1&0\\0&0&1\end{pmatrix}$	$\begin{pmatrix}1&0&0\\0&1&0\\0&0&1\end{pmatrix}$	$\begin{pmatrix}1&0&0\\0&1&0\\0&0&1\end{pmatrix}$
3	$\begin{pmatrix}1&0&0\\0&-1&0\\0&0&1\end{pmatrix}$	$\begin{pmatrix}1&0&0\\0&1&0\\0&0&1\end{pmatrix}$	$\begin{pmatrix}1&0&0\\0&1&0\\0&0&1\end{pmatrix}$
4	$-\begin{pmatrix}1&0&0\\0&1&0\\0&0&1\end{pmatrix}$	$\begin{pmatrix}0&1&0\\0&0&1\\1&0&0\end{pmatrix}$	$\begin{pmatrix}1&0&0\\0&0&1\\0&1&0\end{pmatrix}$
5	$\begin{pmatrix}1&0&0\\0&1&0\\0&0&1\end{pmatrix}$	$\begin{pmatrix}0&0&1\\1&0&0\\0&1&0\end{pmatrix}$	$\begin{pmatrix}1&0&0\\0&1&0\\0&0&1\end{pmatrix}$

FIGURE 3. Auxiliary matrices used in the eigensteps parameterization around $\mathcal{E}$.

Next, we define the functions $v_k(\lambda)$, $w_k(\lambda)$, $W_k(\lambda)$ according to the table in Figure 5. The columns of Φ are then defined by first setting $U_1(U,\lambda) = U$, $\phi_1(U,\lambda) = U_1(U,\lambda)e_1$, and then iterating

$$\begin{aligned} \phi_{k+1}(U,\lambda) &= U_k(U,\lambda)V_kP_k^*v_k(\lambda) \\ U_{k+1}(U,\lambda) &= U_k(U,\lambda)V_kP_k^*W_k(\lambda)Q_k \end{aligned}$$

for $k = 1, 2, 3, 4, 5$. Observing that

$$\mathcal{Q}(\mathcal{E}) \quad = \quad \begin{pmatrix} 1 & 0 & 0 \\ 0 & \sqrt{2}/2 & \sqrt{2}/2 \\ 0 & \sqrt{2}/2 & -\sqrt{2}/2 \end{pmatrix},$$

one may verify that $\Phi(\mathcal{Q}(\mathcal{E}), \lambda(\mathcal{E})) = \mathcal{E}$.

Noting that $\mathcal{E} \in \mathcal{F}^B_{3,6}$ and $\lambda(\mathcal{E}) \in \Delta^\circ_{3,6}$, continuity of Φ around $(U_0, \lambda_0) = (\mathcal{Q}(\mathcal{E}), \lambda(\mathcal{E}))$ implies that there is an open neighborhood $(U_0, \lambda_0) \in \mathcal{U} \subset \mathcal{O}(3) \times \Delta^\circ_{3,6}$ such that $\Phi(U, \mu) \in \mathcal{F}^B_{3,6}$ for all $(U, \mu) \in \mathcal{U}$. Consequently, Ψ is defined and analytic on $\Phi(\mathcal{U})$. Clearly, $\mathcal{Q}(\Phi(U, \mu)) = U$ and $\lambda(\Phi(U, \mu)) = \mu$ by construction, and hence we have that $\Psi \circ \Phi$ is the identity function on $\mathcal{U}$. By analyticity of Φ and Ψ we conclude that the differentials satisfy

$$I_{T_{(U_0,\lambda_0)}\mathcal{O}(3)\times\Delta_{3,6}} = (D(\Psi \circ \Phi))(U_0, \lambda_0) = D\Psi(\Phi(U_0, \lambda_0))D\Phi(U_0, \lambda_0)$$

where $I_{T_{(U_0,\lambda_0)}}$ is the identity operator on the tangent space of $\mathcal{O}(3) \times \Delta_{3,6}$ at (U_0, λ_0). This implies that $D\Phi(U_0, \lambda_0)$ has maximal rank, and therefore we obtain the following proposition:

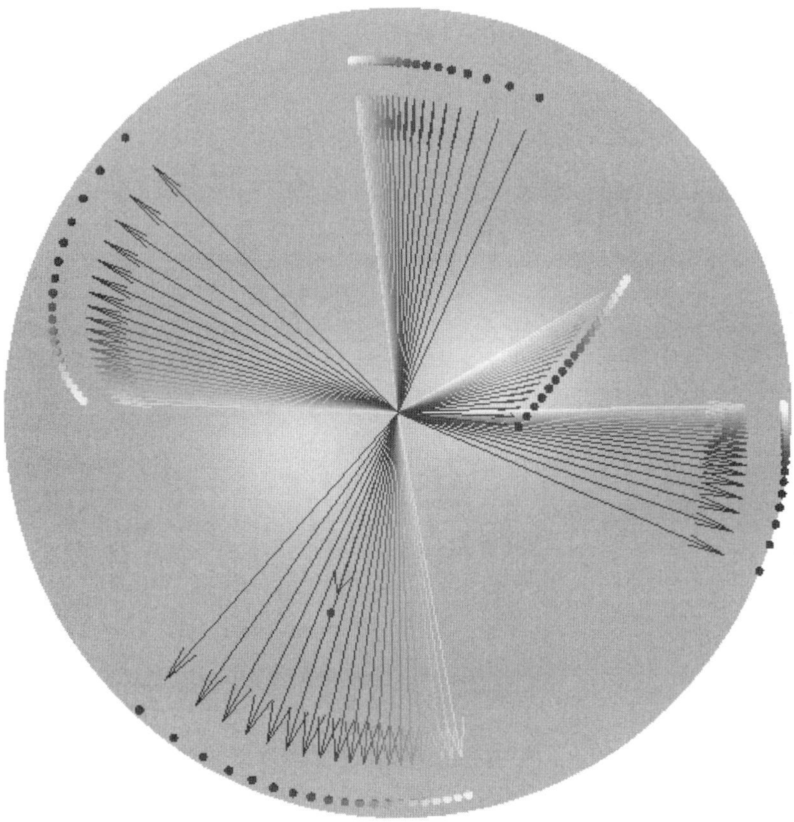

FIGURE 4. The trajectories carved out by the frame vectors while varying the second column of $\lambda(\mathcal{E})$. The resulting trajectory is a synchronized rotation of the two orthonormal bases.

k	$v_k(\lambda)$	$w_k(\lambda)$	$W_k(\lambda)$
1	$\begin{pmatrix} \sqrt{-\frac{(\lambda_{11}-\lambda_{12})(\lambda_{11}-\lambda_{22})(\lambda_{11}-\lambda_{32})}{(\lambda_{11}-\lambda_{21})(\lambda_{11}-\lambda_{31})}} \\ \sqrt{-\frac{(\lambda_{21}-\lambda_{12})(\lambda_{21}-\lambda_{22})}{(\lambda_{21}-\lambda_{11})}} \\ 0 \end{pmatrix}$	$\begin{pmatrix} \sqrt{\frac{(\lambda_{12}-\lambda_{11})(\lambda_{12}-\lambda_{21})(\lambda_{12}-\lambda_{31})}{(\lambda_{12}-\lambda_{22})(\lambda_{12}-\lambda_{32})}} \\ \sqrt{\frac{(\lambda_{22}-\lambda_{11})(\lambda_{22}-\lambda_{21})}{(\lambda_{21}-\lambda_{11})}} \\ 0 \end{pmatrix}$	$\begin{pmatrix} \frac{v_{11}(\lambda)w_{11}(\lambda)}{\lambda_{12}-\lambda_{11}} & \frac{v_{11}(\lambda)w_{21}(\lambda)}{\lambda_{22}-\lambda_{11}} & 0 \\ \frac{v_{21}(\lambda)w_{11}(\lambda)}{\lambda_{12}-\lambda_{21}} & \frac{v_{21}(\lambda)w_{21}(\lambda)}{\lambda_{22}-\lambda_{21}} & 0 \\ 0 & 0 & 1 \end{pmatrix}$
2	$\begin{pmatrix} \sqrt{-\frac{(\lambda_{12}-\lambda_{13})(\lambda_{12}-\lambda_{23})(\lambda_{12}-\lambda_{33})}{(\lambda_{12}-\lambda_{22})(\lambda_{12}-\lambda_{32})}} \\ \sqrt{-\frac{(\lambda_{22}-\lambda_{13})(\lambda_{22}-\lambda_{23})(\lambda_{22}-\lambda_{33})}{(\lambda_{22}-\lambda_{12})(\lambda_{22}-\lambda_{32})}} \\ \sqrt{-\frac{(\lambda_{32}-\lambda_{13})(\lambda_{32}-\lambda_{23})(\lambda_{32}-\lambda_{33})}{(\lambda_{32}-\lambda_{12})(\lambda_{32}-\lambda_{22})}} \end{pmatrix}$	$\begin{pmatrix} \sqrt{\frac{(\lambda_{13}-\lambda_{12})(\lambda_{13}-\lambda_{22})(\lambda_{13}-\lambda_{32})}{(\lambda_{13}-\lambda_{23})(\lambda_{13}-\lambda_{33})}} \\ \sqrt{\frac{(\lambda_{23}-\lambda_{12})(\lambda_{23}-\lambda_{22})(\lambda_{23}-\lambda_{32})}{(\lambda_{23}-\lambda_{13})(\lambda_{23}-\lambda_{33})}} \\ \sqrt{\frac{(\lambda_{33}-\lambda_{12})(\lambda_{33}-\lambda_{22})(\lambda_{33}-\lambda_{32})}{(\lambda_{33}-\lambda_{13})(\lambda_{33}-\lambda_{23})}} \end{pmatrix}$	$\begin{pmatrix} \frac{v_{12}(\lambda)w_{12}(\lambda)}{\lambda_{13}-\lambda_{12}} & \frac{v_{12}(\lambda)w_{22}(\lambda)}{\lambda_{23}-\lambda_{12}} & \frac{v_{12}(\lambda)w_{32}(\lambda)}{\lambda_{33}-\lambda_{12}} \\ \frac{v_{22}(\lambda)w_{12}(\lambda)}{\lambda_{13}-\lambda_{22}} & \frac{v_{22}(\lambda)w_{22}(\lambda)}{\lambda_{23}-\lambda_{22}} & \frac{v_{22}(\lambda)w_{32}(\lambda)}{\lambda_{33}-\lambda_{22}} \\ \frac{v_{32}(\lambda)w_{12}(\lambda)}{\lambda_{13}-\lambda_{32}} & \frac{v_{32}(\lambda)w_{22}(\lambda)}{\lambda_{23}-\lambda_{32}} & \frac{v_{32}(\lambda)w_{32}(\lambda)}{\lambda_{33}-\lambda_{32}} \end{pmatrix}$
3	$\begin{pmatrix} \sqrt{-\frac{(\lambda_{13}-\lambda_{14})(\lambda_{13}-\lambda_{24})(\lambda_{13}-\lambda_{34})}{(\lambda_{13}-\lambda_{23})(\lambda_{13}-\lambda_{33})}} \\ \sqrt{-\frac{(\lambda_{23}-\lambda_{14})(\lambda_{23}-\lambda_{24})(\lambda_{23}-\lambda_{34})}{(\lambda_{23}-\lambda_{13})(\lambda_{23}-\lambda_{33})}} \\ \sqrt{-\frac{(\lambda_{33}-\lambda_{14})(\lambda_{33}-\lambda_{24})(\lambda_{33}-\lambda_{34})}{(\lambda_{33}-\lambda_{13})(\lambda_{33}-\lambda_{23})}} \end{pmatrix}$	$\begin{pmatrix} \sqrt{\frac{(\lambda_{14}-\lambda_{13})(\lambda_{14}-\lambda_{23})(\lambda_{14}-\lambda_{33})}{(\lambda_{14}-\lambda_{24})(\lambda_{14}-\lambda_{34})}} \\ \sqrt{\frac{(\lambda_{24}-\lambda_{13})(\lambda_{24}-\lambda_{23})(\lambda_{24}-\lambda_{33})}{(\lambda_{24}-\lambda_{14})(\lambda_{24}-\lambda_{34})}} \\ \sqrt{\frac{(\lambda_{34}-\lambda_{13})(\lambda_{34}-\lambda_{23})(\lambda_{34}-\lambda_{33})}{(\lambda_{34}-\lambda_{14})(\lambda_{34}-\lambda_{24})}} \end{pmatrix}$	$\begin{pmatrix} \frac{v_{13}(\lambda)w_{13}(\lambda)}{\lambda_{14}-\lambda_{13}} & \frac{v_{13}(\lambda)w_{23}(\lambda)}{\lambda_{24}-\lambda_{13}} & \frac{v_{13}(\lambda)w_{33}(\lambda)}{\lambda_{34}-\lambda_{13}} \\ \frac{v_{23}(\lambda)w_{13}(\lambda)}{\lambda_{14}-\lambda_{23}} & \frac{v_{23}(\lambda)w_{23}(\lambda)}{\lambda_{24}-\lambda_{23}} & \frac{v_{23}(\lambda)w_{33}(\lambda)}{\lambda_{34}-\lambda_{23}} \\ \frac{v_{33}(\lambda)w_{13}(\lambda)}{\lambda_{14}-\lambda_{33}} & \frac{v_{33}(\lambda)w_{23}(\lambda)}{\lambda_{24}-\lambda_{33}} & \frac{v_{33}(\lambda)w_{33}(\lambda)}{\lambda_{34}-\lambda_{33}} \end{pmatrix}$
4	$\begin{pmatrix} \sqrt{-\frac{(\lambda_{24}-\lambda_{15})(\lambda_{24}-\lambda_{25})(\lambda_{24}-\lambda_{35})}{(\lambda_{24}-\lambda_{14})(\lambda_{24}-\lambda_{34})}} \\ \sqrt{-\frac{(\lambda_{34}-\lambda_{15})(\lambda_{34}-\lambda_{25})(\lambda_{34}-\lambda_{35})}{(\lambda_{34}-\lambda_{14})(\lambda_{34}-\lambda_{24})}} \\ 0 \end{pmatrix}$	$\begin{pmatrix} \sqrt{\frac{(\lambda_{25}-\lambda_{24})(\lambda_{25}-\lambda_{34})}{(\lambda_{25}-\lambda_{35})}} \\ \sqrt{\frac{(\lambda_{35}-\lambda_{14})(\lambda_{35}-\lambda_{24})(\lambda_{35}-\lambda_{34})}{(\lambda_{35}-\lambda_{15})(\lambda_{35}-\lambda_{25})}} \\ 0 \end{pmatrix}$	$\begin{pmatrix} \frac{v_{14}(\lambda)w_{14}(\lambda)}{\lambda_{25}-\lambda_{24}} & \frac{v_{14}(\lambda)w_{24}(\lambda)}{\lambda_{35}-\lambda_{24}} & 0 \\ \frac{v_{24}(\lambda)w_{14}(\lambda)}{\lambda_{25}-\lambda_{34}} & \frac{v_{24}(\lambda)w_{24}(\lambda)}{\lambda_{35}-\lambda_{34}} & 0 \\ 0 & 0 & 1 \end{pmatrix}$
5	$\begin{pmatrix} 1 \\ 0 \\ 0 \end{pmatrix}$	$\begin{pmatrix} 1 \\ 0 \\ 0 \end{pmatrix}$	$\begin{pmatrix} 1 & 0 & 0 \\ 0 & 1 & 0 \\ 0 & 0 & 1 \end{pmatrix}$

FIGURE 5. Auxilary functions for the eigensteps parameterization around $\mathcal{E}$.

PROPOSITION 4.4. *The map $\Phi : \mathcal{U} \to \Phi(\mathcal{U})$ is a diffeomorphism.*

Comparing this parameterization with the one obtained through the elimination theoretic arguments, this coordinate system is relatively simple to write down, but the paths are not quite as intuitive. In Figure 4, we simply vary the second column of the eigensteps around $\lambda(\mathcal{E})$. The resulting motion is a synchronized rotation of the two orthonormal bases, a motion that would be harder to describe in the intuitive parameterization. Additionally, the proof of Proposition 4.4 invoked the fact that $\lambda(\mathcal{E}) \in \Delta^{\circ}_{3,6}$ to obtain local analyticity of Ψ. For frames with eigensteps on the boundary of the eigensteps polytope, one might still be able to permute the frame vectors to obtain eigensteps in $\Delta^{\circ}_{3,6}$ to obtain a valid parameterization. However, there is no guarantee that such a permutation exists, which leads to the following open problem.

PROBLEM 4.5. *Prove that for any non-orthodecomposable $F \in \mathcal{F}_{d,N}$ there is a permutation matrix $\Pi \in M_{N,N}$ such that $\lambda(F\Pi) \in \Delta^{\circ}_{d,N}$, or else find a counterexample.*

5. Connectivity and irreducibility of $\mathcal{F}_{d,N}$

5.1. Connectivity. Using the eigensteps parameterizations, we can also show that $\mathcal{F}_{d,N}$ is path-connected whenever $N \geq d+2 \geq 4$. For example, we can lift the path

$$\lambda(t) = (1-t)\lambda(\mathcal{E}) + t\lambda(\Xi)$$

to a continuous path $\gamma(t)$ in $\mathcal{F}_{3,6}$ such that $\gamma(0) = \mathcal{E}$ and $\gamma(1)$ has the same eigensteps as Ξ. It is then easy to see that $\gamma(1)$ must be the union of two orthonormal bases. It then remains for us to show that the set of frames consisting of a union of two orthonormal bases is path connected. To demonstrate this, we need a lemma that provides a general class of "spinning-in-place" motions for FUNTFs.

PROPOSITION 5.1. *Suppose that there is a subspace of $\mathbb{R}^d$ such that orthogonal projection of some subset of the columns of $F \in \mathcal{F}_{d,N}$ forms a tight frame for the subspace. Then continuous rotation of this subset along this subspace is a path in $\mathcal{F}_{d,N}$.*

PROOF. Let P denote the orthogonal projection, and suppose $U : [0,1] \to \mathcal{O}(d)$ is a continuous path in the orthogonal matrices with $U(0) = I_d$, and $U(t)x = U(t)x + (I_d - P)x$ for all $t \in [0,1]$, all $x \in \mathbb{R}^d$. Without loss of generality (since we may always factor paths through a permutation), assume $F = (F_1\ F_2)$ with PF_2 the tight frame for its span. Then it is clear that $F(t) = (F_1\, U(t)F_2)$ has unit norm columns for all $t \in [0,1]$ and

$$\begin{aligned}
F(t)F(t)^* &= \begin{pmatrix} F_1 & U(t)F_2 \end{pmatrix} \begin{pmatrix} F_1^* \\ F_2^* U(t)^* \end{pmatrix} \\
&= F_1F_1^* + U(t)F_2F_2^*U(t)^* \\
&= F_1F_1^* + U(t)(PF_2 + (I_d - P)F_2)(PF_2 + (I_d - P)F_2)^*U(t)^* \\
&= F_1F_1^* + U(t)(cP + (I_d - P)F_2F_2^*(I_d - P)^*)U(t)^* \\
&= F_1F_1^* + cP + (I_d - P)F_2F_2^*(I_d - P)^* \\
&= F_1F_1^* + F_2F_2^* = FF^* \\
&= \frac{N}{d} I_d,
\end{aligned}$$

where c is the tight-frame constant for PF_2 in its span. Thus, the tightness condition is also satisfied for all $t \in [0,1]$ and the proposition follows. $\square$

We will immediately use Proposition 5.1 in the following form: if two vectors in a frame form an orthonormal pair, then continuously rotating this pair in their span while leaving the other frame vectors fixed produces a path in $\mathcal{F}_{d,N}$.

Now, since the orthogonal group has two connected components, this is easy to show if there is a motion which swaps any pair of vectors. While swapping vectors between the two orthonormal bases is as simple as aligning the two vectors by a continuous rotation, and then reversing the rotation while the roles of the vectors are reversed.

If the pair is within the same orthonormal basis, by aligning a pair from the other orthonormal basis in the plane of the targets, this process can be reduced to the two dimensional case. Figure 6 illustrates a continuous motion that swaps the targets and returns the other vectors to their original position.

The main trick here is that we use liftings of the eigensteps map to connect any frame to a particular set of frames or a "hub," and then we show that this "hub" is connected. Generalizing this argument leads to the following theorem.

THEOREM 5.2 (Cahill, Mixon, Strawn, 2014). *If $N \geq d+2 \geq 4$, then $\mathcal{F}_{d,N}$ is path-connected.*

5.2. Irreducibility. By the following proposition, if we can also show that the non-singular points of $\mathcal{F}_{3,6}$ are path-connected and dense in $\mathcal{F}_{3,6}$, then we know that $\mathcal{F}_{3,6}$ is an irreducible algebraic variety.

PROPOSITION 5.3 (Cahill, Mixon, Strawn, 2014). *Suppose V is an algebraic variety such that*

(i) *the set of non-singular points of V is path-connected, and*
(ii) *the set of non-singular points is dense in V.*

Then V is an irreducible algebraic variety.

First, we argue that the non-orthodecomposable frames of $\mathcal{F}_{3,6}$ are dense in $\mathcal{F}_{3,6}$. The following characterization of the orthodecomposable frames makes this argument go through very easily.

PROPOSITION 5.4. *A frame $F \in \mathcal{F}_{3,6}$ is orthodecomposable if and only if there are two distinct frame vectors f_i and f_j that are parallel.*

PROOF. If F is orthodecomposable, then there is a partition of F into F_1 and F_2 so that the linear spans of the vectors in F_1 and F_2 (denoted V_1 and V_2) are non-trivial orthogonal subspaces of $\mathbb{R}^3$. Consequently, either V_1 or V_2 has dimension equal to 1, and hence by the tight frame bound condition either F_1 or F_2 consists of two parallel vectors.

On the other hand, assuming that there are vectors f_i and f_j which are parallel, the tight frame bound condition requires that all other vectors in F are orthogonal to f_i and f_j. Consequently, F is orthodecomposable. $\square$

PROPOSITION 5.5. *If $F \in \mathcal{F}_{3,6}$ is orthodecomposable, then it is a union of two orthonormal bases.*

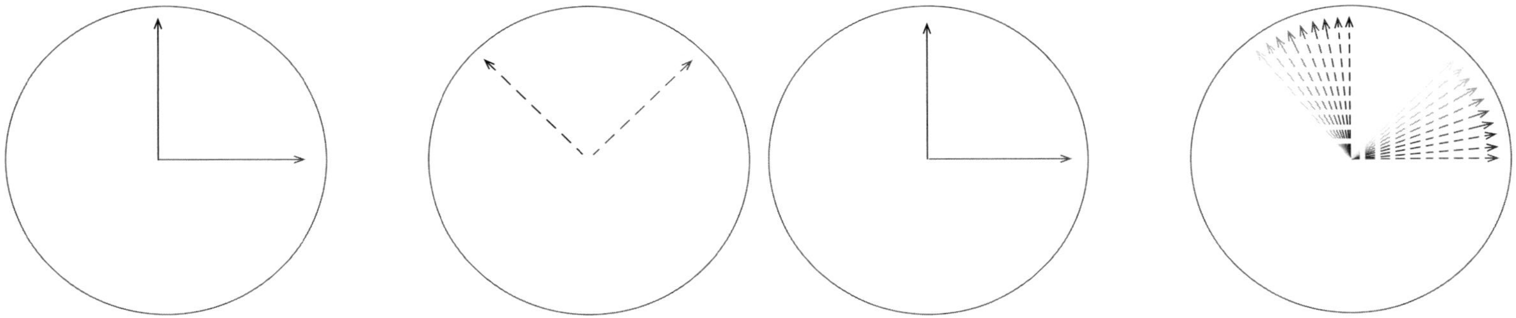

(A) Initial positions

(B) Align a chaperone along the first target

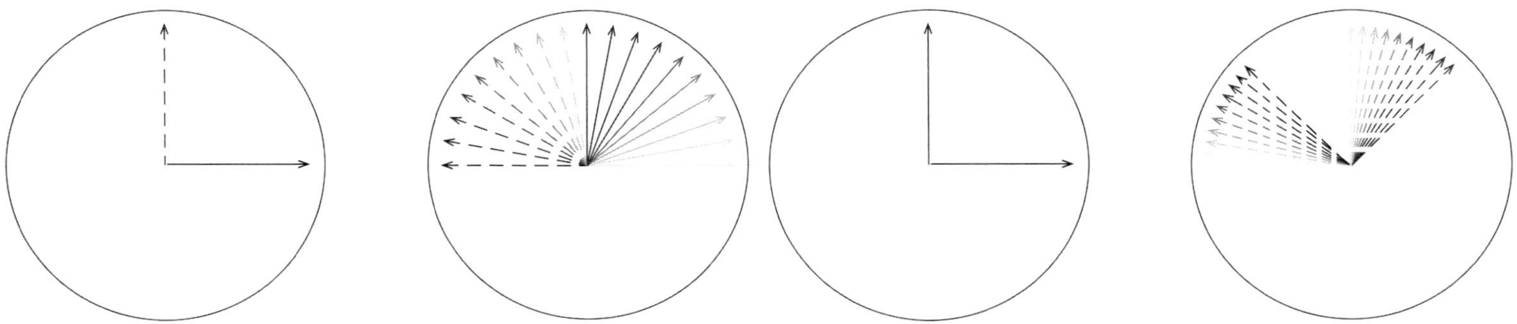

(C) Use the other chaperone to align the first target along the second target

(D) Return the chaperones to their original position

FIGURE 6. Swapping vectors within an orthonormal basis. The vectors targeted for swapping are the solid blue and solid black pair. The dashed pair play the role of chaperones during this transposition.

PROOF. This follows from the fact that all frames in $\mathcal{F}_{2,4}$ consist of a union of two orthonormal bases which in turn follows from Theorem 3.3 of [**B**] and the fact that parallelograms have parallel sides. More explicitly, Theorem 3.3 of [**B**] shows that FUNTFs of $\mathbb{R}^2$ may be identified with unimodular (modulus 1) complex sequences $\{z_i\}_{i=1}^N$ satisfying the condition

$$\sum_{i=1}^{N} z_i^2 = 0.$$

The identification is the most obvious one, where a vector (x_i, y_i) is identified with the complex number $z_i = x_i + \sqrt{-1}y_i$. In the case of $\mathcal{F}_{2,4}$, this means that the squared complex sequence forms a rhombus, and it can be deduced that if z^2 appears in the sum, then $-z^2$ must also appear. Looking back at the identification map, that means that if (x, y) appears in the FUNTF, then $\pm(y, -x)$ must also appear in the FUNTF. □

Now that we have characterized the singularities of $\mathcal{F}_{3,6}$, we verify the second hypothesis of Proposition 5.3.

PROPOSITION 5.6. *The non-singular points of $\mathcal{F}_{3,6}$ are Euclidean dense in $\mathcal{F}_{3,6}$.*

PROOF. Observing that there is always a small rotation of one orthonormal basis that can bring all of its vectors out of alignment with another orthonormal basis (and thus the union is arbitrarily close to a non-orthodecomposable union of two orthonormal bases), and since all unions of two orthonormal bases are in $\mathcal{F}_{3,6}$, we conclude that any orthodecomposable frame in $\mathcal{F}_{3,6}$ is a limit of non-orthodecomposable frames in $\mathcal{F}_{3,6}$ under the Euclidean topology. □

Now, we need to show that the non-singular points of $\mathcal{F}_{3,6}$ form a path-connected set to verify the first hypothesis of Proposition 5.3. During the course of constructing this path, we may need to permute or negate vectors to obtain continuous paths with endpoints having a form that are relatively simple to display. The actual path constructed may be obtained by inverting the permutation or negation over the path.

This path we construct is similar to the one in the connectivity proof, but we need to take great care to avoid the orthodecomposable frames along our path. The path has three parts:

(1) A path connecting to eigensteps such that any frame having such eigensteps is not orthodecomposable (and hence the path avoids the orthodecomposable frames)
(2) A path connecting a frame with these eigensteps to the "hub" of non-orthodcomposable unions of two orthonormal bases of $\mathbb{R}^3$
(3) A sequence of paths that effectively perform arbitrary transpositions of columns so that the final frame consists of two positively-oriented orthonormal bases (a frame of the form $(U\ U')$ where U and U' are fixed orthonormal bases that share no parallel vectors).

For the first part of our path, we lift the eigensteps map to continuously connect to a frame that has the eigensteps

$$\begin{pmatrix} 1 & 3/2 & 3/2 & 2 & 2 & 2 \\ 0 & 1/2 & 3/2 & 3/2 & 2 & 2 \\ 0 & 0 & 0 & 1/2 & 1 & 2 \end{pmatrix}$$

by Theorem 4.3. Note that any frame having such eigensteps must be non-orthodecomposable.

The second part of our path is a little more involved. Noting that the first three vectors of a frame having the above eigensteps effectively form a member of $\mathcal{F}_{2,3}$ (but now in a subspace of $\mathbb{R}^3$), any frame with the above eigensteps can be permuted and negated so that there is a continuous rotation sending it to the block form

$$F = \begin{pmatrix} F_1 & \sqrt{1/3}F_2 \\ \mathbf{0}_3^* & \sqrt{2/3}\mathbf{1}_3^* \end{pmatrix}$$

where

$$F_1 = \begin{pmatrix} 1 & -1/2 & -1/2 \\ 0 & \sqrt{3}/2 & -\sqrt{3}/2 \end{pmatrix} \in \mathcal{F}_{2,3}$$

and $F_2 \in \mathcal{F}_{2,3}$. Note that any frame of this form is non-orthodecomposable. Moreover, since $F_2 \in \mathcal{F}_{2,3}$,

$$\begin{pmatrix} F_1 & \sqrt{1/3}UF_2 \\ \mathbf{0}_3^* & \sqrt{2/3}\mathbf{1}_3^* \end{pmatrix} \in \mathcal{F}_{3,6}$$

for any orthogonal matrix U. This means that we can apply Proposition 5.1 so to continuously rotate the last three columns to have rows of the form

$$F_2 = \begin{pmatrix} 1 & * & * \\ 0 & * & * \end{pmatrix}.$$

Orthogonality of the rows of F_2 with $\mathbf{1}_3$ and the fact that $F_2 \in \mathcal{F}_{2,3}$ therefore fixes the first row to be $(1 \ -1/2 \ -1/2)$. By possibly performing one final permutation, we may therefore assume that $F_1 = F_2$. Now, consider the path

$$\gamma(t) = \begin{pmatrix} \sqrt{1-t^2}F_1 & \sqrt{\frac{1}{3}+t^2}F_1 \\ -t\mathbf{1}_3^* & \sqrt{\frac{2}{3}-t^2}\mathbf{1}_3^* \end{pmatrix}$$

for $t \in [0, 1/\sqrt{3}]$. Observe that

$$\begin{aligned}
\gamma(t)\gamma(t)^* &= \begin{pmatrix} \sqrt{1-t^2}F_1 & \sqrt{\frac{1}{3}+t^2}F_1 \\ -t\mathbf{1}_3^* & \sqrt{\frac{2}{3}-t^2}\mathbf{1}_3^* \end{pmatrix} \begin{pmatrix} \sqrt{1-t^2}F_1^* & -t\mathbf{1}_3 \\ \sqrt{1/3+t^2}F_1^* & \sqrt{\frac{2}{3}-t^2}\mathbf{1}_3 \end{pmatrix} \\
&= \begin{pmatrix} (1-t^2)F_1F_1^* + (1/3+t^2)F_1F_1^* & -t\sqrt{1-t^2}F_1\mathbf{1}_3 + \sqrt{\frac{2}{3}-t^2}\sqrt{\frac{1}{3}+t^2}F_1\mathbf{1}_3 \\ \left(-t\sqrt{(}1-t^2)F_1\mathbf{1}_3 + \sqrt{\frac{2}{3}-t^2}\sqrt{\frac{1}{3}+t^2}F_1\mathbf{1}_3\right)^* & 3t^2 + 3(2/3-t^2) \end{pmatrix} \\
&= \begin{pmatrix} (4/3)(3/2)I_2 & \mathbf{0}_2 \\ \mathbf{0}_2^* & 2 \end{pmatrix} \\
&= 2I_3,
\end{aligned}$$

and noting that the columns of $\gamma(t)$ are always on the unit sphere, we conclude that γ is continuous path in $\mathcal{F}_{3,6}$. In particular, this path does not pass through any orthodecomposable frames since none of the columns can be parallel. Now,

$$\gamma(1/\sqrt{3}) = \begin{pmatrix} \sqrt{\frac{2}{3}}F_1 & \sqrt{\frac{2}{3}}F_1 \\ -\sqrt{\frac{1}{3}}\mathbf{1}_3^* & \sqrt{\frac{1}{3}}\mathbf{1}_3^* \end{pmatrix}$$

and it is easy to verify that

$$\begin{pmatrix} \sqrt{\frac{2}{3}}F_1 \\ -\sqrt{\frac{1}{3}}\mathbf{1}_3^* \end{pmatrix} \text{ and } \begin{pmatrix} \sqrt{\frac{2}{3}}F_1 \\ \sqrt{\frac{1}{3}}\mathbf{1}_3^* \end{pmatrix}$$

FIGURE 7. The motion arriving at a non-orthodecomposable union of two orthonormal bases.

are both orthonormal bases, and $\gamma(1/\sqrt{3})$ is a union of orthonormal bases which is non-orthodecomposable. Figure 7 depicts how the frame vectors move along this path. In particular, the $\mathcal{F}_{2,3}$ both triplets of vectors swing "down" in a synchronous manner until both triplets form orthonormal bases. Consequently, we have landed on the "hub" without any of the vectors becoming parallel. Now that we have arrived at the "hub", we observe that all non-orthodecomposable unions of two orthonormal bases arising from a particular column ordering are path-connected. Thus, the last step of our proof involves showing that there are continuous paths in $\mathcal{F}_{3,6}$ connecting a non-orthodecomposable union of two orthonormal bases to a reordering of its columns.

We now proceed to demonstrate that arbitrary permutations of vectors for non-orthodecomposable unions of two orthonormal bases of $\mathbb{R}^d$ may be obtained via continuous paths that avoid the orthodecomposable unions of two orthonormal bases. Once this is established, the final part of our path may be constructed to complete the proof.

In the proof of path-connectivity of $\mathcal{F}_{3,6}$, we showed how to construct paths that acted as transpositions, and composing transpositions gives us the full permutation group. However, the paths in this case require the alignment of vectors, and hence pass through the orthodecomposable frames. Now that such a path is not an option, we must provide an alternative path that avoids the orthodecomposable frames.

Assume that we begin at the following non-orthodecomposable system, which is a union of two orthonormal bases:

$$\begin{pmatrix} 1 & 0 & 0 & 0 & \frac{\sqrt{2}}{2} & -\frac{\sqrt{2}}{2} \\ 0 & \frac{\sqrt{2}}{2} & -\frac{\sqrt{2}}{2} & 1 & 0 & 0 \\ 0 & \frac{\sqrt{2}}{2} & \frac{\sqrt{2}}{2} & 0 & \frac{\sqrt{2}}{2} & \frac{\sqrt{2}}{2} \end{pmatrix}.$$

Our goal is to swap the second and fifth column without putting any of the vectors in a parallel position. Figure 8(A) depicts this frame, and the second and fifth columns are indicated by the blue and cyan markers, respectively.

During this motion, we shall make extensive use of the fact that continuously rotating any orthonormal pair in their span induces a continuous path through $\mathcal{F}_{3,6}$. As a first application of this idea, we spin the first and fourth columns, effectively moving away from the "hub" of unions of two orthonormal bases. This motion is depicted in Figure 8(B). Next, Figure 9 shows how we spin the four vectors sticking out (essentially inducing a 4-cycle). The remaining motions will essentially perform the action of a 3-cycle on the vectors marked blue, green, and black.

First, we spin the blue and green vectors by $\pi/4$ radians in their span as depicted in Figure 10. After this motion, the vectors marked green and black are now orthogonal and Figure 11 illustrates how we rotate these two $\pi/2$ radians in their span. Lastly, we spin the vectors marked with black and blue by $\pi/4$ radians. If we return the first and third columns to their original position, we see that we have swapped the vectors marked with blue and cyan.

This completes the proof of connectivity of the non-singular points of $\mathcal{F}_{3,6}$, and Proposition 5.3 now implies the following theorem.

THEOREM 5.7. *The algebraic variety $\mathcal{F}_{3,6}$ is irreducible.*

The case when $N = 2d$ is the most complicated part of the general proof of this result for arbitrary N and d. The proof of the general theorem is shown in [**D**]:

THEOREM 5.8 (Cahill, Mixon, Strawn, 2014). *If $N \geq d+2$ and $N > 4$, then $\mathcal{F}_{d,N}$ is an irreducible algebraic variety.*

One implication of this theorem is that a generic (in the algebro-geometric sense) FUNTF is full-spark (see [**A**]), which means that large generic systems of FUNTFs should exhibit optimal phase transition phenomena for compressed sensing (see [**J**]). Another implication is that the singular sets of the FUNTF varieties are not obtained by intersecting irreducible components (except when $d = 2$ and $N = 4$). This means that the singular sets of the FUNTF varieties could be quite complicated, and it should be expected that a full elucidation of the singular sets of the FUNTF varieties could require a large amount of machinery.

6. A final challenge

Now that we have clearly seen the power of the eigensteps construction in finite dimensions, we consider the following problem to be of great interest.

PROBLEM 6.1. *Find an analogue of the eigensteps construction and parameterizations for spaces of finite frames on an infinite dimensional separable Hilbert space.*

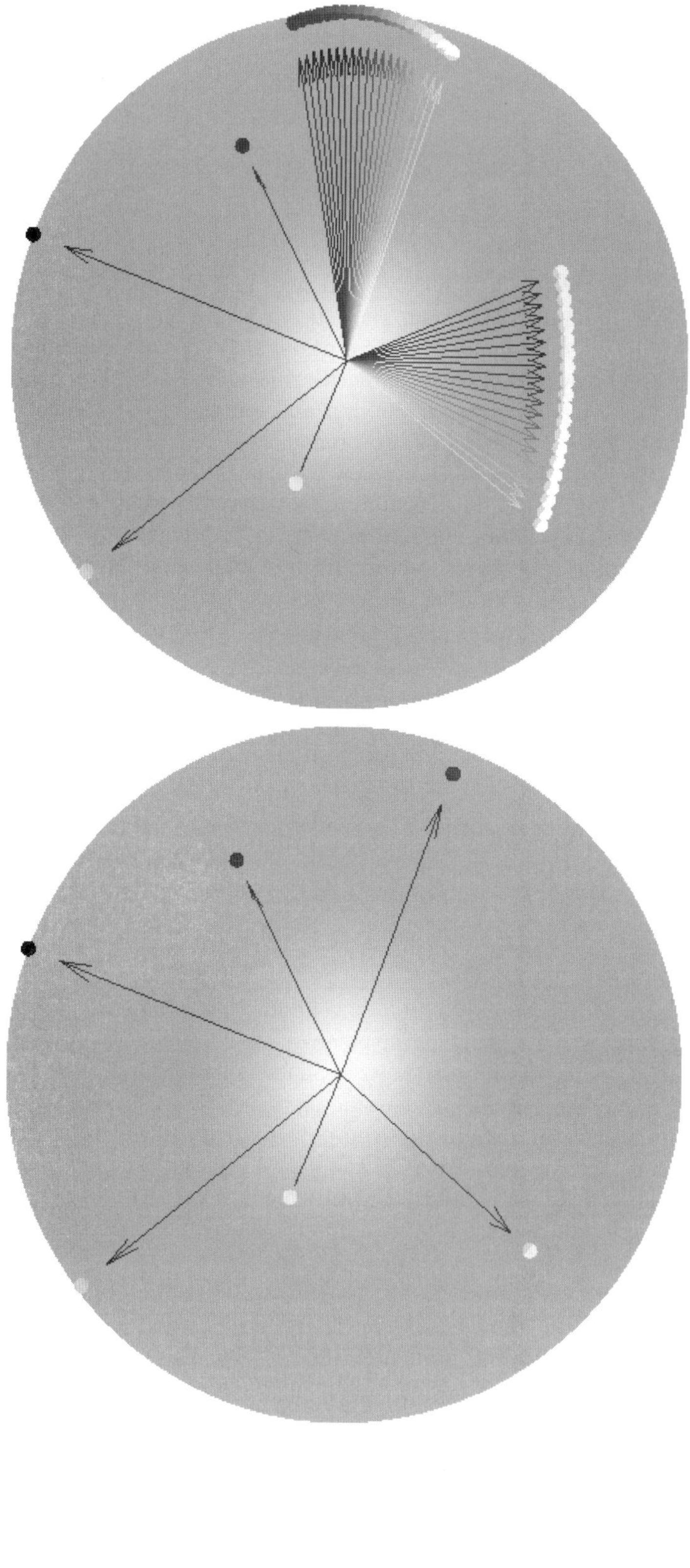

(A) The starting point for the transposition.

(B) This first motion ensures that the ensuing path does not pass through an orthodecompsable frame.

FIGURE 8. Initial positioning for the transposition

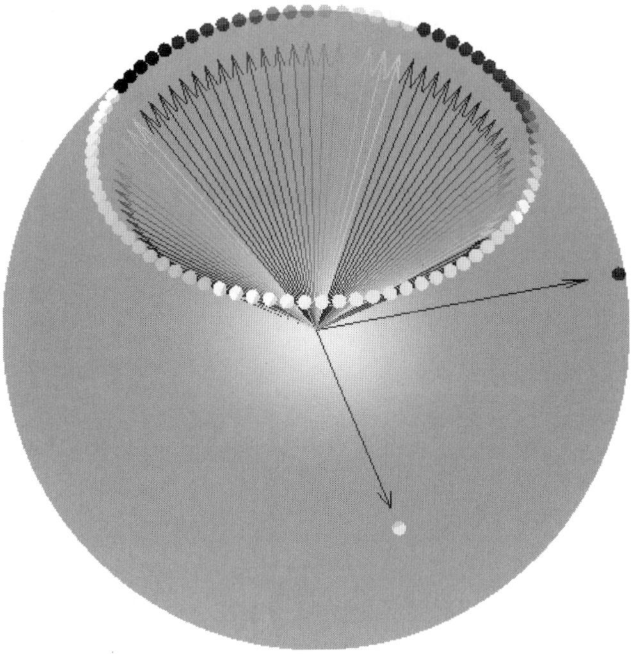

FIGURE 9. Rotation of the four vector that are "sticking out" induces a 4-cycle.

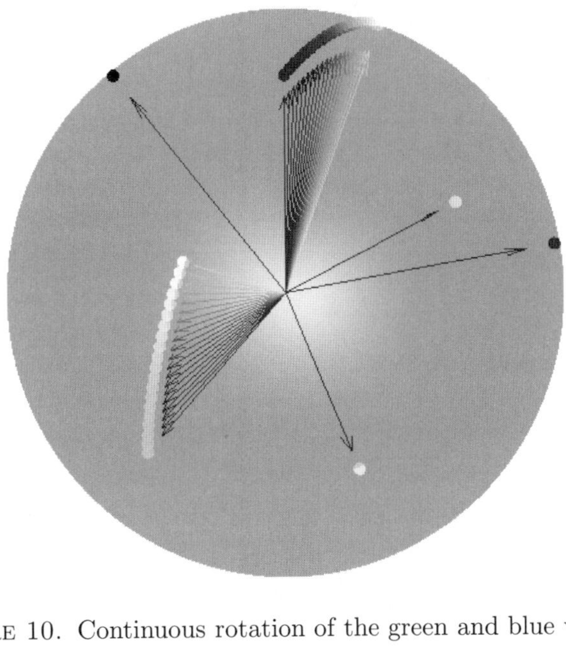

FIGURE 10. Continuous rotation of the green and blue vectors.

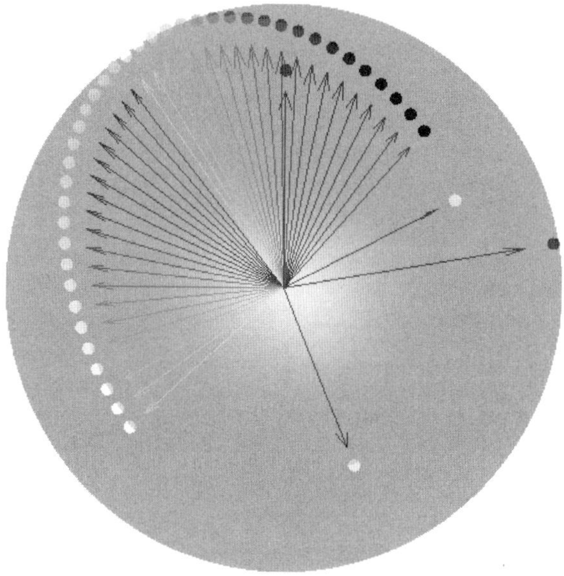

FIGURE 11. Continuous rotation of the green and black vectors.

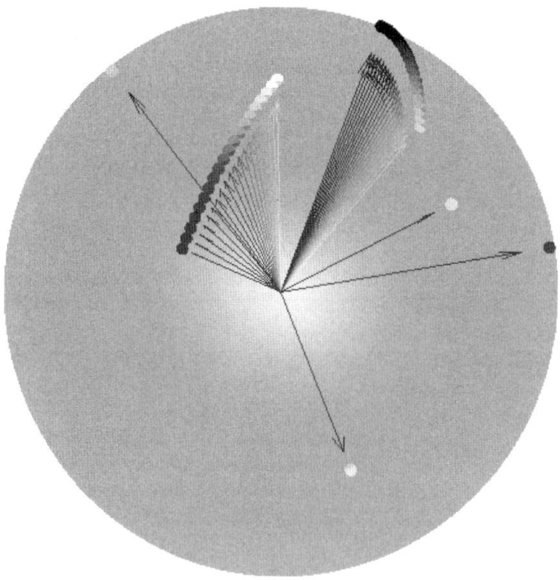

FIGURE 12. Continuous rotation of the blue and black vectors. This completes the 3-cycle on the black, blue, and green vectors, and the net result is a transposition on the blue and cyan points.

References

[A] B. Alexeev, J. Cahill, and D. G. Mixon, *Full spark frames*, J. Fourier Anal. Appl. **18** (2012), no. 6, 1167–1194, DOI 10.1007/s00041-012-9235-4. MR3000979

[B] J. J. Benedetto and M. Fickus, *Finite normalized tight frames*, Adv. Comput. Math. **18** (2003), no. 2-4, 357–385, DOI 10.1023/A:1021323312367. Frames. MR1968126 (2004c:42059)

[C] J. Cahill, M. Fickus, D. G. Mixon, M. J. Poteet, and N. Strawn, *Constructing finite frames of a given spectrum and set of lengths*, Appl. Comput. Harmon. Anal. **35** (2013), no. 1, 52–73, DOI 10.1016/j.acha.2012.08.001. MR3053746

[D] J. Cahill, D. Mixon, and N. Strawn. Connectivity and Irreducibility of Algebraic Varieties of Finite Unit Norm Tight Frames. Available online: arXiv:1311.4748

[E] K. Dykema and N. Strawn, *Manifold structure of spaces of spherical tight frames*, Int. J. Pure Appl. Math. **28** (2006), no. 2, 217–256. MR2228007

[F] V. Guillemin and A. Pollack, *Differential topology*, AMS Chelsea Publishing, Providence, RI, 2010. Reprint of the 1974 original. MR2680546 (2011e:58001)

[G] R. Hartshorne, *Algebraic geometry*, Springer-Verlag, New York-Heidelberg, 1977. Graduate Texts in Mathematics, No. 52. MR0463157 (57 #3116)

[H] R. B. Holmes and V. I. Paulsen, *Optimal frames for erasures*, Linear Algebra Appl. **377** (2004), 31–51, DOI 10.1016/j.laa.2003.07.012. MR2021601 (2004j:42028)

[I] J. Giol, L. V. Kovalev, D. Larson, N. Nguyen, and J. E. Tener, *Projections and idempotents with fixed diagonal and the homotopy problem for unit tight frames*, Oper. Matrices **5** (2011), no. 1, 139–155, DOI 10.7153/oam-05-10. MR2798802 (2012f:47001)

[J] H. Monajemi, S. Jafarpour, M. Gavish, Stat 330/CME 362 Collaboration, and D. L. Donoho, *Deterministic matrices matching the compressed sensing phase transitions of Gaussian random matrices*, Proc. Natl. Acad. Sci. USA **110** (2013), no. 4, 1181–1186, DOI 10.1073/pnas.1219540110. MR3037097

[K] M. Püschel, J. Kovačević. Real, tight frames with maximal robustness to erasures. Proceedings of Data Compression Conference, 63–72, 2005.

[L] N. Strawn, *Finite frame varieties: nonsingular points, tangent spaces, and explicit local parameterizations*, J. Fourier Anal. Appl. **17** (2011), no. 5, 821–853, DOI 10.1007/s00041-010-9164-z. MR2838109 (2012h:42062)

[M] N. Strawn, *Optimization over finite frame varieties and structured dictionary design*, Appl. Comput. Harmon. Anal. **32** (2012), no. 3, 413–434, DOI 10.1016/j.acha.2011.09.001. MR2892742

[N] T. Strohmer and R. W. Heath Jr., *Grassmannian frames with applications to coding and communication*, Appl. Comput. Harmon. Anal. **14** (2003), no. 3, 257–275, DOI 10.1016/S1063-5203(03)00023-X. MR1984549 (2004d:42053)

Department of Mathematics and Statistics, Georgetown University, Washington D.C., 20007

E-mail address: nate.strawn@georgetown.edu

Proceedings of Symposia in Applied Mathematics
Volume **73**, 2016
http://dx.doi.org/10.1090/psapm/073/00630

Preconditioning techniques in frame theory and probabilistic frames

Kasso A. Okoudjou

ABSTRACT. In this chapter we survey two topics that have recently been investigated in frame theory. First, we give an overview of the class of scalable frames. These are (finite) frames with the property that each frame vector can be rescaled in such a way that the resulting frames are tight. This process can be thought of as a preconditioning method for finite frames. In particular, we: (1) describe the class of scalable frames; (2) formulate various equivalent characterizations of scalable frames, and relate the scalability problem to the Fritz John ellipsoid theorem. Next, we discuss some results on a probabilistic interpretation of frames. In this setting, we: (4) define probabilistic frames as a generalization of frames and as a subclass of continuous frames; (5) review the properties of certain potential functions whose minimizers are frames with certain optimality properties.

CONTENTS

1. Introduction
2. Preconditioning techniques in frame theory
3. Probabilistic frames
Acknowledgment
References

1. Introduction

This chapter is devoted to two topics that have been recently investigated within frame theory: (a) frame preconditioning methods investigated under the vocable *scalable frames*, and (b) probabilistic methods in frame theory referred to as *probabilistic frames*. Before getting to the details on each of these topics we briefly review some essential facts on frame theory and refer to [**15**, **42**, **50**, **51**] for more on frames and their applications. In all that follows we restrict ourselves to (finite) frames in $\mathbb{R}^N$.

2010 *Mathematics Subject Classification.* Primary 42C15; Secondary 52A20, 52B11.
Key words and phrases. Parseval frame, Scalable frame, Fritz John theorem, Probabilistic frames, frame potential, continuous frames.

1.1. Review on finite frame theory.

DEFINITION 1.1. A set $\Phi = \{\varphi_k\}_{k=1}^M \subseteq \mathbb{R}^N$ is a *frame* for $\mathbb{R}^N$ if $\exists\, A, B > 0$ such that $\forall x \in \mathbb{R}^N$,

$$A\|x\|^2 \leq \sum_{k=1}^M |\langle x, \varphi_k\rangle|^2 \leq B\|x\|^2.$$

If, in addition, each φ_k is unit-norm, we say that Φ is a *unit-norm frame.* The set of frames for $\mathbb{R}^N$ with M elements will be denoted by $\mathcal{F}(M, N)$, and simply $\mathcal{F}$ if M and N are fixed. In addition, we shall denote the subset of unit-norm frames by $\mathcal{F}_u(M, N)$, i.e.,

$$\mathcal{F}_u(M, N) := \left\{\{\varphi_k\}_{i=1}^M \in \mathcal{F}(M, N) : \|\varphi_k\|_2 = 1 \text{ for } k = 1, \ldots, M\right\}.$$

We shall investigate frames via the *analysis operator*, L, defined by

$$L : \mathbb{R}^N \to \mathbb{R}^M : x \mapsto Lx = \{\langle x, \varphi_k\rangle\}_{k=1}^M.$$

The *synthesis operator* is the adjoint L^* of L and is defined by

$$L^* : \mathbb{R}^M \to \mathbb{R}^N : c = (c_k)_{k=1}^M \mapsto L^*c = \sum_{k=1}^M c_k\varphi_k.$$

It is easily seen that the canonical matrix associated to L^* is the $N \times M$ matrix whose k^{th} column is the frame vector φ_k. As such we shall abuse notation and denote this matrix by Φ again. Consequently, the canonical matrix associated with L is simply Φ^T, the transpose of Φ.

The *frame operator* $S = L^*L$ is given by

$$S : \mathbb{R}^N \to \mathbb{R}^N : x \mapsto Sx = \sum_{k=1}^M \langle x, \varphi_k\rangle\varphi_k,$$

and its matrix will be denoted (again) by S with

$$S = \Phi\Phi^T.$$

The *Gramian (operator)* of the frame is defined by

$$G = LL^* = \Phi^T\Phi.$$

In fact, the Gramian in an $M \times M$ matrix whose $(i, j)^{th}$ entry is $\langle \varphi_j, \varphi_i\rangle$.

$\Phi = \{\varphi_k\}_{k=1}^M \subset \mathbb{R}^M$ is a frame if and only if S is a positive definite matrix on $\mathbb{R}^M$. In this case,

$$\{\tilde{\varphi}_k\}_{k=1}^M = \{S^{-1}\varphi_k\}_{k=1}^M$$

is also a frame, called *the canonical dual frame*, and, for each $x \in \mathbb{R}^N$, we have

$$x = \sum_{k=1}^M \langle x, \varphi_k\rangle\tilde{\varphi}_k = \sum_{k=1}^M \langle x, \tilde{\varphi}_k\rangle\varphi_k. \tag{1.1}$$

A frame Φ is a *tight frame* if we can choose $A = B$. In this case the frame operator is simply a multiple of the identity operator. To any frame $\Phi = \{\varphi_k\}_{k=1}^M \subset \mathbb{R}^M$ is associated a *canonical tight frame* given by

$$\{\varphi_k^\dagger\}_{k=1}^M = \{S^{-1/2}\varphi_k\}_{k=1}^M \subset \mathbb{R}^N$$

such that for every $x \in \mathbb{R}^N$,

$$x = \sum_{k=1}^{M} \langle x, \varphi_k^\dagger \rangle \varphi_k^\dagger. \tag{1.2}$$

If Φ is a tight frame of unit-norm vectors, we say that Φ is a *finite unit-norm tight frame (FUNTF)*. In this case, the reconstruction formula (1.1) reduces to

$$\forall x \in \mathbb{R}^N, \quad x = \tfrac{N}{M} \sum_{k=1}^{M} \langle x, \varphi_k \rangle \varphi_k. \tag{1.3}$$

FUNTFs are one of the most fundamental objects in frame theory with several applications. Other chapters in this volume will delve more into FUNTFs. The reconstruction formulas (1.3) and (1.2) are very reminiscent of the expansion of a vector in an orthonormal basis for $\mathbb{R}^d$. However, due to the redundancy of frames, the coefficients in these reconstruction formulas are not unique. But the "simplicity" of these reconstruction formulas makes the use of tight frames very attractive in many applications. This in turn, spurs the need for methods to characterize and construct tight frames.

A major development in this direction is due to Benedetto and Fickus [**7**] who proved that for each $\Phi = \{\varphi_k\}_{k=1}^M \subset \mathbb{R}^N$, such that $\|\varphi_k\| = 1$ for each k, we have

$$\mathrm{FP}(\Phi) = \sum_{j=1}^{M} \sum_{k=1}^{M} |\langle \varphi_j, \varphi_k \rangle|^2 \geq \tfrac{M}{N} \max(M, N), \tag{1.4}$$

where $\mathrm{FP}(\Phi)$ is the *frame potential.* The bound given in (1.4) is the global minimum of FP and is achieved by orthonormal systems if $M \leq N$, and by tight frames if $M > N$. Casazza, Fickus, Kovačević, Leon and Tremain [**16**] extended this result by removing the condition that the frames are unit norm. In essence, these results suggest that one may effectively search for FUNTFs by minimizing the frame potential. In practice, techniques such as the steepest-descent method can be used to find the minimizers of the frame potential [**17, 59**]. For other related results on the frame potential we refer to [**33, 46**], and [**15**, Chapter 10]. The frame potential and some of its generalization will be considered in Section 3.

Construction of FUNTFs has seen a lot of research activities in recent years and as a result a number of construction methods have been offered. Casazza, Fickus, Mixon, Yang and Zhou [**18**] introduced the *spectral tetrix* method for constructing FUNTFs. This method has been extended to construct all tight frames, [**19, 34, 55**]. There have also been other new insights in the construction of tight frames, leading to methods rooted in differential and algebraic geometry, see [**12, 13, 69, 70**]. Some of these methods will be introduced in some of the other chapters of this volume.

1.2. Scalable frames. While these powerful algebro-geometric methods can construct all tight frames with given vector lengths and frame constant, they have not been able to incorporate any extra requirement. To put it simply, constructing application-specific FUNTFs involves extra construction constraints, which usually makes the problem very difficult. However, one could ask if a (non) tight frame can be (algorithmically) transformed into a tight one. An analogous problem has been investigated for decades in numerical linear algebra. Indeed, preconditioning methods are routinely used to convert large and poorly-conditioned systems of linear equations $Ax = b$, into better conditioned ones [**9, 36, 43**]. For example, a matrix A is (row/column) scalable if there exit diagonal matrices D_1, D_2 with positive diagonal entries such that D_1A, AD_2, or D_1AD_2 have constant row/column sum, [**6, 36, 47, 48, 67**]. Matrix scaling is part of general preconditioning schemes in numerical linear algebra [**9, 43**].

One of the goals of these notes is to survey recent developments in preconditioning methods in finite frame theory. In particular, we describe recent developments in answering the following question:

QUESTION 1.2. Is it possible to (algorithmically) transform a (non-tight) frame into a tight frame?

In Section 2, we outline a convex geometric approach that has been proposed to answer this question. For example, we consider the case of solving Question 1.2 using some classes of scaling matrices to transform a non-tight frame into a tight one. More specifically, we give an overview of recent results addressing the following problem.

QUESTION 1.3. Is it possible to rescale the norms of the vectors in a (non-tight) frame to obtain a tight frame?

Frames that answer positively Question 1.3 are termed *scalable frames*, and were characterized in [**52**], see also [**54**]. This characterization is operator-theoretical and solved the problem in both the finite and the infinite dimensional settings. More precisely, in the finite dimensional setting, the main result of [**52**] characterizes the set of non scalable frames and gives a simple geometric condition for a frame to be non scalable in dimensions 2 and 3, see Section 2.2. Other characterizations of scalable frames using the properties of the so-called diagram vector ([**41**]) appeared in [**25**]. We refer to [**14**] for some other results about scalable frames.

1.3. Probabilistic frames. While frames are intrinsically defined through their spanning properties, in real euclidean spaces, they can also be viewed as distributions of point masses. In this context, the notion of *probabilistic frames* was introduced as a class of probability measures with finite second moment and whose support spans the entire space [**29, 30, 32**]. Probabilistic frames are special cases of continuous frames as introduced by S. T. Ali, J.-P. Antoine, and P.-P. Gazeau [**1**], see also [**35**].

In Section 3, we consider frames from this probabilistic point of view. To begin we note that probabilistic frames are extensions of the notion of frames previously defined. Indeed, consider a frame, $\Phi = \{\varphi_k\}_{k=1}^M$ for $\mathbb{R}^N$ and define the probability measure

$$\mu_\Phi = \tfrac{1}{M}\sum_{k=1}^{M}\delta_{\varphi_k}$$

where δ_x is the Dirac mass at $x \in \mathbb{R}^N$. It is easily seen that the second moment of μ_Φ is finite, i.e.,

$$\int_{\mathbb{R}^N} \|x\|^2 d\mu_\Phi(x) = \tfrac{1}{M}\sum_{k=1}^{M}\|\varphi_k\|^2 < \infty,$$

and the span of the support of μ_Φ is $\mathbb{R}^N$. Thus, each (finite) frame can be associated to a probabilistic frame. We shall present other examples of probabilistic frames associated to Φ in Section 3. By analogy to the theory of finite frame, we shall say that a probability measure on $\mathbb{R}^N$ with finite second moment is a *probabilistic frame* if the linear span of its support is $\mathbb{R}^N$.

It is known that the space $\mathcal{P}_2(\mathbb{R}^N)$ of probability measures with finite second moments can be equipped with the (2-)Wasserstein metric. In this setting, many questions in frame theory can be seen as analysis problems on a subset of the Wasserstein metric space $\mathcal{P}_2(\mathbb{R}^N)$. In Section 3 we introduce this metric space and derive some immediate properties of frames in this setting. Moreover, this probabilistic setting allows one to use non-algebraic tools to investigate questions from frame theory. For instance, tools like convolution of measures has been used in [**32**] to build new (probabilistic) frames from old ones.

One of the main advantages in analyzing frames in the context of the Wasserstein metric lies in the powerful tools available to solve some optimization problems involving frames using the framework of optimal transport. While we will not delve into any details in this chapter, we point out that optimization of functionals like the frame potential can be studied in this setting. For example, C. Wickman recently showed that a potential function that generalizes Benedetto and Fickus's frame potential can be minimized in the Wasserstein space using some optimal transport notions [**82**]. In the last part of the lecture, we shall focus on a family of potentials that generalize the frame potential and present a survey of recent results involving their minimization. In particular, this family includes the *coherence* of a set of vectors, which is important quantity in compressed sensing [**71, 79, 81**], as well as a functional whose minimizers are, in some cases, solutions to the Zauner's conjecture [**63, 84**].

Though we shall not elaborate on these here, it is worth mentioning that probabilistic frames are related to many other areas including: (a) the covariance of matrices multivariate random vectors [**57, 64, 65, 76, 77**]; (b) directional statistics where there are used to test whether certain data are uniformly distributed; see, [**31, 58**]; (c) isotropic measures [**37, 61**], which, as we shall show, are related to the class of *tight probabilistic frames*. We refer to [**32**] for an overview of other relationships between probabilistic frames and other areas.

The rest of this chapter is organized as follows. In Section 2 we give an overview of the recent developments on scalable frames or preconditioning of frames. In Section 3 we deal with probabilistic frames and outline the recently introduced Wasserstein metric tools to deal with frame theory.

2. Preconditioning techniques in frame theory

Scalable frames are frames $\Phi = \{\varphi_k\}_{k=1}^M$ for $\mathbb{R}^N$ for which there exist nonnegative scalars $\{c_k\}_{k=1}^M \subset [0,\infty)$ such that

$$\tilde{\Phi} = \{c_k\varphi_k\}_{k=1}^M$$

is a tight frame. Scalable frames were first introduced and characterized in [**52**]. Both infinite and finite dimensional settings were considered. In this section, we only focus on the latter giving an overview of recent methods developed to understand scalable frames. The results that we shall describe give an exact characterization of the set of scalable frames from various perspectives. However, the important and very practical question of developing algorithms to find the weights $\{c_k\}$ that make the frame tight will not be considered here, and we refer to [**23**] for a sample of results on this topic. Similarly, when a frame fails to be scalable, one could seek to relax the tightness condition and seek an "almost scalable frame". These considerations are sources of ongoing research and will not be taken upon here. Finally, it is worth pointing out a very interesting application of the theory of scalable frames to wavelets constructed from the Laplacian Pyramid scheme [**44**].

The rest of this section is organized as follows. In Section 2.1 we define scalable frames and derive some of their elementary properties. We then outline a characterization of the set of scalable frames in terms of certain convex polytopes in Section 2.2. This characterization is preceded by motivating examples of scalable frames in dimension 2. In Sections 2.3 and 2.4 we give two other equivalent characterizations of scalable frames. The first of these characterizations has geometric interpretation, while the second one is based on Fritz John's ellipsoid theorem.

2.1. Scalable frames: Definition and properties. The following definitions of scalable frames first appeared in [**52, 53**]:

DEFINITION 2.1. Let M, N be integers such that $N \leq M$. A frame $\Phi = \{\varphi_k\}_{k=1}^M$ in $\mathbb{R}^N$ is called *scalable, respectively, strictly scalable,* if there exist nonnegative, respectively, positive, scalars $\{c_k\}_{k=1}^M$ such that $\{c_k\varphi_k\}_{k=1}^M$ is a tight frame for $\mathbb{R}^N$. The set of scalable frames, respectively, strictly scalable frames, is denoted by $\mathcal{SC}(M,N)$, respectively, $\mathcal{SC}_+(M,N)$.

Moreover, given an integer m with $N \leq m \leq M$, $\Phi = \{\varphi_k\}_{k=1}^M$ is said to be *m-scalable*, respectively, *strictly $m-$scalable*, if there exist a subset $\Phi_I = \{\varphi_k\}_{k\in I}$ with $I \subseteq \{1, 2, \ldots, M\}$, $\#I = m$, such that $\Phi_I = \{\varphi_k\}_{k\in I}$ is scalable, respectively, strictly scalable.

We denote the set of m-scalable frames, respectively, strictly m-scalable frames in $\mathcal{F}(M,N)$ by $\mathcal{SC}(M,N,m)$, respectively, $\mathcal{SC}_+(M,N,m)$.

When the integer m is fixed in a given context, we will simply refer to an $m-$scalable frame as a *scalable frame.* The role of the parameter m is especially relevant when dealing with frames of very large redundancy, i.e., when $M/N \gg 1$. In such a case, choosing a "reasonable" m such that the frame is $m-$scalable could potentially lead to sparse representations for signals in $\mathbb{R}^N$. In addition, the problems of finding the weights that make a frame scalable as well as determining the smallest m such that a given frame is $m-$ scalable have been considered in [**14, 21**]. We shall give more details about this question in Section 2.2.

We now point out some special and trivial examples of scalable frames. When $M = N$, a frame Φ is scalable if and only if Φ is an orthogonal set. Moreover, when $M \geq N$, if Φ contains an orthogonal basis, then it is clearly $N-$scalable. Thus, given $M \geq N$, the set $\mathcal{SC}(M,N,N)$ consists exactly of frames that contains an orthogonal basis for $\mathbb{R}^N$.

So from now on we shall assume without loss of generality that $M \geq N+1$, that Φ contains no orthogonal basis, and that $\varphi_k \neq \pm\varphi_\ell$ for $\ell \neq k$.

Given a frame $\Phi \subset \mathbb{R}^N$, assume that $\Phi = \Phi_1 \cup \Phi_2$ where

$$\Phi_1 = \{\varphi_k^{(1)} \in \Phi : \varphi_k^{(1)}(N) \geq 0\}$$

and

$$\Phi_2 = \{\varphi_k^{(2)} \in \Phi : \varphi_k^{(2)}(N) < 0\}.$$

In other words, Φ_1 consists of all frame vectors from Φ whose N^{th} coordinates are nonnegative. Then the frame $\Phi' = \Phi_1 \cup (-\Phi_2) = \{\varphi_k^{(1)}\} \cup \{-\varphi_k^{(2)}\}$ has the same frame operator as Φ. In particular, Φ is a tight frame if and only if Φ' is a tight frame. In addition, Φ is scalable if and only if Φ' is scalable with exactly the same set of weights. Note that the frame vectors in Φ' are all in the upper-half space. Thus, when convenient we shall assume without loss of generality that all the frame vectors are in the upper-half space, that is $\Phi \subset \mathbb{R}^{N-1} \times \mathbb{R}_+$ where $\mathbb{R}_+ = [0, \infty)$.

We note that a frame $\Phi = \{\varphi_k\}_{k=1}^M \subset \mathbb{R}^N$ with $\varphi_k \neq 0$ for each $k = 1, \ldots, M$ is scalable if and only if $\Phi' = \{\frac{\varphi_k}{\|\varphi_k\|}\}_{k=1}^M$ is scalable. Consequently, we might assume in the sequel that we work with frames consisting of unit norm vectors.

We now collect a number of elementary properties of the set of scalable frames in $\mathbb{R}^N$. We refer to [**52, 53**] for details.

PROPOSITION 2.2. Let $M \geq N$, and $m \geq 1$ be integers.

(i) If $\Phi \in \mathcal{F}$ is m-scalable then $m \geq N$.
(ii) For any integers m, m' such that $N \leq m \leq m' \leq M$ we have that

$$\mathcal{SC}(M, N, m) \subset \mathcal{SC}(M, N, m'),$$

and

$$\mathcal{SC}(M, N) = \mathcal{SC}(M, N, M) = \bigcup_{m=N}^{M} \mathcal{SC}(M, N, m).$$

(iii) $\Phi \in \mathcal{SC}(M, N)$ if and only if $T(\Phi) \in \mathcal{SC}(M, N)$ for one (and hence for all) orthogonal transformation(s) T on $\mathbb{R}^N$.
(iv) Let $\Phi = \{\varphi_k\}_{k=1}^{N+1} \in \mathcal{F}(N+1, N) \setminus \{0\}$ with $\varphi_k \neq \pm\varphi_\ell$ for $k \neq \ell$. If $\Phi \in \mathcal{SC}_+(N+1, N, N)$, then $\Phi \notin \mathcal{SC}_+(N+1, N)$.

REMARK 2.3. We point out that part (iii) of Proposition 2.2 is equivalent to saying that Φ is not scalable if one can find an orthogonal transformation T on $\mathbb{R}^N$ such that $T(\Phi)$ is not scalable.

Besides these elementary properties, a study of the topological properties of the set of scalable frames was considered in [**52, 53**]. In particular,

PROPOSITION 2.4. Let $M \geq N \geq 2$.

(i) $\mathcal{SC}(M, N)$ is closed in $\mathcal{F}(M, N)$. Furthermore, for each $N \leq m \leq M$, $\mathcal{SC}(M, N, m)$ is closed in $\mathcal{F}(M, N)$.
(ii) If $M < N(N+1)/2$, then the interior of $\mathcal{SC}(M, N)$ is empty.

Part (i) of proposition 2.4 was proved in [**52**, Corrollary 3.3] and [**53**, Proposition 4.1], while part (ii) first appeared in [**53**, Theorem 4.2].

2.2. Convex polytopes associated to scalable frames. We now proceed to write an explicit formulation of the scalability problem. From this formulation a convex geometric characterization of $\mathcal{SC}(M, N)$ will follow. To start, we recall that Φ denote the synthesis operator associated to the frame $\Phi = \{\varphi_k\}_{k=1}^M$. Φ is (m-) scalable if and only if there are positive numbers $\{x_k\}_{k\in I}$ with $\#I = m \geq N$ such that $\widetilde{\Phi} = \Phi X$ satisfies

$$\widetilde{\Phi}\widetilde{\Phi}^T = \Phi X^2 \Phi^T = \tilde{A} I_N = \frac{\sum_{k\in I} x_k^2 \|\varphi_k\|^2}{N} I_N \tag{2.1}$$

where X is the diagonal matrix with the weights x_k on its main diagonal if $k \in I$ and 0 for $k \notin I$, and I_N is the $N \times N$ identity matrix. Moreover,

$$\|X\|_0 = \#\{k : x_k > 0\} = m \geq N.$$

By rescaling the diagonal matrix X, we can assume that $\tilde{A} = 1$. Thus, (2.1) is equivalent to solving

$$\Phi Y \Phi^T = I_N \tag{2.2}$$

for $Y = \frac{1}{\tilde{A}} X^2$.

To gain some intuition let us consider the two dimensional case with $M \geq 3$. In particular, let us describe when $\Phi = \{\varphi_k\}_{k=1}^M \subset S^1$ is a scalable frame. Without loss of generality, we may assume that $\Phi = \{\varphi_k\}_{k=1}^M \subset \mathbb{R} \times \mathbb{R}_+$, $\|\varphi_k\| = 1$, and $\varphi_\ell \neq \pm\varphi_k$ for $\ell \neq k$. Thus

$$\varphi_k = \begin{pmatrix} \cos\theta_k \\ \sin\theta_k \end{pmatrix} \in S^1$$

with

$$0 = \theta_1 < \theta_2 < \theta_3 < \ldots < \theta_M < \pi.$$

Let $Y = (y_k)_{k=1}^M \subset [0, \infty)$, then (2.2) becomes

$$\begin{pmatrix} \sum_{k=1}^M y_k \cos^2\theta_k & \sum_{k=1}^M y_k \sin\theta_k \cos\theta_k \\ \sum_{k=1}^M y_k \sin\theta_k \cos\theta_k & \sum_{k=1}^M y_k \sin^2\theta_k \end{pmatrix} = \begin{pmatrix} 1 & 0 \\ 0 & 1 \end{pmatrix}. \tag{2.3}$$

This is equivalent to

$$\begin{cases} \sum_{k=1}^M y_k \cos^2\theta_k &= 1 \\ \sum_{k=1}^M y_k \sin^2\theta_k &= 1 \\ \sum_{k=1}^M y_k \sin\theta_k \cos\theta_k &= 0, \end{cases}$$

and using some row operations we arrive at

$$\begin{cases} \sum_{k=1}^M y_k \sin^2\theta_k &= 1 \\ \sum_{k=1}^M y_k \cos 2\theta_k &= 0 \\ \sum_{k=1}^M y_k \sin 2\theta_k &= 0. \end{cases}$$

For Φ to be scalable we must find a nonnegative vector $Y = (y_k)_{k=1}^M$ in the kernel of the matrix whose k^{th} column is $\begin{pmatrix} \cos 2\theta_k \\ \sin 2\theta_k \end{pmatrix}$. Notice that the first equation is just a normalization condition.

We now describe the the subset of the kernel of this matrix that consists of non-trivial nonnegative vectors. Observe that the matrix can be reduced to

$$\begin{pmatrix} 1 & \cos 2\theta_2 & \ldots & \cos 2\theta_M \\ 0 & \sin 2\theta_2 & \ldots & \sin 2\theta_M \end{pmatrix}. \tag{2.4}$$

EXAMPLE 2.5. We first consider the case $M = 3$. In this case, we have $0 = \theta_1 < \theta_2 < \theta_3 < \pi$, and the (2.4) becomes

$$\begin{pmatrix} 1 & \cos 2\theta_2 & \cos 2\theta_3 \\ 0 & \sin 2\theta_2 & \sin 2\theta_3 \end{pmatrix}. \tag{2.5}$$

If there exists an index $k_0 \in \{2, 3\}$ with $\sin 2\theta_{k_0} = 0$, then $\theta_{k_0} = \pi/2$ and the corresponding frame contains an ONB and, hence is scalable.

(i) Moreover, if $k_0 = 2$, then $0 = \theta_1 < \theta_2 = \pi/2 < \theta_3 < \pi$. In this case, the fame is 2− scalable but not 3− scalable, i.e., the frame is in $\mathcal{SC}_+(3,2,2) \setminus \mathcal{SC}(3,2,3)$. This is illustrated by Figure 1.
(ii) If $k_0 = 3$, then $0 = \theta_1 < \theta_2 < \theta_3 = \pi/2$. By symmetry (with respect to the y axis) we conclude again that the fame is 2− scalable but not 3− scalable.

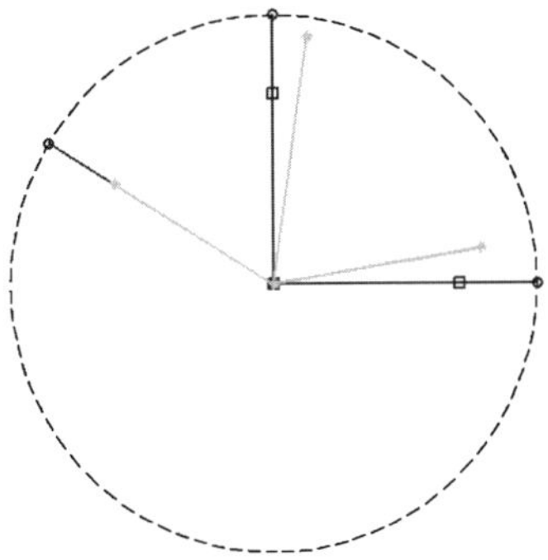

FIGURE 1. A scalable frame (contains an orthonormal basis) with 3 vectors in $\mathbb{R}^2$. The original frame vectors are in blue, the frame vectors obtained by scaling are in red, and for comparison the associated canonical tight frame vectors are in green.

Assume now that $\theta_k \neq \pi/2$ for $k = 2, 3$. If $\theta_3 < \pi/2$, then the frame cannot be scalable. Indeed, $u = (z_1, z_2, z_3)$ belongs to the kernel of (2.5) if and only if

$$\begin{cases} z_1 &= \frac{\sin 2(\theta_3 - \theta_2)}{\sin 2\theta_2} z_3, \\ z_2 &= -\frac{\sin 2\theta_3}{\sin 2\theta_2} z_3, \end{cases} \tag{2.6}$$

where $z_3 \in \mathbb{R}$. The choice of the angles implies that $z_2 z_3 \leq 0$ and $z_1 z_3 \leq 0$ with equality if and only if $z_3 = 0$. This is illustrated by Figure 2. Similarly, if $0 = \theta_1 < \pi/2 < \theta_2 < \theta_3 < \pi$, then the frame cannot be scalable.

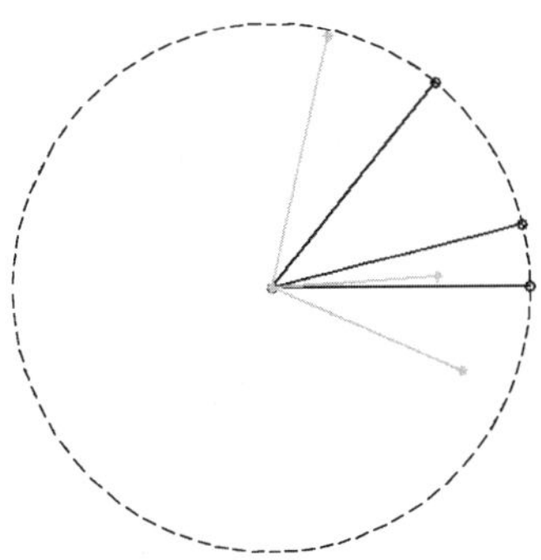

FIGURE 2. A non scalable frame with 3 vectors in $\mathbb{R}^2$. The original frame vectors are in blue, for comparison the associated canonical tight frame vectors are in green.

On the other hand if $0 = \theta_1 < \theta_2 < \pi/2 < \theta_3 < \pi$, then it follows from (2.6) $z_1 > 0$ and $z_2 > 0$ for all $z_3 > 0$ if and only if $\theta_3 - \theta_2 < \pi/2$. Consequently, when $0 = \theta_1 < \theta_2 < \pi/2 < \theta_3 < \pi$ the frame $\Phi \in \mathcal{SC}_+(3, 2, 3)$ if and only if $0 < \theta_3 - \theta_2 < \pi/2$. This is illustrated by Figure 3.

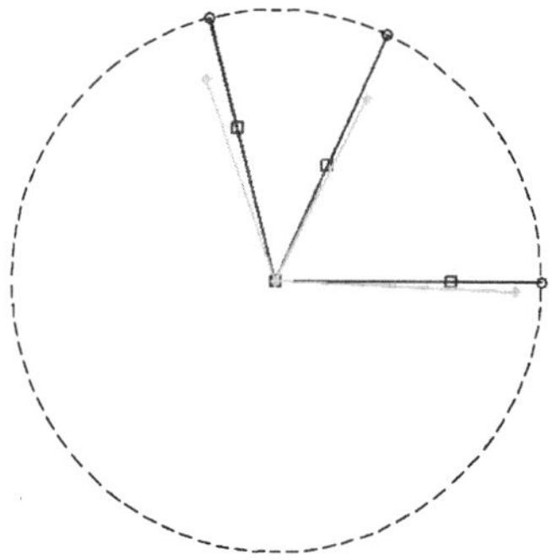

FIGURE 3. A scalable frame with 3 vectors in $\mathbb{R}^2$. The original frame vectors are in blue, the frame vectors obtained by scaling are in red, and for comparison the associated canonical tight frame vectors are in green.

EXAMPLE 2.6. Assume now that $M = 4$. Then we are lead to seek nonnegative non-trivial vectors in the null space of

$$\begin{pmatrix} 1 & \cos 2\theta_2 & \cos 2\theta_3 & \cos 2\theta_4 \\ 0 & \sin 2\theta_2 & \sin 2\theta_3 & \sin 2\theta_4 \end{pmatrix}.$$

If there exists an index $k_0 \in \{2, 3, 4\}$ with $\sin 2\theta_{k_0} = 0$, then $\theta_{k_0} = \pi/2$ and the corresponding frame contains an ONB. Consequently, the frame is scalable. In particular,

(1) When $k_0 = 2$, the null space of the matrix is described by

$$\begin{cases} z_1 & = & z_2 + \frac{\sin 2(\theta_4 - \theta_3)}{\sin 2\theta_3} z_4, \\ z_3 & = & -\frac{\sin 2\theta_4}{\sin 2\theta_3} z_4, \end{cases}$$

where $z_2, z_4 \in \mathbb{R}$. Note that $z_3 \leq 0$, with equality only when $z_4 = 0$, in which case $z_3 = 0$ and the frame will be 2−scalable, but not m−scalable for $m = 3, 4$. This is illustrated by the left figure in Figure 4.

(2) If instead, $k_0 = 3$, then a similar argument shows that

$$\begin{cases} z_1 & = & z_3 + \frac{\sin 2(\theta_4 - \theta_2)}{\sin 2\theta_2} z_4, \\ z_2 & = & -\frac{\sin 2\theta_4}{\sin 2\theta_2} z_4, \end{cases}$$

where $z_3, z_4 \in \mathbb{R}$. Any choice of $z_4 > 0$ will result in $z_2 > 0$. If we choose $\theta_4 - \theta_2 < \pi/2$, then $z_3 \geq 0$ will lead to a 3− scalable frame or a 4−scalable frame. If instead, $\theta_4 - \theta_2 \geq \pi/2$ we can always choose $z_3 > 0$ and large enough to guarantee that $z_1 > 0$, hence Φ will be 4−scalable.

(3) When $k_0 = 4$, then $\Phi \in \mathcal{SC}_+(4, 2, 2) \setminus \mathcal{SC}(4, 2, m)$ for $m = 3, 4$.

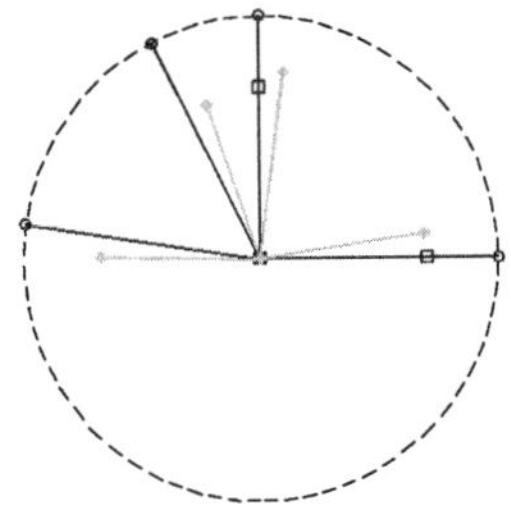

FIGURE 4. A scalable frame (contains an orthonormal basis) with with 4 vectors in $\mathbb{R}^2$. The original frame vectors are in blue, the frame vectors obtained by scaling are in red, and for comparison the associated canonical tight frame vectors are in green.

(4) When $\sin 2\theta_k \neq 0$ for $k \in \{2,3,4\}$ then

$$\begin{cases} z_1 &= \frac{\sin 2(\theta_3-\theta_2)}{\sin 2\theta_2} z_3 + \frac{\sin 2(\theta_4-\theta_2)}{\sin 2\theta_2} z_4, \\ z_2 &= -\frac{\sin 2\theta_3}{\sin 2\theta_2} z_3 - \frac{\sin 2\theta_4}{\sin 2\theta_2} z_4, \end{cases}$$

where $z_3, z_4 \in \mathbb{R}$. A choice of $z_3, z_4 \geq 0$ will lead to a scalable frame if at least $z_1 \geq 0$ or $z_2 \geq 0$. For example, Figure 5 shows a scalable frame.

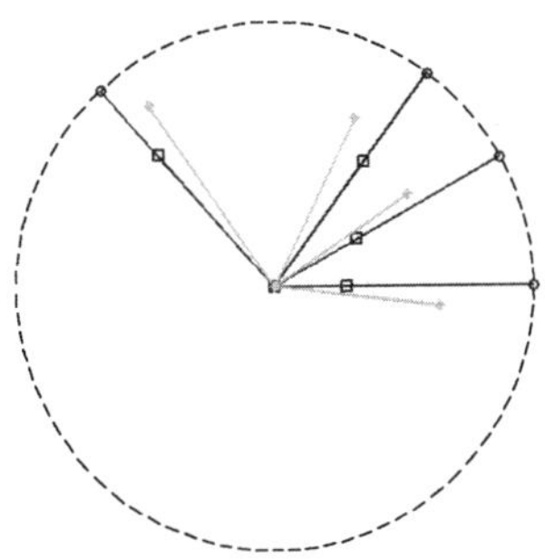

FIGURE 5. A scalable frame with with 4 vectors in $\mathbb{R}^2$. The original frame vectors are in blue, the frame vectors obtained by scaling are in red, and for comparison the associated canonical tight frame vectors are in green.

But in this case we could also get non scalable frame, see Figure 6 . The implication here is that the scalability of the frames depends on the relative position of the frame vectors (hence) the angles θ_k. This will be made rigorous in Section 2.3.

More generally in this two dimensional case, we can continue this analysis of the transformation given by the matrix (2.4) to characterize when $\Phi = \{\varphi_k\}_{k=1}^M$ is scalable. From the figures shown in Examples 2.5 and 2.6, it is clear that some geometric considerations are involved. Before elaborating more on these geometric considerations in Section 2.3, we now consider the general case $M \geq N \geq 2$. In particular, we follow the algebraic approach given in the two-dimensional case, by writing out the equations in (2.2) and collecting all the diagonal terms one the

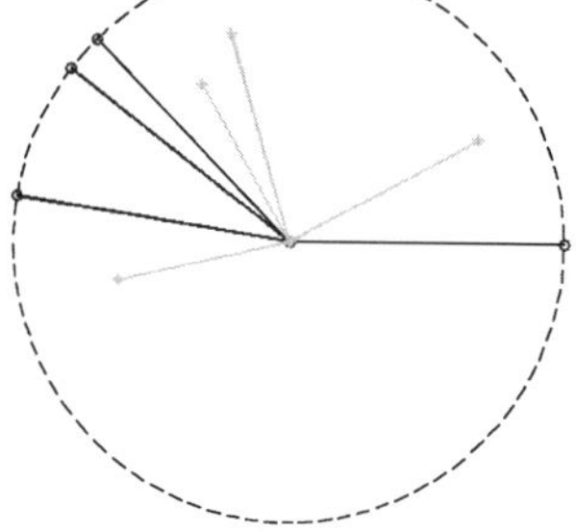

FIGURE 6. A non scalable frame with 4 vectors in $\mathbb{R}^2$. The original frame vectors are in blue, for comparison the associated canonical tight frame vectors are in green.

one hand, and the non-diagonal terms on the other, we see that for a frame to be m-scalable it is necessary and sufficient that there exists $u = (c_1^2, c_2^2, \ldots, c_M^2)^T$ with $\|u\|_0 := \#\{u_k : u_k > 0\} \leq m$ which is a solution of the following linear system of $\frac{N(N+1)}{2}$ equations in the M unknowns $(y_j)_{j=1}^M$:

$$(2.7) \qquad \begin{cases} \sum_{j=1}^{M} \varphi_j(k)^2 y_j &= 1 \quad \text{for } k = 1, \ldots, N, \\ \sum_{j=1}^{M} \varphi_j(\ell)\varphi_j(k) y_j &= 0 \quad \text{for } k > \ell = 1, \ldots, N. \end{cases}$$

We can further reduce this linear system in the following manner. We keep all the equations with homogeneous right-hand sides, i.e., those coming from the non diagonal terms of (2.2). There are $N(N-1)/2$ such equations. The remaining N equations come from the diagonal terms of (2.2), and their right hand-sides are all 1. We can use row operations to reduce these to a new set of N linear equations the first of which will be

$$\sum_{j=1}^{M} \varphi_j(1)^2 y_j = 1.$$

For $k = 2, \ldots, N$, the kth equation is obtained by subtracting row 1 from row k leading to

$$\sum_{j=1}^{M} (\varphi_j(k)^2 - \varphi_j(1)^2) y_j = 0.$$

The condition

$$\sum_{j=1}^{M} \varphi_j(1)^2 y_j = 1$$

is a normalization condition, indicating that if Φ can be scaled with $y = (y_j)_{j=1}^M \subset [0, \infty)$, then it can also be scaled by λy for any $\lambda > 0$. Thus, ignoring this condition and collecting all the remaining equations, we see that Φ is m-scalable if and only if there exists a nonnegative vector $u \in \mathbb{R}^M$ with $\|u\|_0 \leq m$ such that

$$F(\Phi)u = 0,$$

where the $(N-1)(N+2)/2 \times M$ matrix $F(\Phi)$ is given by

$$F(\Phi) = \begin{pmatrix} F(\varphi_1) & F(\varphi_2) & \dots & F(\varphi_M) \end{pmatrix},$$

where $F : \mathbb{R}^N \to \mathbb{R}^d$, $d := (N-1)(N+2)/2$, is defined by

$$(2.8)\quad F(x) = \begin{pmatrix} F_0(x) \\ F_1(x) \\ \vdots \\ F_{N-1}(x) \end{pmatrix}, \qquad F_0(x) = \begin{pmatrix} x_1^2 - x_2^2 \\ x_1^2 - x_3^2 \\ \vdots \\ x_1^2 - x_N^2 \end{pmatrix}, \qquad F_k(x) = \begin{pmatrix} x_k x_{k+1} \\ x_k x_{k+2} \\ \vdots \\ x_k x_N \end{pmatrix},$$

and $F_0(x) \in \mathbb{R}^{N-1}$, $F_k(x) \in \mathbb{R}^{N-k}$, $k = 1, 2, \dots, N-1$.

To summarize, we arrive at the following result that was proved in [**53**, Proposition 3.7]

PROPOSITION 2.7. [**53**, Proposition 3.7] A frame Φ for $\mathbb{R}^N$ is m-scalable, respectively, strictly m-scalable, if and only if there exists a nonnegative $u \in \ker F(\Phi)\setminus\{0\}$ with $\|u\|_0 \leq m$, respectively, $\|u\|_0 = m$, where $\|u\|_0 = \#\{k : u_k > 0\}$.

In the two dimensional case the map F reduces to

$$F\begin{pmatrix} x \\ y \end{pmatrix} = \begin{pmatrix} x^2 - y^2 \\ xy \end{pmatrix}.$$

However, in all the previous examples we considered instead the more geometric map :

$$\widetilde{F}\begin{pmatrix} x \\ y \end{pmatrix} = \begin{pmatrix} x^2 - y^2 \\ 2xy \end{pmatrix}.$$

It is readily seen that $F(\Phi)$ and $\widetilde{F}(\Phi)$ have exactly the same kernel. In fact the map $\widetilde{F}$ carries the following geometric interpretation. Let L_θ be a line through the origin in $\mathbb{R}^2$ which makes an angle θ with the positive $x-$axis with $\theta \in [0, \pi]$. Then the image of L_θ by $\widetilde{F}$ is the line $L_{2\theta}$ that makes an angle 2θ with the positive $x-$axis. That is, $\widetilde{F}$ just rotates counterclockwise the line L_θ by an angle equal to θ.

In the two dimensional case we exploited the geometric meaning of the map F or $\widetilde{F}$ to describe the subset of nonnegative vectors of the nullspace of $\widetilde{F}(\Phi)$. More generally, to find nonnegative vectors in the nullspace of the matrix $F(\Phi)$ we can appeal to one of the formulation of Farkas lemma:

LEMMA 2.8. [**60**, Lemma 1.2.5] *For every real $N \times M$-matrix A exactly one of the following cases occurs:*

(i) *The system of linear equations $Ax = 0$ has a nontrivial nonnegative solution $x \in \mathbb{R}^M$, i.e., all components of x are nonnegative and at least one of them is strictly positive.*
(ii) *There exists $y \in \mathbb{R}^N$ such that $y^T A$ is a vector with all entries strictly positive.*

Applying this in the two dimensional case, we see that for the frame to be scalable, the second alternative in Farkas's lemma should not hold. That is there must exist no vector in $\mathbb{R}^2$ that lies on "one side" of all the vectors $F(\varphi_k)$ for $k = 1, 2, \dots, M$. We illustrate this by the following figures:

We make the following observation about the first alternative of Lemma 2.8. If $\{A_k\}_{k=1}^M \subset \mathbb{R}^N$ represents the column vectors of A, then there exists of a vector

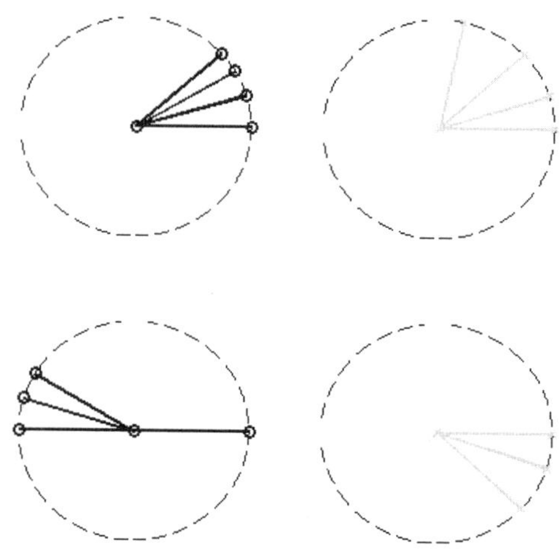

FIGURE 7. Frames with 4 vectors in $\mathbb{R}^2$ (in blue, top and bottom left) and their images by the map F (in green, top and bottom right). Both of these examples result in non scalable frames.

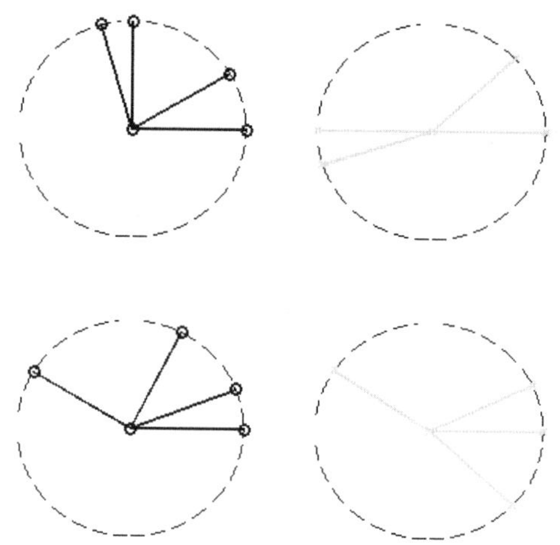

FIGURE 8. Frames with 4 vectors in $\mathbb{R}^2$ (in blue, top and bottom left) and their images by the map F (in green, top and bottom right). Both of these examples result in scalable frames.

$0 \neq x = (x_k)_{k=1}^M$ with $x_k \geq 0$ such that $Ax = 0$ is equivalent to saying that

$$\sum_{k=1}^{M} x_k A_k = 0.$$

Without loss of generality we may assume that $\sum_{k=1}^{M} x_k = 1$, in which case the condition is equivalent to 0 being a convex combination of the column vectors of A. Thus, having a nontrivial nonnegative vector in the null space of A is a statement about the convex hull of the columns of A.

Motivated by the geometric intuition we gained from the two-dimensional setting and to effectively use Farkas's lemma, we introduce a few notions from convex geometry, especially the theory of convex polytopes, and we refer to [**68**, **80**] for more details on these concepts. For a finite set $X = \{x_i\}_{k=1}^{M} \subset \mathbb{R}^N$, the *polytope generated by* X is the convex hull of X, which is a compact convex subset of $\mathbb{R}^N$. In particular, we denote this set by P_X (or $\text{co}(X)$), and we have

$$P_X = \text{co}(X) := \left\{ \sum_{k=1}^{M} \alpha_k x_k : \alpha_k \geq 0, \sum_{k=1}^{M} \alpha_k = 1 \right\}.$$

The *affine hull generated by* X is defined by

$$\text{aff}(X) := \left\{ \sum_{k=1}^{M} \alpha_k x_k : \sum_{k=1}^{M} \alpha_k = 1 \right\}.$$

We have $\text{co}(X) \subset \text{aff}(X)$. The *relative interior* of the polytope $\text{co}(X)$ denoted by $ri\,\text{co}(X)$, is the interior of $\text{co}(X)$ in the topology induced by $\text{aff}(X)$. We have that $ri\,\text{co}(X) \neq \emptyset$ as long as $\#X \geq 2$, and

$$ri\,\text{co}(X) = \left\{ \sum_{k=1}^{M} \alpha_k x_k : \alpha_k > 0, \sum_{k=1}^{M} \alpha_k = 1 \right\}.$$

The *polyhedral cone generated by* X is the closed convex cone $C(X)$ defined by

$$C(X) = \left\{ \sum_{i=k}^{M} \alpha_k x_k : \alpha_k \geq 0 \right\}.$$

The *polar cone* of $C(X)$ is the closed convex cone $C^\circ(X)$ defined by

$$C^\circ(X) := \{x \in \mathbb{R}^N : \langle x, y \rangle \leq 0 \text{ for all } y \in C(X)\}.$$

The cone $C(X)$ is said to be *pointed* if $C(X) \cap (-C(X)) = \{0\}$, and *blunt* if the linear space generated by $C(X)$ is $\mathbb{R}^N$, i.e., $\text{span}\, C(X) = \mathbb{R}^N$.

Using Proposition 2.7, we see that Φ is $(m-)$scalable if there exists $\{\lambda_k\}_{k\in I} \subset [0, \infty)$, $\#I = m$ such that

$$\sum_{k\in I} \lambda_k F(\varphi_k) = 0.$$

This is equivalent to saying that 0 belongs to the polyhedral cone generated by $F(\Phi_I) = \{F(\varphi_k)\}_{k\in I}$. Without loss of generality we can assume that $\sum_{k\in I} \lambda_k = 1$ which implies that 0 belongs to the polytope generated by $F(\Phi_I) = \{F(\varphi_k)\}_{k\in I}$. Putting these observations together with Lemma 2.8 the following results were established in [**53**, theorem 3.9]. In the sequel, we shall denote by $[K]$ the set of integers $\{1, 2, \ldots, K\}$ where $K \in \mathbb{N}$.

THEOREM 2.9. [**53**, Theorem 3.9] *Let $M \geq N \geq 2$, and let m be such that $N \leq m \leq M$. Assume that $\Phi = \{\varphi_k\}_{k=1}^{M} \in \mathcal{F}^*(M, N)$ is such that $\varphi_k \neq \varphi_\ell$ when $k \neq \ell$. Then the following statements are equivalent:*

(i) *Φ is $m-$scalable, respectively, strictly $m-$scalable,*

(ii) *There exists a subset* $I \subset [M]$ *with* $\#I = m$ *such that* $0 \in \text{co}(F(\Phi_I))$, *respectively,* $0 \in ri\,\text{co}(F(\Phi_I))$.
(iii) *There exists a subset* $I \subset [M]$ *with* $\#I = m$ *for which there is no* $h \in \mathbb{R}^d$ *with* $\langle F(\varphi_k), h\rangle > 0$ *for all* $k \in I$, *respectively, with* $\langle F(\varphi_k), h\rangle \geq 0$ *for all* $k \in I$, *with at least one of the inequalities being strict.*

The details of the proof of this result can be found in [**53**, Theorem 3.9]. We point out however, that the equivalence of (i) and (ii) follows from considering $\text{co}(F(\Phi_I))$ which is the polytope in $\mathbb{R}^N$ generated by the vectors $\{F(\varphi_k)\}_{k\in I}$.

By removing the normalization condition that the ℓ^1 norm of the weights making a frame scalable is unity, Theorem 2.9 can be stated in terms of the polyhedral cone $C(F(\Phi))$ generated by $F(\Phi)$. This is the content of the following result which was proved in [**53**, Corollary 3.14]:

COROLLARY 2.10. [**53**, Corollary 3.14] Let $\Phi = \{\varphi_k\}_{k=1}^M \in \mathcal{F}^*$, and let $N \leq m \leq M$ be fixed. Then the following conditions are equivalent:

(i) Φ is strictly m-scalable .
(ii) There exists $I \subset [M]$ with $\#I = m$ such that $C(F(\Phi_I))$ is not pointed.
(iii) There exists $I \subset [M]$ with $\#I = m$ such that $C(F(\Phi_I))^\circ$ is not blunt.
(iv) There exists $I \subset [M]$ with $\#I = m$ such that the interior of $C(F(\Phi_I))^\circ$ is empty.

The map F given in (2.8) is related to the diagram vector of [**41**], and was used in [**25**] to give a different and equivalent characterization of scalable frames which we now present. We start by presenting an interesting necessary condition for scalability in both $\mathbb{R}^N$ and $\mathbb{C}^N$ proved in [**25**, Theorem 3.1]:

THEOREM 2.11. [**25**, Theorem 3.1] *Let* $\Phi = \{\varphi_k\}_{k=1}^M \in \mathcal{F}_u(M,N)$. *If* $\Phi \in \mathcal{SC}(M,N)$, *then there is no unit vector* $u \in \mathbb{R}^N$ *such that* $|\langle u, \varphi_k\rangle| \geq \frac{1}{\sqrt{N}}$ *for all* $k = 1, 2, \ldots, M$ *and* $|\langle u, \varphi_k\rangle| > \frac{1}{\sqrt{N}}$ *for at least one* k.

As pointed out in [**25**] the condition in Theorem 2.11 is also sufficient only when $N = 2$. We wish to compare this result to the following theorem that give a necessary and a (different) sufficient condition for scalability in $\mathbb{R}^N$, and these two conditions are necessary and sufficient only for $N = 2$.

THEOREM 2.12. [**22**, Theorem 4.1] *Let* $\Phi \in \mathcal{F}_u(M,N)$. *Then the following hold:*

(a) (*A necessary condition for scalability*) *If* Φ *is scalable, then*

$$\min_{\|d\|_2=1} \max_k |\langle d, \varphi_k\rangle| \geq \frac{1}{\sqrt{N}}. \tag{2.9}$$

(b) (*A sufficient condition for scalability*) *If*

$$\min_{\|d\|_2=1} \max_k |\langle d, \varphi_k\rangle| \geq \sqrt{\frac{N-1}{N}}, \tag{2.10}$$

then Φ *is scalable.*

Clearly when $N = 2$ the right hand sides of both (2.9) and (2.10) coincide leading to a necessary and sufficient condition.

Observe that (2.9) is equivalent to the fact that for each unit vector $d \in \mathbb{R}^N$, there exists $k = 1, 2, \dots, M$ such that

$$|\langle d, \varphi_k \rangle| \geq \frac{1}{\sqrt{N}}$$

which is different from the condition in theorem 2.11.

We can now present the characterization of scalable obtained in [**25**, Theorem 3.2] and which is based on the Gramian of the diagram vectors. More precisely, for each $v \in \mathbb{R}^N$, we define the diagram vector to be the vector $\tilde{v} \in \mathbb{R}^{N(N-1)}$ given by

$$\tilde{v} = \frac{1}{\sqrt{N-1}} \begin{bmatrix} v(1)^2 - v(2)^2 \\ \vdots \\ v(N-1)^2 - v(N)^2 \\ \sqrt{2N} v(1) v(2) \\ \vdots \\ \sqrt{2N} v(N-1) v(N) \end{bmatrix}, \tag{2.11}$$

where the difference of the squares $v^2(i) - v^2(j)$ and the product $v(i)v(j)$ occur exactly once for $i < j, i = 1, 2, \dots, N-1$. Using this notion, the following result was proved:

THEOREM 2.13. [**25**, Theorem 3.2] *Let $\Phi = \{\varphi_k\}_{k=1}M \in \mathcal{F}_u$ be a frame of unit-norm vectors, and $\widetilde{G} = (\langle \tilde{\varphi}_k, \tilde{\varphi}_\ell \rangle)$ be the Gramian of the diagram vectors $\{\tilde{\varphi}_k\}_{k=1}^M$. Suppose that $\widetilde{G}$ is not invertible. Let $\{v_1, v_2, \dots, v_\ell\}$ be a basis of the nullspace of $\widetilde{G}$ and set*

$$r_i := \begin{bmatrix} v_1(i) \\ \vdots \\ v_\ell(i) \end{bmatrix},$$

for $i = 1, 2, \dots, M$. Then Φ is scalable if and only if $0 \notin \text{co}\{r_1, r_2, \dots, r_M\}$.

When a frame $\Phi = \{\varphi_k\}_{k=1}^M \subset \mathbb{R}^N$ is scalable, then there exist $\{c_k\}_{k=1}^M \subset [0, \infty)$ such that $\{c_k \varphi_k\}_{k=1}^M$ is a tight frame. The nonnegative vector $\omega = \{c_k^2\}_{k=1}^M$ is called a *scaling* of Φ [**14**]. The scaling $\omega = \{c_k^2\}_{k=1}^M \subset [0, \infty)$ is said to be a *minimal scaling* if $\{\varphi_k : c_k^2 > 0\}$ has no proper subset which is scalable. The notion of minimal scalings has recently found some very interesting applications on some structural decomposition of frames; see, [**21**, Section 4] for more details. It turns out that finding the scalings of a scalable frame can be reduced to finding its minimal scalings. More specifically, the following result was proved in [**14**, theorem 3.5]:

THEOREM 2.14. [**14**, Theorem 3.5] *Suppose $\Phi = \{\varphi_k\}_{k=1}^M \in \mathcal{F}_u$ is a scalable frame, and let $\omega = \{\omega_k\}_{k=1}^M \subset [0, \infty)$ be one of its minimal scalings. Then $\{\varphi_k \varphi_k^T : \omega_k > 0\}$ is linearly independent. Furthermore, every scaling of Φ is a convex combination of minimal scalings.*

2.3. A geometric condition for scalability. The two dimensional case we examined earlier (Example 2.5 and Example 2.6) indicates that a frame is not scalable when the frame vectors "cluster" in certain "small" plane regions. In fact, broadly speaking, the frame is not scalable if its vectors lies in a double cone $C \cap (-C)$ with a "small" aperture. This was formalized in theorem 2.9 and Corollary 2.10. We can further exploit these results to give a more formal geometric characterization of scalable frames.

To begin, we rewrite (iii) of Theorem 2.9 in the following form. For $x = (x_k)_{k=1}^N \in \mathbb{R}^N$ and $h = (h_k)_{k=1}^d \in \mathbb{R}^d$, we have that

$$\langle F(x), h\rangle = \sum_{\ell=2}^{N} h_{\ell-1}(x_1^2 - x_\ell^2) + \sum_{k=1}^{N-1}\sum_{\ell=k+1}^{N} h_{k(N-1-(k-1)/2)+\ell-1}x_k x_\ell. \tag{2.12}$$

Consequently, fixing $h \in \mathbb{R}^d$, $\langle F(x), h\rangle$ is a homogeneous polynomial of degree 2 in $x_1, x_2, \ldots, x_N$. Denote the set of all polynomials of this form by $\boldsymbol{P}_2^N$. Then $\boldsymbol{P}_2^N$ can be identified with the subspace of real symmetric $N \times N$ matrices whose trace is 0. Indeed, for each $N \geq 2$, and each $p \in \boldsymbol{P}_2^N$,

$$p(x) = \sum_{\ell=2}^{N} a_{\ell-1}(x_1^2 - x_\ell^2) + \sum_{k=1}^{N-1}\sum_{\ell=k+1}^{N} a_{k(N-(k+1)/2)+\ell-1}x_k x_\ell,$$

we have $p(x) = \langle Q_p x, x\rangle$, where Q_p is the symmetric $N \times N$-matrix with entries

$$Q_p(1,1) = \sum_{k=1}^{N-1} a_k, \qquad Q_p(\ell,\ell) = -a_{\ell-1} \quad \text{for } \ell = 2, 3, \ldots, N$$

and

$$Q_p(k,\ell) = \frac{1}{2}a_{k(N-(k+1)/2)+\ell-1} \quad \text{for } k = 1, \ldots, N-1,\ \ell = k+1, \ldots, N.$$

Thus, $\langle F(x), h\rangle = \langle Q_h x, x\rangle = 0$ defines a quadratic surface in $\mathbb{R}^N$, and condition (iii) in Theorem 2.9 stipulates that for Φ to be scalable, one cannot find such a quadratic surface such that the frame vectors (with index in I) all lie on (only) "one side" of this surface. By taking the contrapositive statement we arrived at the following result that was proved differently in [**52**, Theorem 3.6]. In particular, it provides a characterization of non-scalability of finite frames, and we shall use it to give a very interesting geometric condition on the frame vectors for non-scalable frames.

THEOREM 2.15. [**52**, Theorem 3.6] *Let* $\Phi = \{\varphi_k\}_{k=1}^M \in \mathcal{F}^*$. *Then the following statements are equivalent.*

(i) *Φ is not scalable.*
(ii) *There exists a symmetric matrix $Y \in \mathbb{R}^{N\times N}$ with $Tr(Y) < 0$ such that $\langle Y\varphi_k, \varphi_k\rangle \geq 0$ for all $k = 1, \ldots, M$.*
(iii) *There exists a symmetric matrix $Y \in \mathbb{R}^{N\times N}$ with $Tr(Y) = 0$ such that $\langle Y\varphi_k, \varphi_k\rangle > 0$ for all $k = 1, \ldots, M$.*

To derive the geometric condition for non-scalability we need some set up. It is not difficult to see that each symmetric $N \times N$ matrix Y in (iii) of theorem 2.15 corresponds to a quadratic surface. We call this surface *a conical zero-trace quadric.* The exact definition of such quadratic surface is

DEFINITION 2.16. [**52**, Definition 3.4] Let the *class of conical zero-trace quadrics* $\mathcal{C}_N$ be defined as the family of sets

$$\left\{x \in \mathbb{R}^N : \sum_{k=1}^{N-1} a_k \langle x, e_k\rangle^2 = \langle x, e_N\rangle^2\right\}, \tag{2.13}$$

where $\{e_k\}_{k=1}^N$ runs through all orthonormal bases of $\mathbb{R}^N$ and $(a_k)_{k=1}^{N-1}$ runs through all tuples of elements in $\mathbb{R} \setminus \{0\}$ with $\sum_{k=1}^{N-1} a_k = 1$.

We define the *interior* of the conical zero-trace quadric in (2.13), by

$$\left\{x \in \mathbb{R}^N : \sum_{k=1}^{N-1} a_k \langle x, e_k\rangle^2 < \langle x, e_N\rangle^2\right\},$$

and the *exterior* of the conical zero-trace quadric in (2.13) by

$$\left\{x \in \mathbb{R}^N : \sum_{k=1}^{N-1} a_k \langle x, e_k\rangle^2 > \langle x, e_N\rangle^2\right\}.$$

It is then easy to see that Theorem 2.15 is equivalent to the following result established in [**52**, theorem 3.6]

THEOREM 2.17. [**52**, Theorem 3.6] *Let $\Phi \subset \mathbb{R}^N \setminus \{0\}$ be a frame for $\mathbb{R}^N$. Then the following conditions are equivalent.*

(i) *Φ is not scalable.*
(ii) *All frame vectors of Φ are contained in the interior of a conical zero-trace quadric of $\mathcal{C}_N$.*
(iii) *All frame vectors of Φ are contained in the exterior of a conical zero-trace quadric of $\mathcal{C}_N$.*

The geometric meaning of this result is best illustrated by considering frames in $\mathbb{R}^2$ and $\mathbb{R}^3$, in which case the sets $\mathcal{C}_N$ for $N = 2, 3$, have very simple descriptions given in [**52**]. For our purposes here it suffices to say that each set in $\mathcal{C}_2$ is the boundary surface of a *quadrant cone* in $\mathbb{R}^2$, i.e., the union of two orthogonal one-dimensional subspaces (lines through the origin) in $\mathbb{R}^2$. The sets in $\mathcal{C}_3$ are the boundary surfaces of a particular class of *elliptical cones* in $\mathbb{R}^3$. We give examples of sets in $\mathcal{C}_N$ $N = 2, 3$ in Figure 9 (a) and (b).

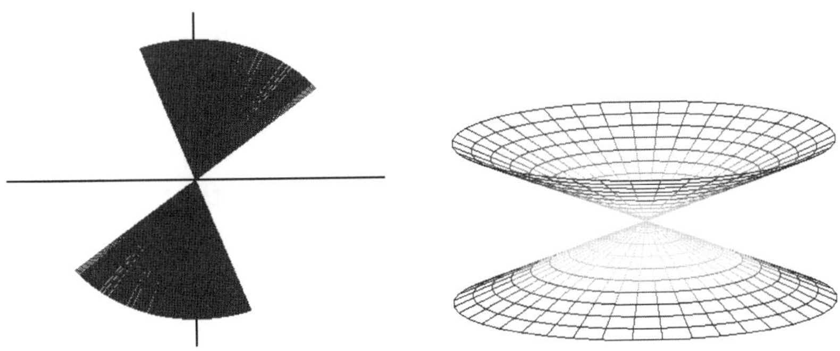

(a) (b)

FIGURE 9. (a) shows a sample region of vectors of a non-scalable frame in $\mathbb{R}^2$. (b) shows example of $\mathcal{C}_3^-$ and $\mathcal{C}_3^+$ which determine sample regions in $\mathbb{R}^3$.

We can now state the following corollary that give a clear geometric insight into the set of non-scalable frames. In particular, the frame vectors cannot lie in a "small cone".

COROLLARY 2.18. [**52**, Corollary 3.8]

(i) A frame $\Phi \subset \mathbb{R}^2 \setminus \{0\}$ for $\mathbb{R}^2$ is not scalable if and only if there exists an open quadrant cone which contains all frame vectors of Φ.

(ii) A frame $\Phi \subset \mathbb{R}^3 \setminus \{0\}$ for $\mathbb{R}^3$ is not scalable if and only if all frame vectors of Φ are contained in the interior of an elliptical conical surface with vertex 0 and intersecting the corners of a rotated unit cube.

2.4. Scalable frames and Fritz John theorem. The last characterization of scalable frames we should discuss is based on Fritz John's ellipsoid theorem. Before we state this theorem, we recall from Section 2.2 that given a set of points $Y = \{y_k\}_{k=1}^L \subset \mathbb{R}^N$, P_Y is the polytope generated by Y.

Given an $N \times N$ positive definite matrix X and a point $c \in \mathbb{R}^N$, we define an N-dimensional ellipsoid centered at c as

$$E(X, c) = c + X^{-1/2}(B) = \{v : \langle X(v-c), (v-c)\rangle \leq 1\},$$

where B is the unit ball in $\mathbb{R}^N$. We recall that the volume of the ellipsoid is given by Volume$(E(X,c)) = (\det(X))^{-1/2}\omega_N$, where ω_N is the volume of the unit ball in $\mathbb{R}^N$.

A convex body $K \subset \mathbb{R}^N$ is a nonempty compact convex subset of $\mathbb{R}^N$. It is well-known that for any convex body K with nonempty interior in $\mathbb{R}^N$ there is a unique ellipsoid of minimal volume containing K; e.g., see [**72**, Chapter 3]. We refer to [**4, 5, 40, 45, 72**] for more on these extremal ellipsoids. Fritz John ellipsoid theorem [**45**] gives a description of this ellipsoid. More specifically:

THEOREM 2.19. [**45**, Section 4] *Let $K \subset B = B_2^N(0,1)$ (unit ball in $\mathbb{R}^N$) be a convex body with nonempty interior. Then B is the ellipsoid of minimal volume containing K if and only if there exist $\{\lambda_k\}_{k=1}^m \subset (0, \infty)$ and $\{u_k\}_{k=1}^m \subset \partial K \cap S^{N-1}$, $m \geq N+1$ such that*

(i) $\sum_{k=1}^m \lambda_k u_k = 0$

(ii) $x = \sum_{k=1}^m \lambda_k \langle x, u_k\rangle u_k, \forall x \in \mathbb{R}^N$

where ∂K is the boundary of K and S^{N-1} is the unit sphere in $\mathbb{R}^N$. In particular, the points u_k are contact points of K and S^{N-1}.

Observe that (ii) of Theorem 2.19 can be written as

$$x = \sum_{k=1}^m \langle x, \sqrt{\lambda_k} u_k\rangle \sqrt{\lambda_k} u_k \quad \textit{for all} \quad x \in \mathbb{R}^N$$

which is equivalently to saying that the vectors $\{u_k\}_{k=1}^m$ form a scalable frame. The difficulty in applying this theorem lies in the fact that determining the contact points u_k and the "multipliers" λ_k is an extremely difficult problem. Nonetheless, we can apply this result to our question since we consider the convex body generated by the frame vectors, in which case these contact points are subset of the frame vectors. In particular, to apply the Fritz John theorem to the scalability problem, we consider a frame $\Phi \in \mathcal{F}_u(M,N)$ of $\mathbb{R}^N$ consisting of unit norm vectors. We define the associated symmetrized frame as

$$\Phi_{\text{Sym}} := \{\varphi_k\}_{k=1}^M \cup \{-\varphi_k\}_{k=1}^M,$$

and we denote the ellipsoid of minimal volume circumscribing the convex hull of the symmetrized frame Φ_{Sym} by E_Φ and refer to it as the *minimal ellipsoid of* Φ.

Its 'normalized' volume is defined by

$$V_\Phi := \frac{\mathrm{Vol}(E_\Phi)}{\omega_N}.$$

Clearly, $V_\Phi \leq 1$, and it is shown in [**22**, theorem 2.11] that equality holds if and only if the frame is scalable. That is, we have

THEOREM 2.20. [**22**, Theorem 2.11] *A frame $\Phi \in \mathcal{F}_u(M,N)$ is scalable if and only if its minimal ellipsoid is the N-dimensional unit ball, in which case $V_\Phi = 1$.*

REMARK 2.21. Given a unit-norm frame Φ, the number V_Φ defined above is one of a few measures of scalability introduced in [**22**]. These are numbers that measure how close to being scalable a frame Φ is. For example, if for a given Φ, $V_\Phi < 1$, then the farther away from 1 it is, the less scalable is Φ. Thus V_Φ along with these other measure of scalability can be used to define "almost" scalable frames. We refer to [**22**] for details.

Using the geometric characterization of scalable frames by V_Φ one can define the following equivalence relation on $\mathcal{F}_u(M,N)$: $\Phi, \Psi \in \mathcal{F}_u(M,N)$ are equivalent if and only if $V_\Phi = V_\Psi$. We denote each equivalence class by the unique volume for all its members. Specifically, for any $0 < a \leq 1$, the class $P[M,N,a]$ consists of all $\Phi \in \mathcal{F}_u(M,N)$ with $V_\Phi = a$. Then, $\mathcal{SC}(M,N) = P[M,N,1]$. This also allows a parametrization of $\mathcal{F}_u(M,N)$:

$$\mathcal{F}_u(M,N) = \bigcup_{a\in(0,1]} P[M,N,a].$$

3. Probabilistic frames

Definition 1.1 introduces frames from a linear algebra perspective through their spanning properties. However, frames can also be viewed as point masses distributed in $\mathbb{R}^d$. In this section we survey a measure theoretical, or more precisely a probabilistic, description of frames. In particular, in Section 3.1 we define probabilistic frames and collect some of their elementary properties. In Section 3.2 We define the probabilistic frame potential and investigate its minimizers. We generalize this notion in Section 3.3 to the concept of pth frame potentials and discuss their minimizers. Probabilistic analogs of these potentials are considered in Section 3.4. This section can be considered as a companion to [**32**] where many of the results we stated below first appeared.

3.1. Definition and elementary properties. Before defining probabilistic frames we first collect some definitions needed in the sequel.

Let $\mathcal{P} := \mathcal{P}(\mathcal{B}, \mathbb{R}^N)$ denote the collection of probability measures on $\mathbb{R}^N$ with respect to the Borel σ-algebra $\mathcal{B}$. Let

$$\mathcal{P}_2 := \mathcal{P}_2(\mathbb{R}^N) = \left\{\mu \in \mathcal{P} : M_2^2(\mu) := \int_{\mathbb{R}^N} \|x\|^2 d\mu(x) < \infty\right\}$$

be the set of all probability measures with finite second moments. Given $\mu, \nu \in \mathcal{P}_2$, let $\Gamma(\mu,\nu)$ be the set of all Borel probability measures γ on $\mathbb{R}^N \times \mathbb{R}^N$ whose marginals are μ and ν, respectively, i.e., $\gamma(A \times \mathbb{R}^N) = \mu(A)$ and $\gamma(\mathbb{R}^N \times B) = \nu(B)$

for all Borel subset A, B in $\mathbb{R}^N$. The space $\mathcal{P}_2$ is equipped with the 2-*Wasserstein metric* given by

$$W_2^2(\mu,\nu) := \min\left\{\int_{\mathbb{R}^N\times\mathbb{R}^N} \|x-y\|^2 d\gamma(x,y), \gamma \in \Gamma(\mu,\nu)\right\}. \tag{3.1}$$

It is known that the minimum defined by (3.1) is attained at a measure $\gamma_0 \in \Gamma(\mu,\nu)$, that is:

$$W_2^2(\mu,\nu) = \int_{\mathbb{R}^N\times\mathbb{R}^N} \|x-y\|^2 d\gamma_0(x,y).$$

We refer to [**2**, Chapter 7], and [**78**, Chapter 6] for more details on the Wasserstein spaces.

DEFINITION 3.1. A Borel probability measure $\mu \in \mathcal{P}$ is a *probabilistic frame* if there exist $0 < A \leq B < \infty$ such that for all $x \in \mathbb{R}^N$ we have

$$A\|x\|^2 \leq \int_{\mathbb{R}^N} |\langle x, y\rangle|^2 d\mu(y) \leq B\|x\|^2. \tag{3.2}$$

The constants A and B are called *lower and upper probabilistic frame bounds*, respectively. When $A = B$, μ is called a *tight probabilistic frame.*

It follows from Definition 3.1 that the upper inequality in (3.2) holds if and only if $\mu \in \mathcal{P}_2$. With a little more work one shows that the lower inequality holds whenever the linear span of the support of the probability measure μ is $\mathbb{R}^N$.

Assume that μ is a tight probabilistic frame, in which case equality holds in (3.2). Hence, choosing $x = e_k$ where $\{e_k\}_{k=1}^N$ is the standard orthonormal basis for $\mathbb{R}^N$ leads to

$$A\|e_k\|^2 = A = \int_{\mathbb{R}^N} \langle e_k, y\rangle^2 d\mu(y).$$

Therefore,

$$NA = \sum_{k=1}^N A\|e_k\|^2 = \int_{\mathbb{R}^N} \sum_{k=1}^N \langle e_k, y\rangle^2 d\mu(y) = \int_{\mathbb{R}^N} \|y\|^2 d\mu(y) = M_2(\mu)^2.$$

Consequently, for a tight probabilistic frame μ, $A = \frac{M_2(\mu)^2}{N}$.

These observations can are summarized in the following result whose proof can be found in [**32**]

THEOREM 3.2. [**32**, Theorem 12.1] *A Borel probability measure $\mu \in \mathcal{P}$ is a* probabilistic frame *if and only if $\mu \in \mathcal{P}_2$ and $E_\mu = \mathbb{R}^N$, where E_μ denotes the linear span of* supp(μ) *in $\mathbb{R}^N$. Moreover, if μ is a tight probabilistic frame, then the frame bound is given by*

$$A = \tfrac{1}{N} M_2^2(\mu) = \tfrac{1}{N}\int_{\mathbb{R}^N} \|y\|^2 d\mu(y).$$

We now consider some examples of probabilistic frames.

EXAMPLE 3.3. (a) A set $\Phi = \{\varphi_k\}_{k=1}^M \subset \mathbb{R}^N$ is a frame if and only if the probability measure $\mu_\Phi = \frac{1}{M}\sum_{k=1}^M \delta_{\varphi_k}$ supported by the set Φ is a probabilistic frame, where δ_φ denotes the Dirac measure supported at $\varphi \in \mathbb{R}^N$.

(b) More generally, let $a = \{a_k\}_{k=1}^M \subset (0, \infty)$ with $\sum_{k=1}^M a_k = 1$. A set $\Phi = \{\varphi_k\}_{k=1}^M \subset \mathbb{R}^N$ is a frame if and only if the probability measure $\mu_{\Phi,a} = \sum_{k=1}^M a_k \delta_{\varphi_k}$ supported by the set Φ is a probabilistic frame.

(c) By symmetry consideration one also shows that the uniform distribution on the unit sphere S^{N-1} in $\mathbb{R}^N$ is a tight probabilistic frame [**30**, Proposition 3.13]. That is, denoting the probability measure on S^{N-1} by $d\sigma$ we have that for all $x \in \mathbb{R}^N$,

$$\tfrac{\|x\|^2}{N} = \int_{\mathbb{R}^N} \langle x, y\rangle^2 d\sigma(y).$$

In the framework of the Wasserstein metric, many properties of probabilistic can be proved. For example, if we denote by $P(A, B)$ the set of probabilistic frames with frame bounds $0 < A \leq B < \infty$, then the following result was proved in [**83**, Proposition 1]:

PROPOSITION 3.4. [**83**, Proposition 1] *$P(A, B)$ is a nonempty, convex, closed subset of $\mathcal{P}_2(\mathbb{R}^N)$.*

Other results including a probabilistic treatment of the frame scalability problem also appeared in [**83**]. Furthermore, in [**82**] an optimal transport approach to minimizing a frame potential that generalizes the Benedetto and Fickus potential was developed. In the process, the smoothness (in the Wasserstein metric) of this potential was derived.

Probabilistic frames can be analyzed in terms of a corresponding analysis operator and its adjoint the synthesis operator. Indeed, let $\mu \in \mathcal{P}$ be a probability measure. The *probabilistic analysis operator* is given by

$$T_\mu : \mathbb{R}^N \to L^2(\mathbb{R}^N, \mu), \quad x \mapsto \langle x, \cdot\rangle.$$

Its adjoint operator is defined by

$$T_\mu^* : L^2(\mathbb{R}^N, \mu) \to \mathbb{R}^N, \quad f \mapsto \int_{\mathbb{R}^N} f(x) x d\mu(x)$$

and is called the *probabilistic synthesis operator*, where the above integral is vector-valued. The *probabilistic frame operator* of μ is

$$S_\mu = T_\mu^* T_\mu,$$

and one easily verifies that

$$S_\mu : \mathbb{R}^N \to \mathbb{R}^N, \qquad S_\mu(x) = \int_{\mathbb{R}^N} \langle x, y\rangle y d\mu(y).$$

If $\{e_j\}_{j=1}^N$ is the canonical orthonormal basis for $\mathbb{R}^N$, then

$$S_\mu e_i = \sum_{j=1}^N m_{i,j}(\mu) e_j,$$

where

$$m_{i,j} = \int_{\mathbb{R}^N} y^{(i)} y^{(j)} d\mu(y)$$

is the (i, j) entry of the matrix of second moments of μ. Thus, the probabilistic frame operator is the matrix of second moments of μ. Consequently, the following results proved in [**32**] follows.

PROPOSITION 3.5. [**32**, Proposition 12.4] Let $\mu \in \mathcal{P}$, then S_μ is well-defined (and hence bounded) if and only if

$$M_2(\mu) < \infty.$$

Furthermore, μ is a probabilistic frame if and only if S_μ is positive definite.

If μ is a probabilistic frame then S_μ is invertible. Let $\tilde{\mu}$ be the *push-forward* of μ through S_μ^{-1} given by

$$\tilde{\mu} = \mu \circ S_\mu.$$

In particular, given any Borel set $B \subset \mathbb{R}^N$ we have

$$\tilde{\mu}(B) = \mu((S_\mu^{-1})^{-1}B) = \mu(S_\mu B).$$

Equivalently, $\tilde{\mu}$ can be defined via integration. Indeed, if f is a continuous bounded function on $\mathbb{R}^N$,

$$\int_{\mathbb{R}^N} f(y)d\tilde{\mu}(y) = \int_{\mathbb{R}^N} f(S_\mu^{-1}y)d\mu(y).$$

In fact, $\tilde{\mu}$ is also a probabilistic frame (with bounds $1/B \leq 1/A$) called the *probabilistic canonical dual frame* of μ. Similarly, when μ is a probabilistic frame, S_μ is positive definite, and its square root exists. The push-forward of μ through $S_\mu^{-1/2}$ is given by

$$\mu^\dagger(B) = \mu(S^{1/2}B)$$

for each Borel set in $\mathbb{R}^N$. The properties of these probability measures are summarized in the following result. We refer to [**32**, Proposition 12.4] and [**32**, Proposition 12.5] for details.

PROPOSITION 3.6. Let $\mu \in \mathcal{P}$ be a probabilistic frame with bounds $0 < A \leq B < \infty$. Then:

(a) $\tilde{\mu}$ is a probabilistic frame with frame bounds $1/B \leq 1/A$.
(b) $\mu^\dagger$ is a tight probabilistic frame.

Consequently, for each $x \in \mathbb{R}^N$ we have:

$$\int_{\mathbb{R}^N} \langle x, y\rangle\, S_\mu y\, d\tilde{\mu}(y) = \int_{\mathbb{R}^N} \langle S_\mu^{-1}x, y\rangle\, y\, d\mu(y) = S_\mu S_\mu^{-1}(x) = x, \tag{3.3}$$

and

$$\int_{\mathbb{R}^N} \langle x, y\rangle\, y\, d\mu^\dagger(y) = \int_{\mathbb{R}^N} \langle S_\mu^{-1/2}x, y\rangle\, S_\mu^{-1/2}y\, d\mu(y) = S_\mu^{-1/2}S_\mu S_\mu^{-1/2}(x) = x. \tag{3.4}$$

It is worth noticing that (3.3) is the analog of the frame reconstruction formula (1.1) while (3.4) is analog of (1.2).

In the context of probabilistic frames, the *probabilistic Gram operator*, or the *probabilistic Gramian* of μ, is the compact integral operator defined on $L^2(\mathbb{R}^N, \mu)$ by

$$G_\mu f(x) = T_\mu T_\mu^* f(x) = \int_{\mathbb{R}^N} K(x,y)f(y)d\mu(y) = \int_{\mathbb{R}^N} \langle x, y\rangle f(y)d\mu(y).$$

It is immediately seen that G_μ is an integral operator with kernel given by $K(x,y) = \langle x, y\rangle$, which is continuous and in $L^2(\mathbb{R}^N \times \mathbb{R}^N, \mu \otimes \mu) \subset L^1(\mathbb{R}^N \times \mathbb{R}^N, \mu \otimes \mu)$, where $\mu \otimes \mu$ is the product measure of μ with itself. Consequently, G_μ is a trace class and Hilbert-Schmidt operator. Moreover, for any $f \in L^2(\mathbb{R}^N, \mu)$, $G_\mu f$ is a uniformly continuous function on $\mathbb{R}^N$. As well-known, G_μ and S_μ have a common spectrum except for the 0. In fact, in the next proposition we collect the properties of G_μ:

PROPOSITION 3.7. [**32**, Proposition 12.4] Let $\mu \in \mathcal{P}$ then G_μ is a trace class and Hilbert-Schmidt operator on $L^2(\mathbb{R}^N)$. The eigenspace corresponding to the eigenvalue 0 has infinite dimension and consists of all functions $0 \neq f \in L^2(\mathbb{R}^N, \mu)$ such that

$$\int_{\mathbb{R}^N} y f(y) d\mu(y) = 0.$$

While new finite frames can be generated from old ones via (linear) algebraic operations, the setting of probabilistic frames allows one to use analytical tools to construct new probability frames from old ones. For example, it was shown in [**32**] when the convolution of a probabilistic frame and a probability measure yields a probabilistic frames. The following is a summary of some of the results proved in [**32**].

PROPOSITION 3.8. [**32**, Theorem 2 & Proposition 2] The following statements hold:

(a) Let $\mu \in \mathcal{P}_2$ be a probabilistic frame and let $\nu \in \mathcal{P}_2$. If $\text{supp}(\mu)$ contains at least $N+1$ distinct vectors, then $\mu * \nu$ is a probabilistic frame.
(b) Let μ and ν be tight probabilistic frames. If $\int_{\mathbb{R}^N} y d\nu(y) = 0$, then $\mu * \nu$ is also a tight probabilistic frame.

3.2. Probabilistic frame potential. One of the motivations of probabilistic frames lies in Benedetto and Fickus's characterization of the FUNTFs as the minimizers of the frame potential (1.4). In describing their results, they motivated it from a physical point of view drawing a parallel to Coulomb's law. It was then clear that the notion of frame potential carries significant information about frames, and can be viewed as describing the interaction of the frame vectors under some "physical force." This in turn partially motivated the introduction of a probabilistic analog to the frame potential in [**29**]. Furthermore, the probabilistic frame potential that we introduce below, can be viewed in the framework of other potential functions, e.g., those investigated by Björck in [**10**]. In this section we review the properties of the probabilistic frame potential investigating in particular its minimizers. The framework of the Wasserstein metric space $(\mathcal{P}_2, W_2)$ also offers the ideal setting to investigate this potential and certain of its generalizations. While we should not report of this analysis here, we shall nevertheless introduce certain generalizations of the probabilistic frame potential whose minimizers are better understood in the context of the Wasserstein metric spaces.

But we first start with the definition of the probabilistic frame potential.

DEFINITION 3.9. The *probabilistic frame potential* is the nonnegative function defined on $\mathcal{P}$ and given by

$$\text{PFP}(\mu) = \iint_{\mathbb{R}^N \times \mathbb{R}^N} |\langle x, y \rangle|^2 \, d\mu(x) \, d\mu(y), \tag{3.5}$$

for each $\mu \in \mathcal{P}$.

The following proposition is an immediate consequence of the above definition:

PROPOSITION 3.10. Let $\mu \in \mathcal{P}$, then $\text{PFP}(\mu)$ is the Hilbert-Schmidt norm of the probabilistic Gramian operator G_μ, that is

$$\|G_\mu\|_{HS}^2 = \iint_{\mathbb{R}^d \times \mathbb{R}^d} \langle x, y \rangle^2 d\mu(x) d\mu(y).$$

Furthermore, if $\mu \in \mathcal{P}_2$, (which is the case when μ is a probabilistic frame) then we have

$$\mathrm{PFP}(\mu) \leq M_2^4(\mu) < \infty.$$

We recall from Definition 3.1 that μ is a tight probabilistic frame if

$$\int_{\mathbb{R}^N} \langle x, y\rangle^2 d\mu(y) = \tfrac{M_2(\mu)^2}{N} \|x\|^2$$

for all $x \in \mathbb{R}^N$. Integrating this equation with respect to x leads to

$$\iint_{\mathbb{R}^N \times \mathbb{R}^N} \langle x, y\rangle^2 d\mu(x)\, d\mu(y) = \mathrm{PFP}(\mu) = \tfrac{M_2^4(\mu)}{N}.$$

It turns out that this value is the absolute lower bound to the probabilistic frame potential.

THEOREM 3.11. [**32**, Theorem 3] *Let $\mu \in \mathcal{P}_2$ be such that $M_2(\mu) = 1$ and set $E_\mu = span(\mathrm{supp}(\mu))$, then the following estimate holds*

$$\mathrm{PFP}(\mu) \geq 1/n \tag{3.6}$$

where n is the number of nonzero eigenvalues of S_μ. Moreover, equality holds if and only if μ is a tight probabilistic frame for E_μ.

In particular, given any probabilistic frame $\mu \in \mathcal{P}_2$ with $M_2(\mu) = 1$, we have

$$\mathrm{PFP}(\mu) \geq 1/N$$

and equality holds if and only if μ is a tight probabilistic frame.

The proof of this result can be found in [**32**, Theorem 3]. Recently, a very simple and elementary proof of the last part of the result was given in [**23**, Theorem 5]. Furthermore, in [**82**] an optimal transport approach to minimizing a modification of the probabilistic frame potential was considered and showed great promise to analyze other potential functions in frame theory. Moreover, this approach has a natural numerical part that could be used as a gradient descent-type method to numerically find the minimizers of the PFP and its generalization.

3.3. The pth frame potentials. The techniques used to prove Theorem 3.11 can be used to investigate the minimizers of other related potential functions, especially when there are defined for probability measures supported on compact sets, such as on the unit sphere. In this section, we define a family of (deterministic) potentials and describe their minimizers. The probabilistic analogs of these results will follow in the next section.

To motivate our definition, we recall the following result due to Strohmer and Heath [**71**], and we refer to [**81**] for historical perspectives on this result.

THEOREM 3.12. [**71**, Theorem 2.3] *For any frame $\Phi = \{\varphi_k\}_{k=1}^M \subset \mathbb{R}^N$ with $\|\varphi_k\| = 1$, we have*

$$\max_{k \neq \ell} |\langle \varphi_k, \varphi_\ell\rangle| \geq \sqrt{\tfrac{M-N}{N(M-1)}}, \tag{3.7}$$

and equality hold if and only if Φ is a FUNTF such that

$$|\langle \varphi_k, \varphi_\ell\rangle| = \sqrt{\tfrac{M-N}{N(M-1)}} \tag{3.8}$$

when $k \neq \ell$. Furthermore, equality can hold only when $M \leq \frac{N(N+1)}{2}$.

A FUNTF that satisfies (3.8) is termed an *equiangular tight frame (ETF)*. Note that the left-hand side of (3.7) can be viewed as a potential function of Φ. Indeed, this is the so-called *coherence* of Φ that we shall defined for reasons to be evident later as

$$\text{FP}_{\infty,M}(\Phi) = \max_{k\neq\ell} |\langle\varphi_k, \varphi_\ell\rangle| \tag{3.9}$$

for $\Phi = \{\varphi_k\}_{k=1}^M \subset S^{N-1}$. In fact, $\text{FP}_{\infty,N}(\Phi)$ as well as the frame potential FP given in (1.4) are members of the family of the pth *frame potentials* defined by:

DEFINITION 3.13. Let M be a positive integer, and $0 < p < \infty$. Given a collection of unit vectors $\Phi = \{\varphi_k\}_{k=1}^M \subset S^{N-1}$, the *p-frame potential* is the functional

$$\text{FP}_{p,M}(\Phi) = \sum_{k,\ell=1}^{M} |\langle\varphi_k, \varphi_\ell\rangle|^p. \tag{3.10}$$

When, $p = \infty$, the definition reduces to

$$\text{FP}_{\infty,M}(\Phi) = \max_{k\neq\ell} |\langle\varphi_k, \varphi_\ell\rangle|.$$

It is clear that $\text{FP}_{p,M}$ and its minimizers are also functions of N, the dimension of the underlying space. However, to keep the notations simple, we shall not make explicit the dependence on N, unless it is necessary.

The case $p = 2$ corresponds to the frame potential FP given in (1.4). As mentioned above, $\text{FP}_{\infty,M}(\Phi)$ is the coherence of Φ and plays a key role in compressed sensing [**3**, **27**, **28**, **38**, **74**]. Moreover, for fixed M, the minimizers of $\text{FP}_{\infty,M}$ are called Grassmanian frames [**8**, **71**]. By using continuity and compactness arguments one can show that given M, N, $\text{FP}_{\infty,M}$ always has a minimum [**8**, Appendix]. The challenge is the construction of these Grassmanian. In [**8**] constructions of Grassmanians were considered for $N = 2$ and $M \geq 2$, and for $N = 3$ when $M \in \{3, 4, 5, 6\}$. The ideas used in these constructions are based on analytical interpretation of some geometric results obtained in [**73**]. The general question of constructing the minimizers of $\text{FP}_{\infty,M}$ for $N \geq 3$ and $M \geq 6$ is still a mostly open question.

Even more, minimizing $\text{FP}_{p,M}$ is an extremely difficult problem as one needs to deal with both p, M, and the ambient dimension N. Some results on the minimizers as well as the value of the minimum as a function of the involved parameters were proved in [**29**]. We refer to [**62**] for earlier results on minimizing the pth frame potential. Before summarizing some of these results we consider the special case where $M = 3$, $N = 2$ and seek the minimizers of

$$\text{FP}_{p,3}(\Phi) = \sum_{k,\ell=1}^{3} |\langle\varphi_k, \varphi_\ell\rangle|^p$$

when $p \in (0, \infty)$ with the usual modification when $p = \infty$, and $\Phi = \{\varphi_k\}_{k=1}^3 \subset S^1$.

When $p = 2$,

$$\text{FP}_{2,3}(\Phi) = \sum_{k,\ell=1}^{3} |\langle\varphi_k, \varphi_\ell\rangle|^2 \geq 9/2$$

with equality if and only if $\Phi = \{\varphi_k\}_{k=1}^3 \subset S^1$ is a FUNTF. A minimizer of $\mathrm{FP}_{2,3}$ is the MB-frame which is pictured below:

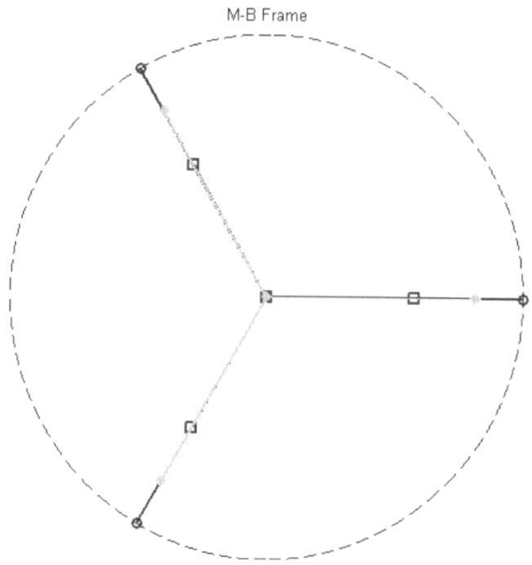

FIGURE 10. An example of Equiangular FUNTF: the MB-frame.

When $p = \infty$,

$$\mathrm{FP}_{\infty,3}(\Phi) = \max_{k \neq \ell} |\langle \varphi_k, \varphi_\ell \rangle| \geq 1/\sqrt{2}$$

with equality if and only if $\Phi = \{\varphi_k\}_{k=1}^3 \subset S^1$ is an ETF. But what happens for other values of $2 \neq p \in (0, \infty)$? This was partially answered in [**29**] for $0 < p \leq 2$ and recently the case $p \geq 2$ was settled [**85**]. Before giving more details on this case, we first collect a number of generic results about the minimizers of $\mathrm{FP}_{p,M}$ when $M \geq N \geq 2$ and $p \in (0, \infty]$.

PROPOSITION 3.14. Let $p \in (0, \infty]$, M, N be positive integers. Let $\Phi = \{\varphi_k\}_{k=1}^M \subset S^{N-1}$ we have:

(a) If $M \geq N$ and $2 < p < \infty$, then

$$\mathrm{FP}_{p,M}(\Phi) \geq M(M-1)\big(\tfrac{M-N}{N(M-1)}\big)^{p/2} + N,$$

and equality holds if and only if Φ is an ETF.

(b) Let $0 < p < 2$ and assume that $M = kN$ for some positive integer k. Then the minimizers of the p-frame potential are exactly the k copies of any orthonormal basis modulo multiplications by ± 1. The minimum of (3.10) over all sets of $M = kN$ unit norm vectors is $k^2 N$.

(c) Assume that $M = N + 1$ and set $p_0 = \frac{\log(\frac{N(N+1)}{2})}{\log(N)} = \frac{\log(\frac{M(M-1)}{2})}{\log(M-1)}$. Assume that $\mathrm{FP}_{p_0,N}(\Phi) \geq N + 3$, with equality holding if and only if $\Phi = \{\varphi_k\}_{k=1}^{N+1}$ is an orthonormal basis plus one repeated vector or an equiangular FUNTF. Then,

(1) for $0 < p < p_0$, for any $\Phi = \{\varphi_k\}_{k=1}^{N+1} \subset S^{N-1}$, we have $\mathrm{FP}_{p,N+1}(\Phi) \geq N + 3$, and equality holds if and only if $\Phi = \{\varphi_k\}_{k=1}^{N+1}$ is an orthonormal basis plus one repeated vector,

(2) for $p_0 < p < 2$, for any $\Phi = \{\varphi_k\}_{k=1}^{N+1} \subset S^{N-1}$, we have $\mathrm{FP}_{p,N+1}(\Phi) \geq 2^{\frac{p}{p_0}}\left((N+1)N\right)^{1-\frac{p}{p_0}} + N + 1$, and equality holds if and only if $\Phi = \{\varphi_k\}_{k=1}^{N+1}$ is an equiangular FUNTF.

In the special case when $N = 2$, part (c) of the proposition becomes:

COROLLARY 3.15. For $N = 2$, $M = 3$, and $p_0 = \frac{\log(3)}{\log(2)}$, the hypothesis of (c) above holds. That is, for any $\Phi = \{\varphi_k\}_{k=1}^{3} \subset S^1$,

$$\mathrm{FP}_{p_0,3}(\Phi) \geq 5,$$

and equality holds if and only if $\Phi = \{\varphi_k\}_{k=1}^{3}$ is an orthonormal basis plus one repeated vector or an equiangular FUNTF.

However, when $N \geq 3$, it is still unknown if the hypothesis of proposition 3.14 (c) holds, and it was conjectured in in [**29**] that with p_0 given in (c),

$$\mathrm{FP}_{p_0,N+1}(\Phi) \geq N + 3$$

with equality if and only if $\Phi = \{\varphi_k\}_{k=1}^{N+1}$ is an orthonormal basis plus one repeated vector or an equiangular FUNTF.

Using Corollary 3.15 one can compute

$$\mu_{p,3,2} = \min\{\mathrm{FP}_{p,2}(\Phi) : \Phi = \{\varphi_k\}_{k=1}^{3} \subset S^1\}$$

for all $p \in (0, \infty]$ leading to

$$\mu_{p,3,2} = \begin{cases} 5 & \text{for } p \in (0, \frac{\log(3)}{\log(2)}] \\ 3 + 6e^{-p\log 2} & \text{for } p \geq \frac{\log(3)}{\log(2)}. \end{cases}$$

The graph of $\mu_{p,3,2}$ when $p \in (0, \infty)$ is given in Figure 11.

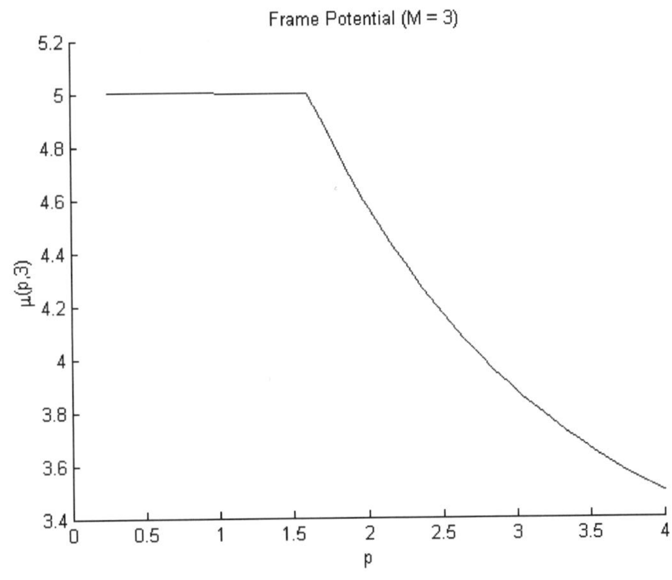

FIGURE 11. Graph of $\mu_{p,3,2}$ when $p \in (0, 4)$.

One can ask about of $\mu_{p,M,2}$ for other values of M. It follows from proposition 3.14 (b) that $\mu_{p,M,2} = 2k^2$ for all $p \in (0,2]$ whenever $M = 2k$ is an even integer. For $p > 2$ or odd M some numerical simulations were considered in [**85**]. For example, the following graphs (Figures 12 and 13) of $\mu_{p,M,2}$ for $M \in \{4,6\}$ were obtained.

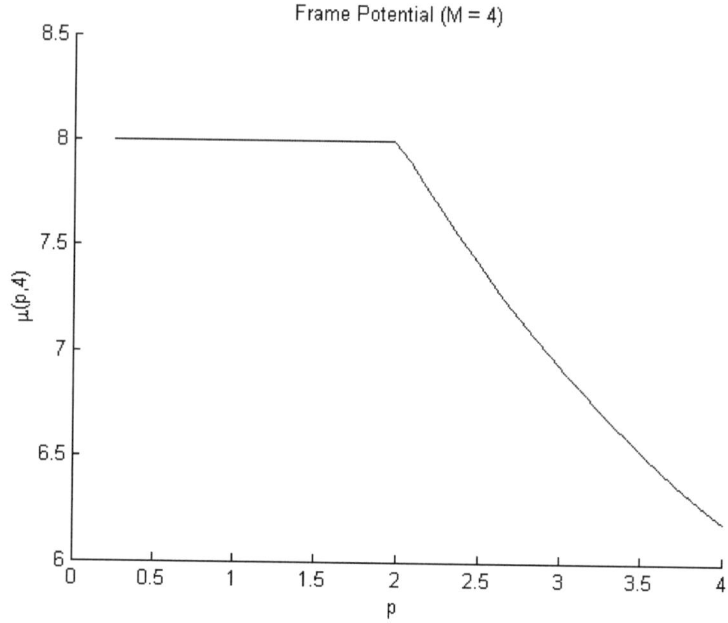

FIGURE 12. Graph of $\mu_{p,4,2}$ when $p \in (0,4)$.

For $M = 5$ the numerical results suggest that the graph of $\mu_{p,5,2}$ is as given in 14. Finally, we plot the behavior of $\mu_{p,M,2}$ as a sequence in M when $p \in (0,4)$ is shown in Figure 15.

For integer values of p, the minimizers of $\mathrm{FP}_{p,M}$ have been investigated in connection with the theory of spherical designs [**26**, **75**].

DEFINITION 3.16. Let t be a positive integer. A *spherical t-design* is a finite subset $\{x_i\}_{i=1}^M$ of the unit sphere S^{N-1} in $\mathbb{R}^N$, such that,

$$\frac{1}{M}\sum_{i=1}^{M} h(x_i) = \int_{S^{N-1}} h(x)d\sigma(x),$$

for all homogeneous polynomials h of total degree equals or less than t in N variables and where σ denotes the uniform surface measure on S^{N-1} normalized to have mass one.

It is easy to see that any spherical $t-$design is also a spherical $t'-$design for all positive integers $t' \leq t$. Spherical $2-$designs are exactly FUNTFs whose center of mass is at the origin. More precisely we have:

PROPOSITION 3.17. $\Phi = \{\varphi_k\}_{k=1}^M \subset S^{N-1}$ is a spherical 2-design if and only if Φ is a FUNTF and $\sum_{k=1}^M \varphi_k = 0$.

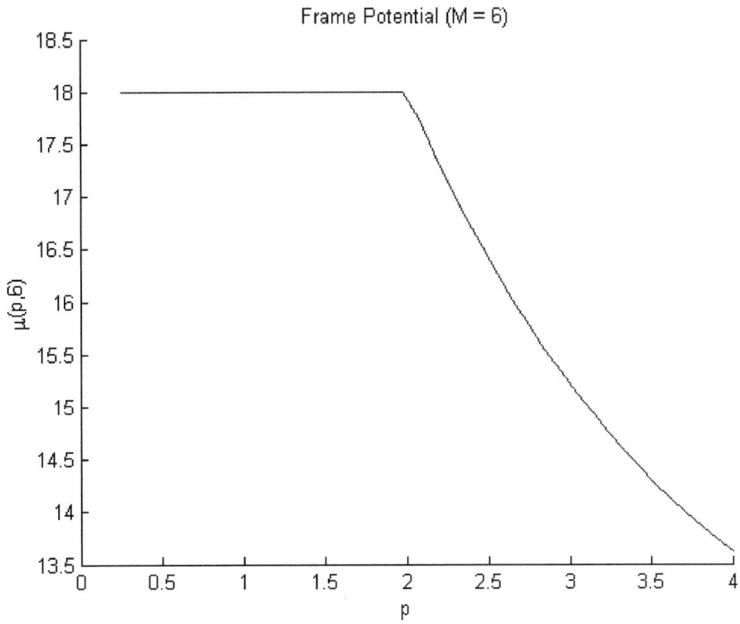

FIGURE 13. Graph of $\mu_{p,6,2}$ when $p \in (0, 4)$.

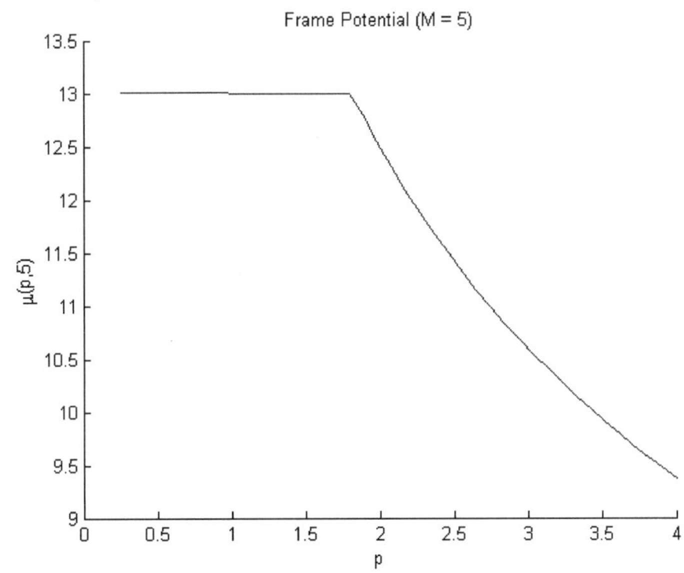

FIGURE 14. Graph of $\mu_{p,5,2}$ when $p \in (0, 4)$.

We refer to [**26**], [**75**, Theorem 3.2] for details on the proof of the above proposition. Recalling that FUNTFs minimize the frame potential, it is not surprising that spherical t-designs also minimize a potential. In particular,

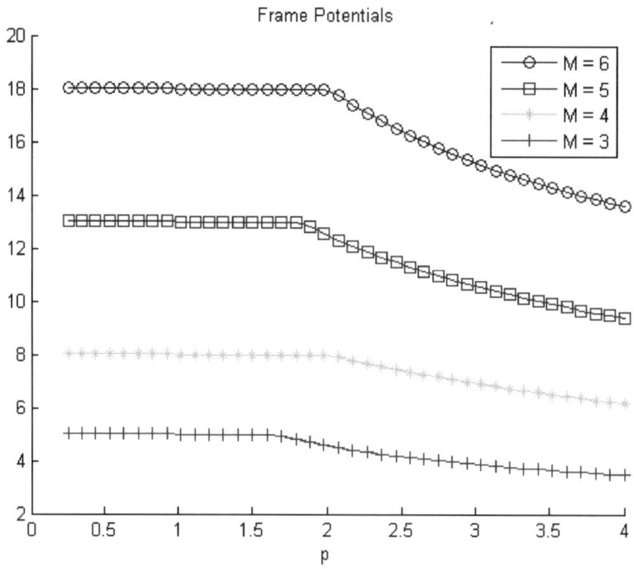

FIGURE 15. Graph of $\mu_{p,M,2}$ when $p \in (0,4)$, $M \in \{3,4,5,6\}$.

THEOREM 3.18. [**75**, Theorem 8.1] *Let $p = 2k$ be an even integer and $\{x_i\}_{i=1}^M = \{-x_i\}_{i=1}^M \subset S^{N-1}$, then*

$$\mathrm{FP}_{p,M}(\{x_i\}_{i=1}^M) \geq \frac{1 \cdot 3 \cdot 5 \cdots (p-1)}{N(N+2)\cdots(N+p-2)} M^2,$$

and equality holds if and only if $\{x_i\}_{i=1}^M$ is a spherical p-design.

3.4. Probabilistic p frame potential. The pth frame potential can be viewed in light of mass distributions on the unit sphere. It is therefore natural to look at it from a probabilistic point of view. This motivates the the introduction of the larger family of potential called *probabilistic p-frame potential.*

For $p \in (0, \infty)$ set

$$\mathcal{P}_p = \{\mu \in \mathcal{P} : M_p^p(\mu) = \int_{\mathbb{R}^N} \|y\|^p d\mu(y) < \infty\}.$$

DEFINITION 3.19. For each $p \in (0,\infty)$, the *probabilistic p−frame potential* is given by

$$\mathrm{PFP}(\mu, p) = \iint_{\mathbb{R}^N \times \mathbb{R}^N} |\langle x, y\rangle|^p \, d\mu(x)\, d\mu(y). \tag{3.11}$$

When μ is a purely atomic measure with atoms on the unit sphere, that is when $\mathrm{supp}(\mu) = \Phi = \{\varphi_k\}_{k=1}^M \subset S^{N-1}$, $\mathrm{PFP}(\mu,p)$ reduces to $\mathrm{FP}_{p,M}$ given in (3.10).

This class of potentials is related to the potentials considered by G. Bjřock [**10**]. More precisely, suppose $F \subset \mathbb{R}^N$ is compact and let $\lambda > 0$. Björck considered the question of maximizing the functional

$$I_F(\mu) = \iint_{F\times F} \|x-y\|^\lambda d\mu(x) d\mu(y)$$

where μ ranges over all positive Borel measures with $\mu(F) = 1$. It turns out that the techniques used in [**10**] to maximize $I_F(\mu)$ can be extended to understand the minimizers of $\mathrm{PFP}(\mu)$ when μ is restricted to a probability measure on the unit sphere S^{N-1} in $\mathbb{R}^N$. In particular, it was proved in [**29**, theorem 4.9] that when restricted to probability measures μ supported on the unit sphere of $\mathbb{R}^N$ and when $0 < p < 2$, then the minimizers of $\mathrm{PFP}(\mu, p)$ are discrete probability measures. Furthermore, the support of such minimizers contains an orthonormal basis B and is contained in the set $\pm B$. More specifically we have:

THEOREM 3.20. [**29**, Theorem 4.9] *Let $0 < p < 2$, then the minimizers of* (3.11) *over all the probability measures supported on the unit sphere S^{N-1} are exactly those probability measures μ that satisfy*

(i) *there is an orthonormal basis $\{e_1, \dots, e_N\}$ for $\mathbb{R}^N$ such that*

$$\{e_1, \dots, e_N\} \subset \mathrm{supp}(\mu) \subset \{\pm e_1, \dots, \pm e_N\}$$

(ii) *there is $f : S^{N-1} \to \mathbb{R}$ such that $\mu(x) = f(x)\nu_{\pm x_1,\dots,\pm x_N}(x)$ and*

$$f(x_i) + f(-x_i) = \frac{1}{N},$$

where the measure $\nu_{\pm x_1,\dots,\pm x_N}(x)$ represent the counting measure of the set $\{\pm x_i : i = 1, \dots, N\}$.

Theorem 3.18 shows that the minimizers of $\mathrm{FP}_{p,M}$ when $p = 2k$ is an even integer, are exactly spherical $p-$designs. In view of this fact, one can ask whether the minimizers of PFP have some special "approximation" properties. This partially motivates that following definition in which we denote by $\mathcal{M}(S^{N-1}, \mathcal{B})$ the space of all Borel probability measures supported on S^{N-1}.

DEFINITION 3.21. [**29**, Definition 4.1] For $0 < p < \infty$, we call $\mu \in \mathcal{M}(S^{N-1}, \mathcal{B})$ a *probabilistic p-frame* for $\mathbb{R}^N$ if and only if there are constants $A, B > 0$ such that

$$A\|y\|^p \leq \int_{S^{N-1}} |\langle x, y\rangle|^p d\mu(x) \leq B\|y\|^p, \quad \forall y \in \mathbb{R}^N. \tag{3.12}$$

We call μ a *tight probabilistic p-frame* if and only if we can choose $A = B$.

By symmetry considerations, it is not difficult to show that the uniform surface measure σ on S^{N-1} is always a tight probabilistic p-frame, for each $0 < p < \infty$. In addition, observe that we can always take $B = 1$ in (3.12). Thus to determine if a probability measure μ on S^{N-1} is a probabilistic $p-$frame one must focus on establishing the lower bound in the above definition. When $p = 2$ this definition reduces to that of probabilistic frames introduced earlier. In fact, more is true:

LEMMA 3.22. [**29**, Lemma 4.5] *If μ is probabilistic frame, then it is a probabilistic p-frame for all $1 \leq p < \infty$. Conversely, if μ is a probabilistic p-frame for some $1 \leq p < \infty$, then it is a probabilistic frame.*

The analogy between tight probabilistic p-frames and spherical $t-$designs can now be made explicitly as one can show the following result which is an analog of theorem 3.18. More specifically, the result below shows that tight probabilistic p-frames are the minimizers of the probabilistic frame potential (3.11) when restricted to probability measures supported on S^{N-1}, and when p is an even integer:

THEOREM 3.23. [**29**, Theorem 4.10] *Let p be an even integer. For any probability measure* μ *on* S^{N-1},

$$\mathrm{PFP}(\mu, p) = \int_{S^{N-1}} \int_{S^{N-1}} |\langle x, y\rangle|^p d\mu(x) d\mu(y) \geq \frac{1 \cdot 3 \cdot 5 \cdots (p-1)}{N(N+2)\cdots(N+p-2)},$$

and equality holds if and only if μ *is a tight probabilistic p-frame.*

By combining Theorem 3.18 and Theorem 3.23 we can conclude that when $p = 2k$ there exists a one-to-one correspondence between the class of spherical $p-$designs and the class of discrete tight probabilistic $p-$frames. More specifically, every spherical $p-$design supports a discrete measure μ which is a tight probabilistic p-frame. This is summarized in the following proposition:

PROPOSITION 3.24. Let $p = 2k$ be an even positive integer. A set $\Phi = \{\varphi_k\}_{k=1}^M \subset S^{N-1}$ is a spherical $p-$design if and only if the probability measure $\mu_\Phi = \frac{1}{M}\sum_{k=1}^N \delta_{\varphi_k}$ is a tight probabilistic $p-$frame.

The question then becomes how to construct tight probabilistic p-frames. When restricted to discrete measures and when $p = 2k$ is an even integer, this problem is equivalent to constructing spherical $p-$designs. This is a difficult problem with known solutions only for certain values of p, M, and N. Of course, and as shown in Section 2 the special case when $p = 2$ leads to the FUNTFs. The analytics methods developed in [**82**] are new promising techniques that could be used to investigated in general the minimizers of $\mathrm{PFP}(\mu)$ when μ ranges over the probability measures on S^{N-1} and $p > 0$.

Acknowledgment

This work was partially supported by a grant from the Simons Foundation (#319197 to Kasso Okoudjou). The author would like to thank Chae Clark and Matthew Begué for their helpful discussions.

References

[1] S. T. Ali, J.-P. Antoine, and J.-P. Gazeau, *Continuous frames in Hilbert space*, Ann. Physics **222** (1993), no. 1, 1–37, DOI 10.1006/aphy.1993.1016. MR1206084 (94e:81107)

[2] L. Ambrosio, N. Gigli, and G. Savaré, *Gradient flows in metric spaces and in the space of probability measures*, Lectures in Mathematics ETH Zürich, Birkhäuser Verlag, Basel, 2005. MR2129498 (2006k:49001)

[3] W. U. Bajwa, R. Calderbank, and D. G. Mixon, *Two are better than one: fundamental parameters of frame coherence*, Appl. Comput. Harmon. Anal. **33** (2012), no. 1, 58–78, DOI 10.1016/j.acha.2011.09.005. MR2915706

[4] K. Ball, *Ellipsoids of maximal volume in convex bodies*, Geom. Dedicata **41** (1992), no. 2, 241–250, DOI 10.1007/BF00182424. MR1153987 (93k:52006)

[5] K. Ball, *An elementary introduction to modern convex geometry*, Flavors of geometry, Math. Sci. Res. Inst. Publ., vol. 31, Cambridge Univ. Press, Cambridge, 1997, pp. 1–58, DOI 10.2977/prims/1195164788. MR1491097 (99f:52002)

[6] V. Balakrishnan and S. Boyd, *Existence and uniqueness of optimal matrix scalings*, SIAM J. Matrix Anal. Appl. **16** (1995), no. 1, 29–39, DOI 10.1137/S0895479892235393. MR1311416 (95j:65043)

[7] J. J. Benedetto and M. Fickus, *Finite normalized tight frames*, Adv. Comput. Math. **18** (2003), no. 2-4, 357–385, DOI 10.1023/A:1021323312367. Frames. MR1968126 (2004c:42059)

[8] J. J. Benedetto, J. D. Kolesar, *Geometric properties of Grassmannian frames for* $\mathbb{R}^2$ *and* $\mathbb{R}^3$, EURASIP J. Applied Signal Processing, **2006**, pp. 1–17.

[9] M. Benzi, *Preconditioning techniques for large linear systems: a survey*, J. Comput. Phys. **182** (2002), no. 2, 418–477, DOI 10.1006/jcph.2002.7176. MR1941848 (2003j:65026)

[10] G. Björck, *Distributions of positive mass, which maximize a certain generalized energy integral*, Ark. Mat. **3** (1956), 255–269. MR0078470 (17,1198b)

[11] J. Bourgain, *On high-dimensional maximal functions associated to convex bodies*, Amer. J. Math. **108** (1986), no. 6, 1467–1476, DOI 10.2307/2374532. MR868898 (88h:42020)

[12] J. Cahill, M. Fickus, D. G. Mixon, M. J. Poteet, and N. Strawn, *Constructing finite frames of a given spectrum and set of lengths*, Appl. Comput. Harmon. Anal. **35** (2013), no. 1, 52–73, DOI 10.1016/j.acha.2012.08.001. MR3053746

[13] J. Cahill and N. Strawn, *Algebraic geometry and finite frames*, Finite frames, Appl. Numer. Harmon. Anal., Birkhäuser/Springer, New York, 2013, pp. 141–170, DOI 10.1007/978-0-8176-8373-3_4. MR2964009

[14] J. Cahill and X. Chen, *A note on scalable frames*, Proceedings of SampTA 2013.

[15] P. G. Casazza and G. Kutyniok, "Finite Frame Theory," Eds., Birkhaüser, Boston, 2012.

[16] P. G. Casazza, M. Fickus, J. Kovačević, M. T. Leon, and J. C. Tremain, *A physical interpretation of tight frames*, Harmonic analysis and applications, Appl. Numer. Harmon. Anal., Birkhäuser Boston, Boston, MA, 2006, pp. 51–76, DOI 10.1007/0-8176-4504-7_4. MR2249305 (2007d:42053)

[17] P. G. Casazza, M. Fickus, and D. G. Mixon, *Auto-tuning unit norm frames*, Appl. Comput. Harmon. Anal. **32** (2012), no. 1, 1–15, DOI 10.1016/j.acha.2011.02.005. MR2854158

[18] P. G. Casazza, M. Fickus, D. G. Mixon, Y. Wang, and Z. Zhou, *Constructing tight fusion frames*, Appl. Comput. Harmon. Anal. **30** (2011), no. 2, 175–187, DOI 10.1016/j.acha.2010.05.002. MR2754774 (2012c:42069)

[19] P. G. Casazza, A. Heinecke, K. Kornelson, Y. Wang, and Z. Zhou, *Necessary and sufficient conditions to perform spectral tetris*, Linear Algebra Appl. **438** (2013), no. 5, 2239–2255, DOI 10.1016/j.laa.2012.10.030. MR3005287

[20] P. G. Casazza and J. Kovačević, *Equal-norm tight frames with erasures*, Adv. Comput. Math. **18** (2003), no. 2-4, 387–430, DOI 10.1023/A:1021349819855. Frames. MR1968127 (2004e:42046)

[21] A. Z. -Y. Chan, M. S. Copenhaver, S. K. Narayan, L. Stokols, and A. Theobold, *On structural decompositions of finite frames,* arXiv:1411.6138, (2014).

[22] X. Chen, G. Kutyniok, K. A. Okoudjou, F. Philipp, and R. Wang, *Measures of scalability*, IEEE Trans. Inform. Theory **61** (2015), no. 8, 4410–4423, DOI 10.1109/TIT.2015.2441071. MR3372361

[23] C. A. Clark and K. A. Okoudjou, *On Optimal Frame Conditioners,* 2015 International Conference on Sampling Theory and Applications (SampTA), 148–152 (DOI: 10.1109/SAMPTA.2015.7148869).

[24] H. Cohn and A. Kumar, *Universally optimal distribution of points on spheres*, J. Amer. Math. Soc. **20** (2007), no. 1, 99–148, DOI 10.1090/S0894-0347-06-00546-7. MR2257398 (2007h:52009)

[25] M. S. Copenhaver, Y. H. Kim, C. Logan, K. Mayfield, S. K. Narayan, M. J. Petro, and J. Sheperd, *Diagram vectors and tight frame scaling in finite dimensions*, Oper. Matrices **8** (2014), no. 1, 73–88, DOI 10.7153/oam-08-02. MR3202927

[26] P. Delsarte, J. M. Goethals, and J. J. Seidel, *Spherical codes and designs*, Geometriae Dedicata **6** (1977), no. 3, 363–388. MR0485471 (58 #5302)

[27] D. L. Donoho and M. Elad, *Optimally sparse representation in general (nonorthogonal) dictionaries via l^1 minimization*, Proc. Natl. Acad. Sci. USA **100** (2003), no. 5, 2197–2202 (electronic), DOI 10.1073/pnas.0437847100. MR1963681 (2004c:94068)

[28] M. Elad and A. M. Bruckstein, *A generalized uncertainty principle and sparse representation in pairs of bases*, IEEE Trans. Inform. Theory **48** (2002), no. 9, 2558–2567, DOI 10.1109/TIT.2002.801410. MR1929464 (2003h:15002)

[29] M. Ehler and K. A. Okoudjou, *Minimization of the probabilistic p-frame potential*, J. Statist. Plann. Inference **142** (2012), no. 3, 645–659, DOI 10.1016/j.jspi.2011.09.001. MR2853573

[30] M. Ehler, *Random tight frames*, J. Fourier Anal. Appl. **18** (2012), no. 1, 1–20, DOI 10.1007/s00041-011-9182-5. MR2885555

[31] M. Ehler and J. Galanis, *Frame theory in directional statistics*, Statist. Probab. Lett. **81** (2011), no. 8, 1046–1051, DOI 10.1016/j.spl.2011.02.027. MR2803742 (2012e:62176)

[32] M. Ehler and K. A. Okoudjou, *Probabilistic frames: an overview*, Finite frames, Appl. Numer. Harmon. Anal., Birkhäuser/Springer, New York, 2013, pp. 415–436, DOI 10.1007/978-0-8176-8373-3_12. MR2964017

[33] M. Fickus, B. D. Johnson, K. Kornelson, and K. A. Okoudjou, *Convolutional frames and the frame potential*, Appl. Comput. Harmon. Anal. **19** (2005), no. 1, 77–91, DOI 10.1016/j.acha.2005.02.005. MR2147063 (2006d:42050)

[34] M. Fickus, D. G. Mixon, M. J. Poteet, and N. Strawn, *Constructing all self-adjoint matrices with prescribed spectrum and diagonal*, Adv. Comput. Math. **39** (2013), no. 3-4, 585–609, DOI 10.1007/s10444-013-9298-z. MR3116042

[35] M. Fornasier and H. Rauhut, *Continuous frames, function spaces, and the discretization problem*, J. Fourier Anal. Appl. **11** (2005), no. 3, 245–287, DOI 10.1007/s00041-005-4053-6. MR2167169 (2006g:42053)

[36] G. E. Forsythe and E. G. Straus, *On best conditioned matrices*, Proc. Amer. Math. Soc. **6** (1955), 340–345. MR0069585 (16,1054j)

[37] A. A. Giannopoulos and V. D. Milman, *Extremal problems and isotropic positions of convex bodies*, Israel J. Math. **117** (2000), 29–60, DOI 10.1007/BF02773562. MR1760584 (2001e:65031)

[38] A. C. Gilbert, S. Muthukrishnan, and M. J. Strauss, *Approximation of functions over redundant dictionaries using coherence*, Proceedings of the Fourteenth Annual ACM-SIAM Symposium on Discrete Algorithms (Baltimore, MD, 2003), ACM, New York, 2003, pp. 243–252. MR1974925

[39] V. K. Goyal, M. Vetterli, and N. T. Thao, *Quantized overcomplete expansions in* $\mathbf{R}^N$*: analysis, synthesis, and algorithms*, IEEE Trans. Inform. Theory **44** (1998), no. 1, 16–31, DOI 10.1109/18.650985. MR1486646 (99a:94004)

[40] O. Güler, *Foundations of optimization*, Graduate Texts in Mathematics, vol. 258, Springer, New York, 2010. MR2680744 (2011e:90002)

[41] D. Han, K. Kornelson, D. Larson, and E. Weber, *Frames for undergraduates*, Student Mathematical Library, vol. 40, American Mathematical Society, Providence, RI, 2007. MR2367342 (2010e:42044)

[42] C. Heil, *What is ... a frame?*, Notices Amer. Math. Soc. **60** (2013), no. 6, 748–750, DOI 10.1090/noti1011. MR3076247

[43] N. J. Higham, *Accuracy and stability of numerical algorithms*, 2nd ed., Society for Industrial and Applied Mathematics (SIAM), Philadelphia, PA, 2002. MR1927606 (2003g:65064)

[44] Y. Hur and K. A. Okoudjou, *Scaling Laplacian pyramids*, SIAM J. Matrix Anal. Appl. **36** (2015), no. 1, 348–365, DOI 10.1137/140988231. MR3327355

[45] F. John, *Extremum problems with inequalities as subsidiary conditions*, Studies and Essays Presented to R. Courant on his 60th Birthday, January 8, 1948, Interscience Publishers, Inc., New York, N. Y., 1948, pp. 187–204. MR0030135 (10,719b)

[46] B. D. Johnson and K. A. Okoudjou, *Frame potential and finite abelian groups*, Radon transforms, geometry, and wavelets, Contemp. Math., vol. 464, Amer. Math. Soc., Providence, RI, 2008, pp. 137–148, DOI 10.1090/conm/464/09081. MR2440134 (2009k:42074)

[47] C. R. Johnson and R. Reams, *Scaling of symmetric matrices by positive diagonal congruence*, Linear Multilinear Algebra **57** (2009), no. 2, 123–140, DOI 10.1080/03081080600872327. MR2492099 (2009m:15024)

[48] L. Khachiyan and B. Kalantari, *Diagonal matrix scaling and linear programming*, SIAM J. Optim. **2** (1992), no. 4, 668–672, DOI 10.1137/0802034. MR1186169 (93h:90058)

[49] B. Klartag and G. Kozma, *On the hyperplane conjecture for random convex sets*, Israel J. Math. **170** (2009), 253–268, DOI 10.1007/s11856-009-0028-7. MR2506326 (2010c:60031)

[50] J. Kovačević and A. Chebira, *Life Beyond Bases: The Advent of Frames (Part I)*, Signal Processing Magazine, IEEE Volume **24**, Issue 4, July 2007, 86–104

[51] J. Kovačević and A. Chebira, *Life Beyond Bases: The Advent of Frames (Part II)*, Signal Processing Magazine, IEEE Volume **24**, Issue 5, Sept. 2007, 115–125.

[52] G. Kutyniok, K. A. Okoudjou, F. Philipp, and E. K. Tuley, *Scalable frames*, Linear Algebra Appl. **438** (2013), no. 5, 2225–2238, DOI 10.1016/j.laa.2012.10.046. MR3005286

[53] G. Kutyniok, K. A. Okoudjou, and F. Philipp, *Scalable frames and convex geometry*, Operator methods in wavelets, tilings, and frames, Contemp. Math., vol. 626, Amer. Math. Soc., Providence, RI, 2014, pp. 19–32, DOI 10.1090/conm/626/12507. MR3329091

[54] G. Kutyniok, K. A. Okoudjou, and F. Philipp, *Perfect preconditioning of frames by a diagonal operator,* Proceedings of the 10th International Conference on Sampling Theory and Applications pp. 85-88.

[55] J. Lemvig, C. Miller, and K. A. Okoudjou, *Prime tight frames*, Adv. Comput. Math. **40** (2014), no. 2, 315–334, DOI 10.1007/s10444-013-9309-0. MR3194708

[56] E. Levina, P. Bickel, *The Earth Mover's distance is the Mallows distance: some insights from statistics,* Eighth IEEE International Conference on Computer Vision, **2** (2001), 251–256.

[57] E. Levina and R. Vershynin, *Partial estimation of covariance matrices*, Probab. Theory Related Fields **153** (2012), no. 3-4, 405–419, DOI 10.1007/s00440-011-0349-4. MR2948681

[58] K. V. Mardia and P. E. Jupp, *Directional statistics*, Wiley Series in Probability and Statistics, John Wiley & Sons, Ltd., Chichester, 2000. Revised reprint of *Statistics of directional data* by Mardia [MR0336854 (49 #1627)]. MR1828667 (2003b:62004)

[59] P. Massey and M. Ruiz, *Minimization of convex functionals over frame operators*, Adv. Comput. Math. **32** (2010), no. 2, 131–153, DOI 10.1007/s10444-008-9092-5. MR2581231 (2011b:42109)

[60] J. Matoušek, *Lectures on discrete geometry*, Graduate Texts in Mathematics, vol. 212, Springer-Verlag, New York, 2002. MR1899299 (2003f:52011)

[61] V. D. Milman and A. Pajor, *Isotropic position and inertia ellipsoids and zonoids of the unit ball of a normed n-dimensional space*, Geometric aspects of functional analysis (1987–88), Lecture Notes in Math., vol. 1376, Springer, Berlin, 1989, pp. 64–104, DOI 10.1007/BFb0090049. MR1008717 (90g:52003)

[62] O. Oktay, *Frame quantization theory and equiangular tight frames*, ProQuest LLC, Ann Arbor, MI, 2007. Thesis (Ph.D.)–University of Maryland, College Park. MR2711357

[63] J. M. Renes, R. Blume-Kohout, A. J. Scott, and C. M. Caves, *Symmetric informationally complete quantum measurements*, J. Math. Phys. **45** (2004), no. 6, 2171–2180, DOI 10.1063/1.1737053. MR2059685 (2004m:81043)

[64] M. Rudelson, *Approximate John's decompositions*, Geometric aspects of functional analysis (Israel, 1992), Oper. Theory Adv. Appl., vol. 77, Birkhäuser, Basel, 1995, pp. 245–249. MR1353463 (96j:46007)

[65] M. Rudelson, *Contact points of convex bodies*, Israel J. Math. **101** (1997), 93–124, DOI 10.1007/BF02760924. MR1484871 (99d:52006)

[66] J. J. Seidel, *Definitions for spherical designs*, J. Statist. Plann. Inference **95** (2001), no. 1-2, 307–313, DOI 10.1016/S0378-3758(00)00297-4. Special issue on design combinatorics: in honor of S. S. Shrikhande. MR1829118 (2002b:05030)

[67] J. Stoer and C. Witzgall, *Transformations by diagonal matrices in a normed space*, Numer. Math. **4** (1962), 158–171. MR0150151 (27 #154)

[68] J. Stoer and C. Witzgall, *Convexity and optimization in finite dimensions. I*, Die Grundlehren der mathematischen Wissenschaften, Band 163, Springer-Verlag, New York-Berlin, 1970. MR0286498 (44 #3707)

[69] N. Strawn, *Finite frame varieties: nonsingular points, tangent spaces, and explicit local parameterizations*, J. Fourier Anal. Appl. **17** (2011), no. 5, 821–853, DOI 10.1007/s00041-010-9164-z. MR2838109 (2012h:42062)

[70] N. Strawn, *Optimization over finite frame varieties and structured dictionary design*, Appl. Comput. Harmon. Anal. **32** (2012), no. 3, 413–434, DOI 10.1016/j.acha.2011.09.001. MR2892742

[71] T. Strohmer and R. W. Heath Jr., *Grassmannian frames with applications to coding and communication*, Appl. Comput. Harmon. Anal. **14** (2003), no. 3, 257–275, DOI 10.1016/S1063-5203(03)00023-X. MR1984549 (2004d:42053)

[72] N. Tomczak-Jaegermann, *Banach-Mazur distances and finite-dimensional operator ideals*, Pitman Monographs and Surveys in Pure and Applied Mathematics, vol. 38, Longman Scientific & Technical, Harlow; copublished in the United States with John Wiley & Sons, Inc., New York, 1989. MR993774 (90k:46039)

[73] L. Fejes Tóth, *Distribution of points in the elliptic plane* (English, with Russian summary), Acta Math. Acad. Sci. Hungar **16** (1965), 437–440. MR0184139 (32 #1612)

[74] J. A. Tropp, *Greed is good: algorithmic results for sparse approximation*, IEEE Trans. Inform. Theory **50** (2004), no. 10, 2231–2242, DOI 10.1109/TIT.2004.834793. MR2097044 (2005e:94036)

[75] B. Venkov, *Réseaux et designs sphériques* (French, with English and French summaries), Réseaux euclidiens, designs sphériques et formes modulaires, Monogr. Enseign. Math., vol. 37, Enseignement Math., Geneva, 2001, pp. 10–86. MR1878745 (2002m:11061)

[76] R. Vershynin, *John's decompositions: selecting a large part*, Israel J. Math. **122** (2001), 253–277, DOI 10.1007/BF02809903. MR1826503 (2002c:46017)

[77] R. Vershynin, *How close is the sample covariance matrix to the actual covariance matrix?*, J. Theoret. Probab. **25** (2012), no. 3, 655–686, DOI 10.1007/s10959-010-0338-z. MR2956207

[78] C. Villani, *Optimal transport*, Grundlehren der Mathematischen Wissenschaften [Fundamental Principles of Mathematical Sciences], vol. 338, Springer-Verlag, Berlin, 2009. Old and new. MR2459454 (2010f:49001)

[79] S. Waldron, *Generalized Welch bound equality sequences are tight frames*, IEEE Trans. Inform. Theory **49** (2003), no. 9, 2307–2309, DOI 10.1109/TIT.2003.815788. MR2004787 (2005a:94024)

[80] R. Webster, *Convexity*, Oxford Science Publications, The Clarendon Press, Oxford University Press, New York, 1994. MR1443208 (98h:52001)

[81] L. R. Welch, *Lower bounds on the maximum cross correlation of signals,* IEEE Trans. Inform. Theory, vol. IT-**20** (1974), 397–399.

[82] C. G. Wickman, *An Optimal Transport Approach to Some Problems in Frame Theory*, ProQuest LLC, Ann Arbor, MI, 2014. Thesis (Ph.D.)–University of Maryland, College Park. MR3259951

[83] C. Wickman Lau and K. A. Okoudjou, *Scalable probabilistic frames,* arXiv:1501.07321 (2015).

[84] G. Zauner, "Quantum designs—Foundations of non-commutative theory of designs" (in German), Ph.D. thesis, University of Vienna, 1999. Available online at http://www.math.univie.ac.at/ neum/papers.html.

[85] MAPS-REU: http://www-math.umd.edu/maps-reu.html

Department of Mathematics &, Norbert Wiener Center, University of Maryland, College Park, Maryland 20742

E-mail address: kasso@math.umd.edu

Proceedings of Symposia in Applied Mathematics
Volume **73**, 2016
http://dx.doi.org/10.1090/psapm/073/00631

Quantization, finite frames, and error diffusion

Alexander Dunkel, Alexander M. Powell, Anneliese H. Spaeth,
and Özgur Yılmaz

ABSTRACT. This chapter gives a brief expository introduction to quantization for finite frame coefficients. Topics discussed include memoryless scalar quantization (MSQ), the uniform noise model, first order Sigma-Delta ($\Sigma\Delta$) quantization, and the role of error diffusion in quantization. The main concepts are illustrated through numerical examples. Stability bounds and basic error bounds are presented for a generalization of $\Sigma\Delta$ quantization which includes the standard first order $\Sigma\Delta$ algorithm and a noise shaping quantization algorithm based on error diffusion.

1. Introduction

Finite frames are a well-studied generalization of orthonormal bases which use redundancy to provide stable and robust signal representations [**6**]. A finite frame allows one to decompose an arbitrary signal $x \in \mathbb{R}^d$ as a linear combination

$$x = \sum_{n=1}^{N} c_n e_n \tag{1.1}$$

using frame vectors $\{e_n\}_{n=1}^N \subset \mathbb{R}^d$ and real coefficients $\{c_n\}_{n=1}^N \subset \mathbb{R}^d$. When $N > d$, the redundancy in frames offers increased design flexibility and can mitigate the effects of noise and other data loss [**1**, **5**, **10**, **17**]. This chapter gives an introductory discussion on noise that is associated with round-off errors in the coefficients c_n and, more generally, with the process of digitally encoding or *quantizing* the coefficients c_n.

2. Finite Frames

A finite collection of vectors, $\{e_n\}_{n=1}^N \subseteq \mathbb{R}^d$ is a *frame* for $\mathbb{R}^d$ with *frame bounds* $0 < A \leq B < \infty$, if

$$\forall x \in \mathbb{R}^d, \qquad A\|x\|^2 \leq \sum_{n=1}^{N} |\langle x, e_n\rangle|^2 \leq B\|x\|^2. \tag{2.1}$$

2010 *Mathematics Subject Classification.* Primary 94A99, 65T99.
Key words and phrases. Analog-to-digital conversion, error diffusion, Sigma-Delta algorithms.

Here, $\|\cdot\|$ denotes the Euclidean norm on $\mathbb{R}^d$. If the frame bounds satisfy $A = B$, then the frame is said to be *tight*. If $\|e_n\| = 1$ for each $n = 1, 2, \cdots, N$, then the frame is said to be *unit-norm*.

If $\{e_n\}_{n=1}^N$ is a frame for $\mathbb{R}^d$, then there exists at least one associated *dual frame* $\{f_n\}_{n=1}^N \subseteq \mathbb{R}^d$ which gives the frame expansions

$$\forall x \in \mathbb{R}^d, \quad x = \sum_{n=1}^{N} \langle x, f_n \rangle \, e_n = \sum_{n=1}^{N} \langle x, e_n \rangle \, f_n. \tag{2.2}$$

Note that if $N > d$, then $\{e_n\}_{n=1}^N$ is a redundant spanning set for $\mathbb{R}^d$ and cannot be a basis. Hence, the choice of dual frame and the choice of coefficients $\{c_n\}_{n=1}^N$ in (1.1) are generally not unique.

It is often convenient to formulate frame theoretic concepts in matrix form. Let $E = \{e_n\}_{n=1}^N \subseteq \mathbb{R}^d$ be a frame for $\mathbb{R}^d$. With slight abuse of notation, E will also denote the $d \times N$ matrix whose j^{th} column is e_j. If one represents $x \in \mathbb{R}^d$ as a $d \times 1$ column vector, then the frame inequality (2.1) can be rewritten as

$$\forall x \in \mathbb{R}^d, \quad A \|x\|^2 \leq \|E^* x\|^2 \leq B \|x\|^2, \tag{2.3}$$

where E^* is the adjoint of E. Moreover, if $F = \{f_n\}_{n=1}^N \subset \mathbb{R}^d$ is a dual frame associated to E, and F denotes the $d \times N$ matrix whose j^{th} column is f_j, then the frame expansions (2.2) can be expressed in matrix form as

$$EF^* = FE^* = I, \tag{2.4}$$

where I is the $d \times d$ identity matrix. This shows that F is a dual frame associated to E if and only if F is a left inverse of E^*.

Dual frames are generally not unique, but there is an important canonical choice. Given a frame E, the *frame operator* is the $d \times d$ matrix $S = EE^*$. The frame operator is invertible, positive, and self-adjoint, e.g. [**8**], and hence

$$I = S^{-1}S = S^{-1}EE^* = (EE^*)^{-1}EE^*. \tag{2.5}$$

Equation (2.5) shows that $F = (EE^*)^{-1}E$, the so-called Moore-Penrose pseudoinverse, is a left inverse to E^*, and hence F defines a dual frame associated to E. The dual frame $F = (EE^*)^{-1}E = S^{-1}E$ is called the *canonical dual frame* associated to E. Rewriting (2.5) in vector form leads to

$$\forall x \in \mathbb{R}^d, \quad x = \sum_{n=1}^{N} \langle x, e_n \rangle \left(S^{-1} e_n\right) = \sum_{n=1}^{N} \left\langle x, \left(S^{-1} e_n\right) \right\rangle e_n. \tag{2.6}$$

Thus, in vector form, the canonical dual frame is simply $F = \{S^{-1}e_n\}_{n=1}^N$.

Unit-norm tight frames for $\mathbb{R}^d$ lead to especially simple frame expansions. If $E = \{e_n\}_{n=1}^N$ is a unit-norm tight frame for $\mathbb{R}^d$, then it can be shown, e.g. [**8**], that the canonical dual frame is given by $F = \{\frac{d}{N} e_n\}_{n=1}^N$, and one has

$$\forall x \in \mathbb{R}^d, \quad x = \frac{d}{N} \sum_{n=1}^{N} \langle x, e_n \rangle \, e_n. \tag{2.7}$$

EXAMPLE 2.1. Fix $N \geq 2$, and define $E_N = \{e_n^N\}_{n=1}^N \subset \mathbb{R}^2$ by

$$e_n^N = \begin{pmatrix} \cos(2\pi n/N) \\ \sin(2\pi n/N) \end{pmatrix}. \tag{2.8}$$

The collection $\{e_n^N\}_{n=1}^N$ is a unit-norm tight frame for $\mathbb{R}^2$, e.g., [**10**], and in particular,

$$\forall x \in \mathbb{R}^2, \quad x = \frac{2}{N}\sum_{n=1}^{N}\langle x, e_n^N\rangle e_n^N.$$

The vectors $\{e_n^N\}_{n=1}^N$ may be viewed geometrically as the Nth roots of unity, and for this reason we will refer to the unit-norm tight frame E_N as the Nth *roots of unity frame.*

3. Quantization

Many physical quantities or signals of interest have analog properties which are physically challenging to measure, represent, and store. *Quantization* is the process that approximates analog signals (or even signals which take on an unwieldy number of values) using a finite set of values which will be more manageably representable in the digital domain. It is desirable to implement quantization methods of low complexity which achieve small error between the quantized, approximate signal, and the original signal.

We consider the problem of quantization for finite frame coefficients. Let $E = \{e_n\}_{n=1}^N \subset \mathbb{R}^d$ be a frame for $\mathbb{R}^d$, with an associated dual frame $F = \{f_n\}_{n=1}^N$. Given a signal $x \in \mathbb{R}^d$, the frame coefficients $E^*x = \{\langle x, e_n\rangle\}_{n=1}^N$ losslessly represent x through the frame expansion

$$x = \sum_{n=1}^{N}\langle x, e_n\rangle f_n.$$

However, the frame coefficients $\langle x, e_n\rangle$ do not "digitally" represent x, since the coefficients $\langle x, e_n\rangle$ take values that lie in the continuum. To obtain a digitally useable representation, the quantization step aims to represent each frame coefficient using finitely many bits. The quantization step is typically lossy.

Let $\mathcal{A} \subset \mathbb{R}$ be a finite set, which we shall refer to as a *quantization alphabet.* The quantization problem for finite frame coefficients seeks a *quantization encoding* $\mathcal{E} : \mathbb{R}^N \longrightarrow \mathcal{A}^N$ where $\{\langle x, e_n\rangle\}_{n=1}^N \longmapsto \{q_n\}_{n=1}^N$ and each $q_n \in \mathcal{A}$. Since the quantization alphabet $\mathcal{A}$ is finite, the quantized coefficients $\{q_n\}_{n=1}^N$ can be digitally stored and processed, and hence can serve as a digital proxy for the actual frame coefficients. To recover an approximation $\widetilde{x} \in \mathbb{R}^d$ to the original signal $x \in \mathbb{R}^d$ from the quantized coefficients $\{q_n\}_{n=1}^N \subset \mathcal{A}$, one needs a suitable *quantization decoding* map $\mathcal{D} : \mathcal{A}^N \longrightarrow \mathbb{R}^d$ with $\mathcal{D} : \{q_n\}_{n=1}^N \longmapsto \widetilde{x}$. It is desirable to have a quantization encoder/decoder pairing with small error

$$\|x - \mathcal{D}(\mathcal{E}(E^*x))\|,$$

but other typical considerations might include:

- The quantization encoder is often implemented as an iterative algorithm that should act sequentially on the frame coefficients.
- Computational complexity and memory usage will constrain the types of quantization algorithms that can be used in practice.
- Circuit implementations of quantization methods should be robust against nonideal circuit elements, and should balance performance with physical factors such as cost and power usage.

- The quantization encoder may only have partial or even no knowledge of the frame $\{e_n\}_{n=1}^N$ that was used to compute the coefficients $\{\langle x, e_n\rangle\}_{n=1}^N$. In some cases, full knowledge of the frame will only be available at the quantization decoding step.

Below, we focus on two classes of quantizers and discuss (the limits of) their effectiveness in quantizing frame expansions while remaining within the design constraints that we have outlined above. Specifically, Section 4 introduces and analyzes memoryless scalar quantizers (MSQ), which are based on quantizing every coefficient independently, e.g., [**10**, **13**, **17**]. In Sections 5 and 6, we focus on $\Sigma\Delta$ quantizers (of order 1) that are examples of *noise-shaping quantizers* [**7**].

4. Memoryless Scalar Quantization

Scalar quantization is a fundamental building block in the quantization process. Given a finite set, $\mathcal{A} \subseteq \mathbb{R}$, which we will refer to as a *quantization alphabet*, the associated *scalar quantizer*, $Q : \mathbb{R} \to \mathcal{A}$, is defined by

$$Q(u) = \arg\min_{q \in \mathcal{A}} |u - q| \,. \tag{4.1}$$

In other words, Q rounds u to the nearest element of $\mathcal{A}$. If u is equidistant from two distinct elements of $\mathcal{A}$, we adopt the convention that Q rounds u to the larger of the two alphabet elements.

A simple approach to quantization is given by quantizing the frame coefficients $\langle x, e_n\rangle$ through

$$q_n = Q(\langle x, e_n\rangle). \tag{4.2}$$

This process is referred to as *memoryless scalar quantization (MSQ)*, or *pulse code modulation (PCM)*. As the name suggests, each quantization or "rounding" that is performed depends only upon a single coefficient under consideration, and hence the quantizer can be said to have no memory of the quantization that occurred in prior coefficients. Unfortunately, as we shall see, memoryless scalar quantization fails to take sufficient advantage of the redundancy of frames, though it is relatively uncomplicated and hence fairly easy to implement.

We shall restrict our attention to the quantization decoder given by *linear reconstruction*

$$\widetilde{x} = \sum_{n=1}^{N} q_n f_n, \tag{4.3}$$

where $\{f_n\}_{n=1}^N$ is a fixed dual frame associated to $\{e_n\}_{n=1}^N$. Moreover, to study the behavior of MSQ with linear reconstruction, we further restrict our attention to the *midrise quantization alphabet*, given by

$$\mathcal{A}_K^\delta = \left\{ \left(-K + \frac{1}{2}\right)\delta, \left(-K + \frac{3}{2}\right)\delta, \dots, \left(-\frac{1}{2}\right)\delta, \left(\frac{1}{2}\right)\delta, \dots, \left(K - \frac{1}{2}\right)\delta \right\}, \tag{4.4}$$

where $\delta > 0$, and $K \in \mathbb{N}$.

Let $\{e_n\}_{n=1}^N \subset \mathbb{R}^d$ be a frame for $\mathbb{R}^d$. If $x \in \mathbb{R}^d$ satisfies

$$\|x\| < \frac{K}{\delta}\left(\max_{1 \le n \le N} \|e_n\|\right)^{-1},$$

then the frame coefficients $\langle x, e_n \rangle$ satisfy

$$|\langle x, e_n \rangle| \leq \|x\| \|e_n\| < \frac{K}{\delta},$$

and the coefficients q_n produced by MSQ in (4.2) satisfy

$$|q_n - \langle x, e_n \rangle| \leq \frac{\delta}{2}. \tag{4.5}$$

From this, and (2.2) and (4.3), it follows that

$$\|x - \widetilde{x}\| \leq \sum_{n=1}^{N} |\langle x, e_n \rangle - q_n| \, \|f_n\| \leq \frac{\delta}{2} \sum_{n=1}^{N} \|f_n\|. \tag{4.6}$$

In the special case when $\{e_n\}_{n=1}^N$ is a unit-norm tight frame for $\mathbb{R}^d$ and $f_n = \frac{d}{N} e_n$ is the canonical dual frame, (4.6) reduces to

$$\|x - \widetilde{x}\| \leq \frac{d\delta}{2}. \tag{4.7}$$

As expected, the error bounds (4.6) and (4.7) improve when a finer quantization alphabet (with smaller δ) is used. However, these error bounds do not show an improvement when a more redundant frame (with larger N) is used.

In spite of the pessimistic bounds (4.6) and (4.7), MSQ typically exhibits some error reduction as a function of oversampling when the frame size N increases. To illustrate this, it will be convenient to briefly discuss the uniform noise model for coefficient quantization errors in MSQ. Equation (4.5) shows that the individual coefficient quantization errors, $q_n - Q(\langle x, e_n \rangle)$, for MSQ lie in the interval $[-\delta/2, \delta/2]$. The uniform noise model goes one step further than this and states that under suitable conditions, the coefficient quantization errors can be well approximated by independent random variables on $[-\delta/2, \delta/2]$. The following theorem [**13**] provides mathematical justification for the uniform noise model.

THEOREM 4.1. *Suppose that x is an absolutely continuous random vector in $\mathbb{R}^d$ and suppose that the vectors $\{e_n\}_{n=1}^N \subset \mathbb{R}^d$ are pairwise linearly independent. The entries of the normalized vector of quantization errors*

$$\frac{1}{\delta} [\langle x, e_1 \rangle - Q(\langle x, e_1 \rangle), \cdots, \langle x, e_N \rangle - Q(\langle x, e_N \rangle)] \tag{4.8}$$

converge in distribution to the uniform distribution on $[-1/2, 1/2]$. Moreover, the entries of (4.8) *are pairwise asymptotically independent as $\delta \to 0$.*

Roughly speaking, Theorem 4.1 says that the approximation for MSQ given by

$$Q(\langle x, e_n \rangle) \approx \langle x, e_n \rangle + \epsilon_n,$$

is accurate in the high resolution limit $\delta \to 0$. In other words, the uniform noise model treats MSQ as an additive source of uniform noise. Since the scalar quantizer $Q(\cdot)$ is a nonlinear function of its input, a rigorous and exact analysis of MSQ is nonlinear and difficult. However, the uniform noise model has the pleasant consequence of linearizing and simplifying the analysis of MSQ. The remainder of this section discusses implications of the uniform noise model, and shows that the uniform noise model is empirically reasonable (which is perhaps surprising from a purely mathematical point of view without knowledge of Theorem 4.1).

For the remainder of this section, let

$$q_n = \langle x, e_n \rangle + \epsilon_n, \tag{4.9}$$

where $\{\epsilon_n\}_{n=1}^N$ are i.i.d. uniform random variables on $[-\delta/2, \delta/2]$. This is done with the uniform noise model in mind, and with the hope that by Theorem 4.1, $\{q_n\}_{n=1}^N$ will serve as a good approximation for $\{Q(\langle x, e_n \rangle)\}_{n=1}^N$. If one now considers the linear reconstruction (4.3), with q_n given by (4.9), then

$$\widetilde{x} - x = \sum_{n=1}^N (q_n - \langle x, e_n \rangle) f_n = \sum_{n=1}^N \epsilon_n f_n,$$

and one may compute the mean squared error (MSE):

$$\begin{aligned}
\mathbb{E}\|\widetilde{x} - x\|^2 &= \mathbb{E}\left[\left\langle \sum_{n=1}^N \epsilon_n f_n, \sum_{m=1}^N \epsilon_m f_m \right\rangle\right] \\
&= \sum_{n=1}^N \sum_{m=1}^N \mathbb{E}[\epsilon_n \epsilon_m] \langle f_n, f_m \rangle \\
&= \sum_{k=1}^N \mathbb{E}|\epsilon_k|^2 \, \|f_k\|^2 \\
&= \frac{\delta^2}{12} \sum_{k=1}^N \|f_k\|^2. \qquad (4.10)
\end{aligned}$$

Here we have used that $\mathbb{E}(\epsilon_n) = 0$ and $\mathbb{E}[\epsilon_n^2] = \delta^2/12$. In the special case when $\{e_n\}_{n=1}^N$ is a unit-norm tight frame for $\mathbb{R}^d$ and $f_n = \frac{d}{N} e_n$ is the associated canonical dual frame, (4.10) reduces to

$$\mathbb{E}\|\widetilde{x} - x\|^2 = \frac{d^2 \delta^2}{12N}. \tag{4.11}$$

The next several examples will compare the actual performance of MSQ against the error predicted by the uniform noise model in (4.11).

EXAMPLE 4.2. Let $E_N = \{e_n^N\}_{n=1}^N \subset \mathbb{R}^2$ be the unit-norm tight frame for $\mathbb{R}^2$ defined by (2.8). In this example we restrict our attention to the case when $N \geq 11$ is odd. When N is odd the frame vectors $\{e_n^N\}_{n=1}^N$ are pairwise linearly independent, and the frame hypothesis of Theorem 4.1 holds.

Using (4.4), let $\delta = .0001$ and $K = 10000$. For each fixed odd N, we randomly select 200 vectors $\{x_j\}_{j=1}^{200} \subset \mathbb{R}^2$ independently according to the uniform distribution on the unit circle $\mathbb{S}^1 = \{x \in \mathbb{R}^2 : \|x\| = 1\}$. For each of these vectors $x_j \in \mathbb{R}^2$, we quantize the frame coefficients $\{\langle x_j, e_n^N \rangle\}_{n=1}^N$ with MSQ by $Q(\langle x_j, e_n^N \rangle)$, and reconstruct linearly using

$$\widetilde{x}_{j,N} = \frac{2}{N} \sum_{n=1}^N Q\left(\langle x_j, e_n^N \rangle\right) e_n^N.$$

We then compute the average squared error

$$\text{MSE}(N) = \frac{1}{200} \sum_{j=1}^{200} \|x_j - \widetilde{x}_{j,N}\|^2.$$

FIGURE 1. Mean squared error from Example 4.2 with $\delta = .0001$ and N odd.

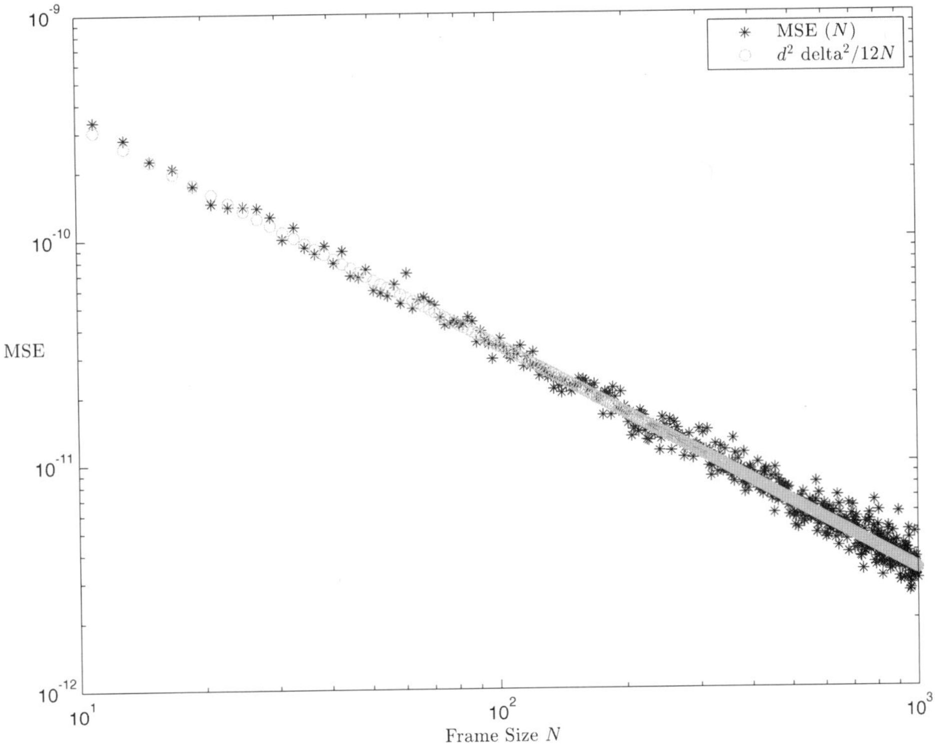

Figure 1 shows a log-log plot of MSE(N) against N (in other words, the figure shows shows a plot of log(MSE(N)) against log(N)). For comparison with (4.11), Figure 1 also shows a log-log plot of $\frac{d^2\delta^2}{12N}$ against N (recall that $d = 2$ in this example). Note that error predicted by (4.11) agrees closely with the actual quantization error.

EXAMPLE 4.3. In this example, we repeat the experiment in Example 4.2, except with the larger (coarser) value of $\delta = .005$ and the value $K = 200$. Figure 2 shows a log-log plot of MSE(N) against N, and also shows a log-log plot of $\frac{d^2\delta^2}{12N}$ against N. Note that the error predicted by (4.11) does not agree as closely with the actual quantization error as in Example 4.2. This can be explained by the fact that Theorem 4.1 requires a sufficiently small value of δ for the uniform noise model to be accurate.

EXAMPLE 4.4. In this example, we repeat the experiment in Example 4.2 with the same values $\delta = .0001$ and $K = 10000$, but instead use the frames E_N in the case when $N \geq 12$ is *even.* Note that when $N > 2$ is even, the elements of the frame $E_N = \{e_n^N\}_{n=1}^N$ are not pairwise linearly independent. Figure 3 shows a log-log plot of MSE(N) against N, and also shows a log-log plot of $\frac{d^2\delta^2}{12N}$ against N. Note that the error predicted by (4.11) does not agree as closely with the actual quantization error as in Example 4.2. This can be explained by the fact that when N is even, the

FIGURE 2. Mean squared error from Example 4.3 with $\delta = .005$ and N odd.

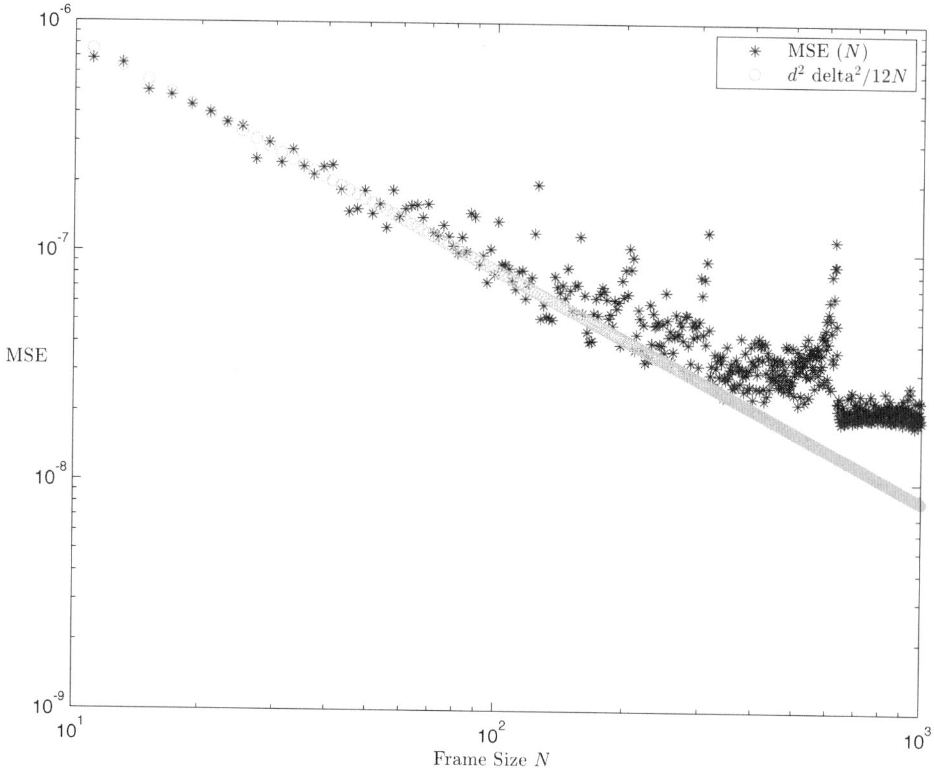

frame E_N does not satisfy the hypotheses of Theorem 4.1 needed for the uniform noise model to be accurate.

EXAMPLE 4.5. Consider the frame $E_5 = \{e_n^5\}_{n=1}^5$ defined by (2.8) as

$$e_n^5 = \begin{pmatrix} \cos(2\pi n/5) \\ \sin(2\pi n/5) \end{pmatrix}.$$

Let $\{x_j\}_{j=1}^{10000} \subset \mathbb{R}^2$ be drawn independently at random according to the uniform distribution on the unit-circle $\mathbb{S}^1$.

Let

$$v_n^\delta(x) = \frac{1}{\delta}\left(\langle x, e_n\rangle - Q_\delta(\langle x, e_n\rangle)\right),$$

and

$$\overline{v}_n^\delta = \frac{1}{10000}\sum_{j=1}^{10000} v_n^\delta(x_j).$$

Let C_δ be the 5×5 sample covariance matrix whose $(k,l)^{\text{th}}$ entry is

$$C_\delta(k,l) = \frac{1}{10000}\sum_{m=1}^{10000}\left(v_k^\delta(x_m) - \overline{v}_k^\delta\right)\left(v_l^\delta(x_m) - \overline{v}_l^\delta\right).$$

FIGURE 3. Mean squared error from Example 4.4 with $\delta = .0001$ and N even.

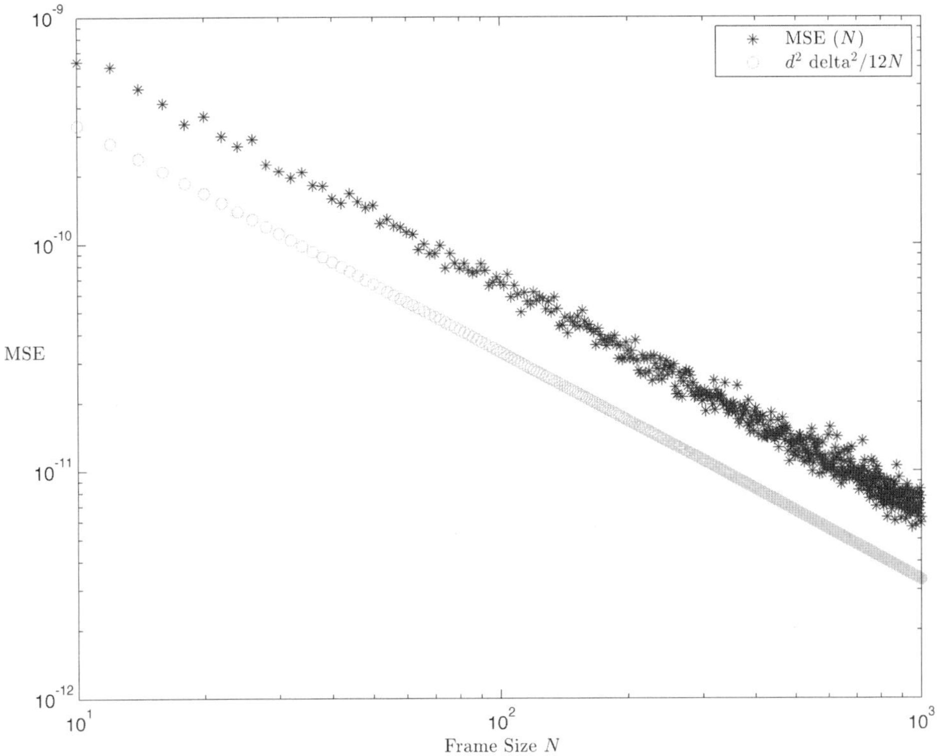

For $\delta = .1$, we have

$$C_\delta = \begin{pmatrix} 0.0931 & -0.0101 & -0.0070 & -0.0048 & -0.0119 \\ -0.0101 & 0.0935 & -0.0111 & -0.0072 & -0.0065 \\ -0.0070 & -0.0111 & 0.0939 & -0.0103 & -0.0075 \\ -0.0048 & -0.0072 & -0.0103 & 0.0930 & -0.0121 \\ -0.0119 & -0.0065 & -0.0075 & -0.0121 & 0.0942 \end{pmatrix}.$$

For $\delta = .001$, we have

$$C_\delta = \begin{pmatrix} 0.0833 & -0.0004 & -0.0008 & -0.0002 & -0.0012 \\ -0.0004 & 0.0838 & 0.0005 & -0.0003 & -0.0021 \\ -0.0008 & 0.0005 & 0.0837 & -0.0009 & -0.0000 \\ -0.0002 & -0.0003 & -0.0009 & 0.0827 & 0.0005 \\ -0.0012 & -0.0021 & -0.0000 & 0.0005 & 0.0840 \end{pmatrix}.$$

For $\delta = .000001$, we have

$$C_\delta = \begin{pmatrix} 0.0837 & 0.0001 & -0.0004 & -0.0012 & -0.0006 \\ 0.0001 & 0.0844 & -0.0001 & 0.0008 & -0.0005 \\ -0.0004 & -0.0001 & 0.0842 & 0.0006 & -0.0000 \\ -0.0012 & 0.0008 & 0.0006 & 0.0834 & 0.0004 \\ -0.0006 & -0.0005 & -0.0000 & 0.0004 & 0.0831 \end{pmatrix}.$$

Note that as $\delta \to 0$, the off-diagonal elements of the sample covariance matrices C_δ become small in magnitude, which is consistent with the entries of (4.8) being uncorrelated as predicted by Theorem 4.1. Moreover, the diagonal elements of C_δ take values near $1/12 \approx .083333$ which is the expected value of a uniform random variable on $[-1/2, 1/2]$; this is consistent with Theorem 4.1.

5. First Order $\Sigma\Delta$ Algorithms

This section introduces the standard and generalized first order $\Sigma\Delta$ algorithms, and discusses their relation to error diffusion. $\Sigma\Delta$ quantization is one of the most common methods of analog-to-digital (A/D) conversion in practice [**9, 12, 16**]. For example, in the context of audio signals – modeled as bandlimited functions – $\Sigma\Delta$ algorithms rely on obtaining redundant representations of bandlimited signals by oversampling. This redundancy is then utilized by an iterative procedure to push the quantization error into an unoccupied part of the signal spectrum [**7, 17**]. This so-called "noise-shaping" formulation of $\Sigma\Delta$ quantization can be generalized to the problem of quantization of arbitrary frame expansions, and provides algorithms that outperform MSQ for various classes of frames, e.g., [**1–3, 19**]. Finally, $\Sigma\Delta$ quantizers can also be used (and again outperform MSQ) in the context of compressive sampling [**11, 14, 18**].

5.1. The standard first order $\Sigma\Delta$ quantizer.

DEFINITION 5.1. The standard *first order* $\Sigma\Delta$ *quantizer* takes a sequence $\{x_n\}_{n=1}^N \subset \mathbb{R}$ as its input, and produces the quantized output $\{q_n\}_{n=1}^N \subset \mathcal{A}_K^\delta$ by initializing $u_0 = 0$ and then iterating the following for $n = 1, \cdots, N$

$$\begin{aligned} q_n &= Q(u_{n-1} + x_n) \\ u_n &= u_{n-1} + x_n - q_n. \end{aligned} \tag{5.1}$$

The numbers $\{u_n\}_{n=0}^N \subset \mathbb{R}$ are called *state variables* of the algorithm.

Let $E = \{e_n\}_{n=1}^N$ be a frame for $\mathbb{R}^d$ and let $F = \{f_n\}_{n=1}^N$ be a dual frame to E. Given $x \in \mathbb{R}^d$, denote its frame coefficients by $x_n = \langle x, e_n \rangle$, so that

$$x = \sum_{n=1}^N x_n f_n. \tag{5.2}$$

Suppose that the quantized coefficients $\{q_n\}_{n=1}^N$ are the output of the first order $\Sigma\Delta$ algorithm (5.1) when the frame coefficients $\{x_n\}_{n=1}^N$ are used as input. The $\Sigma\Delta$ quantized coefficients $\{q_n\}_{n=1}^N$ provide a digital encoding of the frame coefficients $\{\langle x, e_n \rangle\}_{n=1}^N$. To recover a signal from the quantized coefficients $\{q_n\}_{n=1}^N$ we focus on the decoding step given by linear reconstruction

$$\widetilde{x} = \sum_{n=1}^N q_n f_n. \tag{5.3}$$

EXAMPLE 5.2. This numerical example shows the individual steps of the $\Sigma\Delta$ algorithm (5.1) to develop intuition, cf. [**1**]. We work with the quantization alphabet $\mathcal{A}_1^2 = \{-1, 1\}$ and the 7th roots of unity frame $E_7 = \{e_n^7\}_{n=1}^7 \subset \mathbb{R}^2$, given by (2.8).

FIGURE 4. 7th roots of unity

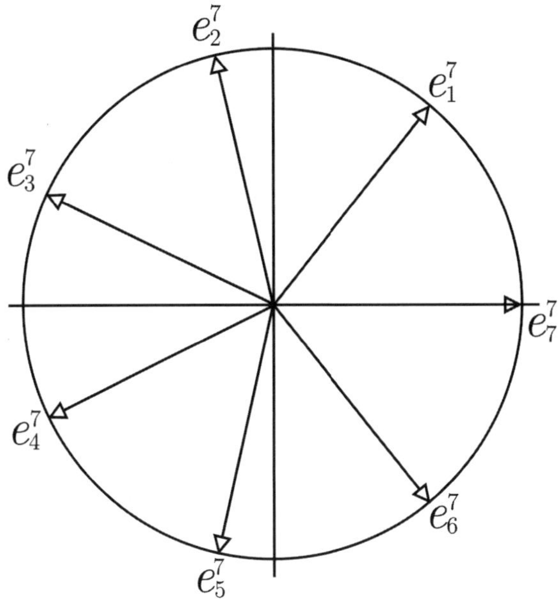

The performance of $\Sigma\Delta$ quantization depends closely on the ordering of the input sequence of frame coefficients. If the roots of unity frame (2.8) is taken in the counterclockwise sequential order,

$$e_1^N, e_2^N, \cdots, e_N^N,$$

then we shall refer to this as the *standard* or *natural ordering* of the Nth roots of unity frame. A visualization of the 7th roots of unity frame taken in the standard ordering is given in Figure 4.

Let $x = \begin{pmatrix} 1/4 \\ -2/3 \end{pmatrix}$. The frame coefficients $x_n = \langle x, e_n \rangle$ of the vector x are approximately given by:

$$\begin{aligned} x_1 &\approx -0.3653 \\ x_2 &\approx -0.7056 \\ x_3 &\approx -0.5145 \\ x_4 &\approx 0.0640 \\ x_5 &\approx 0.5943 \\ x_6 &\approx 0.6771 \\ x_7 &\approx 0.2500. \end{aligned}$$

Now we may calculate $\{q_n\}_{n=1}^7$ and $\{u_n\}_{n=1}^7$ according to (5.1).

$$\begin{aligned}
q_1 &= Q(0 - 0.3653) = -1 & u_1 &\approx 0 - 0.3653 + 1 = 0.6347\\
q_2 &= Q(0.6347 - 0.7056) = -1 & u_2 &\approx 0.6347 - 0.7056 + 1 = 0.9291\\
q_3 &= Q(0.9291 - 0.5145) = 1 & u_3 &\approx 0.9291 - 0.5145 - 1 = -0.5854\\
q_4 &= Q(-0.5854 + 0.0640) = -1 & u_4 &\approx -0.5854 + 0.0640 + 1 = 0.4786\\
q_5 &= Q(0.4786 + 0.5943) = 1 & u_5 &\approx 0.4786 + 0.5943 - 1 = 0.0729\\
q_6 &= Q(0.0729 + 0.6771) = 1 & u_6 &\approx 0.0729 + 0.6771 - 1 = -0.2500\\
q_7 &= Q(-0.2500 + 0.2500) = -1 & u_7 &\approx -0.2500 + 0.2500 + 1 = 1.0000.
\end{aligned}$$

Because $\{e_n^7\}_{n=1}^7$ is a unit-norm tight frame for $\mathbb{R}^2$, the canonical dual frame is given by $f_n = \frac{2}{N}e_n^N$, and with the canonical dual frame, linear reconstruction in (5.3) is given by

$$\widetilde{x}_{\Sigma\Delta} = \frac{2}{7}\sum_{n=1}^{7} q_n e_n^7 \approx \begin{pmatrix} -0.2857\\ -0.7559 \end{pmatrix}. \tag{5.4}$$

The error for $\Sigma\Delta$ quantization is given by

$$\|x - \widetilde{x}_{\Sigma\Delta}\| \approx 0.5431. \tag{5.5}$$

For comparison, if MSQ is used to quantize the frame coefficients $\{x_n\}_{n=1}^7$ then

$$\begin{aligned}
q'_1 &= Q(x_1) = -1\\
q'_2 &= Q(x_2) = -1\\
q'_3 &= Q(x_3) = -1\\
q'_4 &= Q(x_4) = 1\\
q'_5 &= Q(x_5) = 1\\
q'_6 &= Q(x_6) - 1\\
q'_7 &= Q(x_7) = 1.
\end{aligned}$$

Linear reconstruction from MSQ quantized coefficients $\{q'_n\}_{n=1}^7$ gives

$$\widetilde{x}_{\text{MSQ}} = \frac{2}{7}\sum_{n=1}^{7} q'_n e_n^7 \approx \begin{pmatrix} 0.2857\\ -1.2518 \end{pmatrix}.$$

The error for MSQ is given by

$$\|x - \widetilde{x}_{\text{MSQ}}\| \approx 0.5862. \tag{5.6}$$

It is helpful to visualize the outcome of the quantization and reconstruction. To do so, note that $\widetilde{x}_{\Sigma\Delta}$ and $\widetilde{x}_{\text{MSQ}}$ are two particular elements of the following finite set of all possible quantized expansions

$$\Gamma = \left\{ \frac{2}{7}\sum_{n=1}^{7} \epsilon_n e_n^7 : \epsilon_n \in \mathcal{A}_1^2 \right\}. \tag{5.7}$$

Figure 5 shows the set Γ, along with x and the respective approximations $\widetilde{x}_{\Sigma\Delta}$ and $\widetilde{x}_{\text{MSQ}}$, using $\Sigma\Delta$ quantization and MSQ. Figure 5 shows that neither $\Sigma\Delta$ quantization nor MSQ selects the closest vector to x from Γ. In other words, neither $\Sigma\Delta$ quantization nor MSQ performs optimally in this example. However, the first order $\Sigma\Delta$ algorithm does outperform MSQ.

FIGURE 5. Comparing various algorithms for $x = \left(\frac{1}{4}, \frac{-2}{3}\right)$

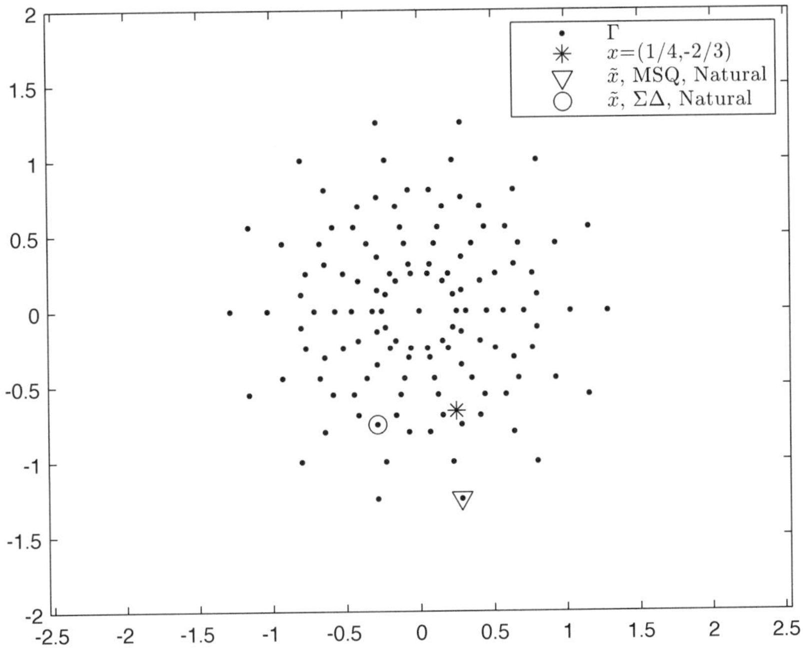

5.2. Generalized $\Sigma\Delta$ quantization and error diffusion. The standard first order $\Sigma\Delta$ algorithm outperformed MSQ in Example 5.2, but there is still much room for improvement. In this section, we consider a generalized first order $\Sigma\Delta$ algorithm, studied by Boufounos and Oppenheim in [**4**], which is designed to improve upon the performance of the standard first order $\Sigma\Delta$ algorithm.

DEFINITION 5.3. Fix $\{c_n\}_{n=1}^N \subset \mathbb{R}$. The *generalized first order $\Sigma\Delta$ algorithm* takes a sequence of real inputs $\{x_n\}_{n=1}^N \subset \mathbb{R}$, and produces the quantized output $\{q_n\}_{n=1}^N \subset \mathcal{A}_K^\delta$ by initializing $u_0 = 0$ and iterating the following for $n = 1, \cdots, N$

$$\begin{aligned} q_n &= Q(u_{n-1}c_n + x_n) \\ u_n &= u_{n-1}c_n + x_n - q_n. \end{aligned} \tag{5.8}$$

Note that the generalized first order $\Sigma\Delta$ algorithm in Definition 5.3 reduces to the standard first order $\Sigma\Delta$ algorithm when $c_n = 1$, for $n = 1, \cdots, N$.

The generalized $\Sigma\Delta$ algorithm (and indeed also the standard $\Sigma\Delta$ algorithm) is sensitive to the order in which frame coefficients are entered as input. We did not emphasize this point in Section 5.1, but will do so here. To capture the dependence of the generalized $\Sigma\Delta$ algorithm on the ordering of the input sequence, let p be a permutation of $\{1, \cdots, N\}$, and assume that the algorithm (5.3) is applied to the input sequence of frame coefficients $x_n = \langle x, e_{p(n)}\rangle, 1 \leq n \leq N$, to obtain the quantized output $\{q_n\}_{n=1}^N$. Taking into account the permation p, one has the frame expansion

$$x = \sum_{n=1}^{N} x_n f_{p(n)}. \tag{5.9}$$

Linear reconstruction will be used to recover $\widetilde{x} \in \mathbb{R}^d$ from the quantized output coefficients $\{q_n\}_{n=1}^N$ by

$$\widetilde{x} = \sum_{n=1}^{N} q_n f_{p(n)}. \tag{5.10}$$

The performance of the generalized $\Sigma\Delta$ algorithm (5.8) depends on the choice of constants $\{c_n\}_{n=1}^N \subset \mathbb{R}$, in addition to the ordering of the frame. We now motivate an important choice of $\{c_n\}_{n=1}^N$ for (5.8), see [**4**]. Let $\{e_n\}_{n=1}^N \subset \mathbb{R}^d$ be a frame for $\mathbb{R}^d$ with dual frame $\{f_n\}_{n=1}^N$. Suppose that $x \in \mathbb{R}^d$, and that the frame coefficients $x_n = \langle x, e_{p(n)} \rangle$ are used as input to (5.8) to obtain the quantized output $\{q_n\}_{n=1}^N$. If $\widetilde{x}$ is linearly reconstructed using (5.10), then the error $x - \widetilde{x}$ satisfies

$$x - \widetilde{x} = \sum_{n=1}^{N} (x_n - q_n) f_{p(n)}.$$

Define $c^\star = \{c_n^\star\}_{n=1}^N \subset \mathbb{R}$ by

$$c_1^\star = 1, \quad \text{and } c_n^\star = \left\langle f_{p(n-1)}, \frac{f_{p(n)}}{\|f_{p(n)}\|^2} \right\rangle, \quad \text{for } 2 \leq n \leq N. \tag{5.11}$$

We now examine the steps in the algorithm in Definition 5.3, when it is implemented with the particular choice of $\{c_n\}_{n=1}^N$ given by $c_n = c_n^\star$ in (5.11).

<u>**Step** $n = 1$</u> in (5.8) performs:

$$\begin{aligned} q_1 &= Q(u_0 c_1^\star + x_1) = Q(x_1) \\ u_1 &= u_0 + x_1 - q_1 = x_1 - q_1. \end{aligned}$$

Intuition:

- The $n = 1$ term in the frame expansion (5.9) is $x_1 f_{p(1)}$.
- The error vector from quantizing $x_1 f_{p(1)}$ to $q_1 f_{p(1)}$ at step $n = 1$ in (5.3) is given by $(x_1 - q_1) f_{p(1)} = u_1 f_{p(1)}$.
- Since the dual frame $\{f_{p(n)}\}_{n=1}^N$ is a redundant spanning set, part of the error $u_1 f_{p(1)}$ may lie in the span of the next dual frame vector $f_{p(2)}$.
- The algorithm (5.8) attempts to eliminate part of the error $u_1 f_{p(1)}$ from step $n = 1$ by diffusing or projecting it forward to step $n = 2$.

<u>**Step** $n = 2$</u> in (5.8) performs:

$$\begin{aligned} q_2 &= Q(u_1 c_2^\star + x_2) \\ u_2 &= u_1 c_2^\star + x_2 - q_2. \end{aligned}$$

Intuition:

- The $n = 2$ term in the frame expansion (5.9) is $x_2 f_{p(2)}$.
- The projection onto $\mathrm{span}\{f_{p(2)}\}$ of the error $u_1 f_{p(1)}$ from step $n = 1$ is given by

$$\left\langle u_1 f_{p(1)}, \frac{f_{p(2)}}{\|f_{p(2)}\|} \right\rangle \frac{f_{p(2)}}{\|f_{p(2)}\|} = u_1 c_2^\star f_{p(2)}.$$

- MSQ would simply quantize $x_2 f_{p(2)}$ to $Q(x_2)f_{p(2)}$. Instead, the step $n=2$ in (5.8) also compensates for the error $u_1 c_2^\star$ that was diffused from step $n=1$, by quantizing $(u_1 c_2^\star + x_2)f_{p(2)}$ to $q_2 f_{p(2)} = Q(u_1 c_2^\star + x_2)f_{p(2)}$.
- The total error vector from quantizing $x_1 f_{p(1)} + x_2 f_{p(2)}$ to $q_1 f_{p(1)} + q_2 f_{p(2)}$ after steps $n = 1, 2$ is given by

$$u_1 f_{p(1)} + (u_2 - c_2^\star u_1) f_{p(2)} = u_2 f_{p(2)} + u_1 (f_{p(1)} - c_2^\star f_{p(2)}). \tag{5.12}$$

  It would be desirable to compensate for the error (5.12) by diffusing all of it to the next step of the algorithm. This would require that the algorithm have memory of both steps u_1 and u_2 of the algorithm at step $n=3$.
- As a practical compromise, the first order algorithm (5.8) only diffuses forward the portion $u_2 f_{p(2)}$ of the error (5.12) that is associated with the previous state variable u_2.

<u>**Step** $n=k$</u> in (5.8) performs:

$$\begin{aligned} q_k &= Q(u_{k-1} c_k^\star + x_k) \\ u_k &= u_{k-1} c_k^\star + x_k - q_k. \end{aligned}$$

Intuition:

- The $n=k$ term in the frame expansion (5.2) is $x_k f_{p(k)}$.
- The total vector error from quantizing $\sum_{j=1}^{k-1} x_j f_{p(j)}$ to $\sum_{j=1}^{k-1} q_j f_{p(j)}$ in the previous steps $n = 1, 2, \cdots, (k-1)$ is given by

$$\begin{aligned} \sum_{j=1}^{k-1} (x_j - q_j) f_{p(j)} &= \sum_{j=1}^{k-1} (u_j - c_j^\star u_{j-1}) f_{p(j)} \\ &= u_{k-1} f_{p(k-1)} + \sum_{j=1}^{k-2} u_j (f_{p(j)} - c_{j+1}^\star f_{p(j+1)}). \end{aligned} \tag{5.13}$$

- The algorithm (5.8) only diffuses the portion $u_{k-1} f_{p(k-1)}$ of the error (5.13) to the current step. For comparison, diffusing the entire error in (5.13) would require the algorithm to have memory of $u_1, u_2, \cdots, u_{k-1}$ at step $n=k$; this would require the system memory to grow with the iteration count. We will not consider higher order algorithms in this chapter, but simply note that greater system memory and increased algorithmic complexity can be used to achieve smaller quantization error.
- The projection onto $\operatorname{span}\{f_{p(k)}\}$ of the portion of the error $u_{k-1} f_{p(k-1)}$ diffused from step $k-1$ is given by

$$\left\langle u_{k-1} f_{p(k-1)}, \frac{f_{p(k)}}{\|f_{p(k)}\|} \right\rangle \frac{f_{p(k)}}{\|f_{p(k)}\|} = u_{k-1} c_k^\star f_{p(k)}. \tag{5.14}$$

- The algorithm quantizes $(u_{k-1} c_k^\star + x_k) f_{p(k)}$ to $Q(u_{k-1} c_k^\star + x_k) f_{p(k)}$.

To summarize the above discussion, if one defines $\{c_n\}_{n=1}^N$ by $c_n = c_n^\star$ in (5.11), then the generalized $\Sigma\Delta$ algorithm of Definition 5.3 can be viewed geometrically as an error diffusion algorithm. When x_n is quantized to q_n, the resulting error is not forgotten, but is partially diffused to the next iteration, where the algorithm will attempt to compensate for the error. This motivates the following definition.

DEFINITION 5.4. Let $E = \{e_n\}_{n=1}^N$ be a frame for $\mathbb{R}^d$ and let $F = \{f_n\}_{n=1}^N$ be an associated dual frame. Let p be a permutation of $\{1, 2, \cdots, N\}$. If $\{c_n\}_{n=1}^N$ is defined by $c_n = c_n^\star$ in (5.11), then we refer to the algorithm in Definition 5.3 as a *first order* $\Sigma\Delta$ *error diffusion algorithm*, or simply, a *first order* $\Sigma\Delta$ *projection algorithm.*

We shall see in Section 6.2 that the first order $\Sigma\Delta$ error diffusion algorithm in Definition 5.4 is able to offer a quantifiable improvement over the standard first order $\Sigma\Delta$ algorithm (5.1). However, this comes at a cost: the choice of $\{c_n^\star\}_{n=1}^N$ in (5.11) requires the error diffusion algorithm to have knowledge of the dual frame $F = \{f_n\}_{n=1}^N$. Moreover, since the error diffusion algorithm implicitly depends on the frame $E = \{e_n\}_{n=1}^N$, the algorithm changes when a different frame is used to compute frame coefficients.

In contrast to the $\Sigma\Delta$ error diffusion algorithm, the standard $\Sigma\Delta$ algorithm (5.1) is universal in the sense that its implementation requires no knowledge of the frame or dual frame. The standard first order $\Sigma\Delta$ algorithm corresponds to the universal choice $c_n = 1$ for all $1 \leq n \leq N$ in Definition 5.3, regardless of the frame or dual frame being used. $\Sigma\Delta$ algorithms are known to be especially suitable for problems involving a "smoothly ordered" dual frame $\{f_n\}_{n=1}^N$, e.g., [**2, 15**]. Roughly speaking, if the dual frame $\{f_n\}_{n=1}^N$ is uniform in norm, and smoothly ordered so that each $\|f_n - f_{n+1}\| \approx 0$ is sufficiently small, then one has $\left\langle f_n, \frac{f_{n+1}}{\|f_{n+1}\|^2} \right\rangle \approx \left\langle \frac{f_n}{\|f_n\|}, \frac{f_{n+1}}{\|f_{n+1}\|} \right\rangle \approx 1$. In other words, if the dual frame has suitable smoothness properties, then the choice $c_n = 1$ used in the standard $\Sigma\Delta$ quantizer often serves as a sufficiently good approximation for the more sophisticated choice $\{c_n^\star\}_{n=1}^N$ from (5.11), used in the $\Sigma\Delta$ error diffusion algorithm.

The geometric intuition for the $\Sigma\Delta$ error diffusion algorithm shows that the ordering of the frame and dual frame strongly influence the performance of the algorithm. In particular, (5.14) shows that the amount of quantization error diffused forward at each step of the algorithm depends on the size of $c_n^\star = \left\langle f_{n-1}, \frac{f_n}{\|f_n\|^2} \right\rangle$. A larger value of $c_n^\star$ is intuitively preferable since this corresponds to diffusing more quantization error forward, and hence being able to compensate for more of the error. Section 6.2 includes results which quantify how the performance of the algorithm depends on the frame ordering. However, for the moment, it will be convenient to consider intuitive orderings which greedily maximize the amount of diffused error at each step.

DEFINITION 5.5. Let $F = \{f_n\}_{n=1}^N \subset \mathbb{R}^d$ and let p be a permutation of $\{1, 2, \cdots, N\}$. The permutation p will be called a *greedy permutation* for F if

$$\forall\ 2 \leq n \leq N, \qquad \left| \left\langle f_{p(n-1)}, \frac{f_{p(n)}}{\|f_{p(n)}\|^2} \right\rangle \right| = \max_{n \leq k \leq N} \left| \left\langle f_{p(n-1)}, \frac{f_{p(k)}}{\|f_{p(k)}\|^2} \right\rangle \right|.$$

If p is a greedy permutation for F, we say that p gives a *greedy ordering* of F.

EXAMPLE 5.6. Recall that the natural ordering of the 7th roots of unity frame $E_7 = \{e_n^7\}_{n=1}^7$ is given by $e_1^7, e_2^7, e_3^7, e_4^7, e_5^7, e_6^7, e_7^7$, see Example 5.2 and Figure 4. It can be verified that the permutation

$$p(1) = 1, p(2) = 5, p(3) = 2, p(4) = 6, p(5) = 3, p(6) = 7, p(7) = 4,$$

gives a greedy ordering

$$e_{p(1)}, e_{p(2)}, e_{p(3)}, e_{p(4)}, e_{p(5)}, e_{p(6)}, e_{p(7)},$$

of the 7th roots of unity frame, see Figure 6.

FIGURE 6. 7th roots of unity with a greedy ordering

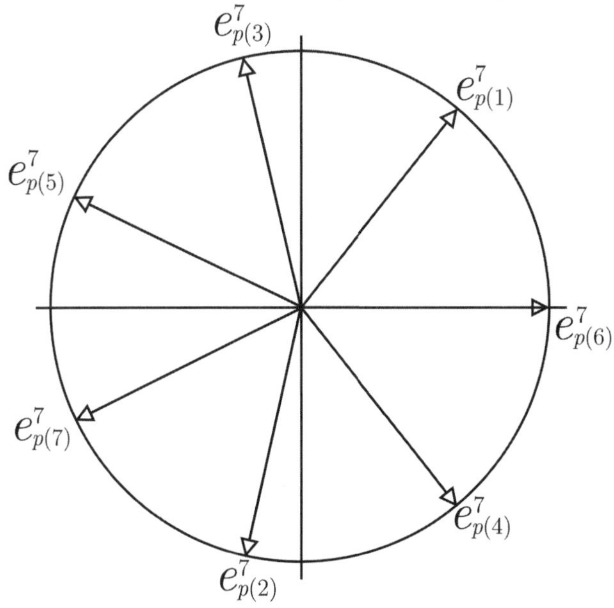

Consider the first order $\Sigma\Delta$ error diffusion algorithm in Definition 5.4 with quantization alphabet $\mathcal{A}_1^2$. We shall repeat the experiment in Example 5.2 but also compare the performance of the $\Sigma\Delta$ error diffusion algorithm, using the natural and greedy orderings.

Let $x = \begin{pmatrix} 1/4 \\ -2/3 \end{pmatrix}$. When the frame coefficients in their natural ordering, $x_n^{\text{nat}} = \langle x, e_n \rangle, 1 \leq n \leq N$, are used as input to the $\Sigma\Delta$ error diffusion algorithm we denote the quantized output as $\{q_n^{\text{nat}}\}_{n=1}^N$. When the frame coefficients in their greedy ordering, $x_n^{\text{gre}} = \langle x, e_{p(n)} \rangle, 1 \leq n \leq N$, are used as input to the $\Sigma\Delta$ error diffusion algorithm, we denote the quantized output as $\{q_n^{\text{gre}}\}_{n=1}^N$. We use linear reconstruction to recover $\widetilde{x}_{\text{nat}}$ and $\widetilde{x}_{\text{gre}}$ from the quantized coefficients by

$$\widetilde{x}_{\text{nat}} = \frac{2}{7} \sum_{n=1}^{7} q_n^{\text{nat}} e_n^7 \quad \text{and} \quad \widetilde{x}_{\text{gre}} = \frac{2}{7} \sum_{n=1}^{7} q_n^{\text{gre}} e_{p(n)}^7.$$

Figure 7 shows the reconstructed signals $\widetilde{x}_{\text{nat}}$ and $\widetilde{x}_{\text{gre}}$, along with $\widetilde{x}_{\text{MSQ}}$ and $\widetilde{x}_{\Sigma\Delta}$ from Example 5.2. The set Γ from (5.7) is also shown for perspective. Note that the $\Sigma\Delta$ error diffusion algorithm with greedy ordering selects the closest element of Γ. Interestingly, the $\Sigma\Delta$ error diffusion algorithm is outperformed by the standard first-order $\Sigma\Delta$ algorithm when run with the natural ordering.

FIGURE 7. Comparing various algorithms for $x = \left(\frac{1}{4}, \frac{-2}{3}\right)$

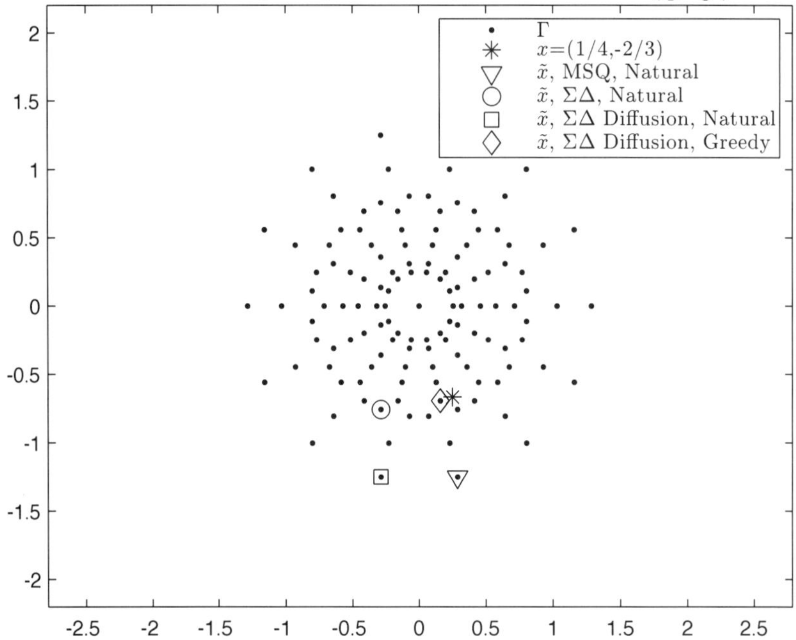

We repeat this experiment with the two different choices $x = \begin{pmatrix} 1/e \\ 1/\sqrt{3} \end{pmatrix}$ and $x = \begin{pmatrix} 1/3 \\ 1/2 \end{pmatrix}$, see Figure 8 and Figure 9 respectively. For both of these choices of x, the $\Sigma\Delta$ error diffusion algorithm in the greedy ordering outperforms the other methods.

6. Stability and Error Bounds

In this section, we prove stability bounds and error bounds for the generalized $\Sigma\Delta$ algorithm in Definition 5.3.

6.1. Stability. The notion of stability for $\Sigma\Delta$ algorithms refers to having quantitative control on the size of the state variables u_n in terms of the size of the input variables x_n. Stability is an important step towards obtaining mathematically rigorous error bounds, but also has practical implications for implementing $\Sigma\Delta$ algorithms in circuitry.

PROPOSITION 6.1. *Fix* $\{c_n\}_{n=1}^N \subset \mathbb{R}$ *and consider the generalized first order* $\Sigma\Delta$ *algorithm* (5.8) *with quantization alphabet* $\mathcal{A}_K^\delta$. *Suppose that* $M = \max_{1\leq n\leq N} |c_n| \leq 2K$. *If*

$$\forall n = 1, ..., N, \qquad |x_n| < \left(K - \frac{M}{2}\right)\delta, \tag{6.1}$$

then

$$\forall n = 0, ..., N, \qquad |u_n| \leq \frac{\delta}{2}. \tag{6.2}$$

FIGURE 8. Comparing various algorithms for $x = \left(\frac{1}{e}, \frac{1}{\sqrt{3}}\right)$

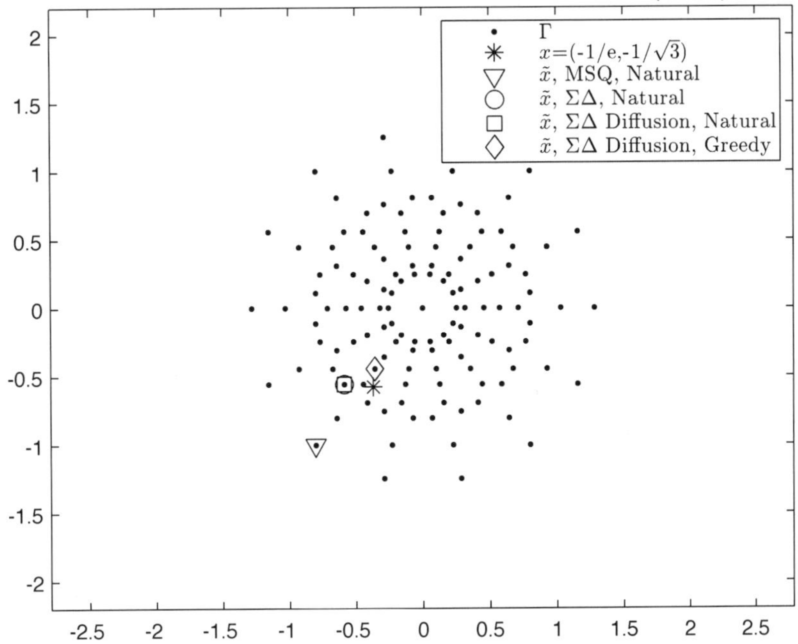

FIGURE 9. Comparing various algorithms for $x = \left(\frac{1}{3}, \frac{1}{2}\right)$

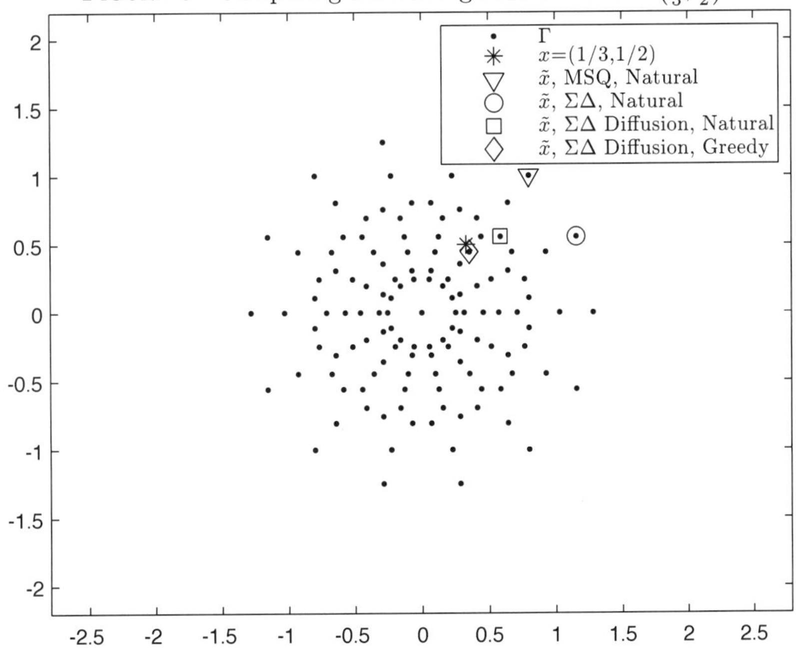

PROOF. The proof of (6.2) proceeds by induction. The base case $n = 0$ holds by the assumption in Definition 5.3 that $u_0 = 0$. Next, for the inductive step, suppose that $|u_{n-1}| \leq \frac{\delta}{2}$. By (6.1),

$$|u_{n-1}c_n + x_n| < \frac{M\delta}{2} + \left(K - \frac{M}{2}\right)\delta = K\delta.$$

Since Q is the scalar quantizer associated to $\mathcal{A}_K^\delta$ it follows that

$$|u_n| = |u_{n-1}c_n + x_n - q_n| = |u_{n-1}c_n + x_n - Q(u_{n-1}c_n + x_n)| \leq \frac{\delta}{2}.$$

□

In the case of standard first order $\Sigma\Delta$ quantization (when each $c_n = 1$), Proposition 6.1 recovers stability bounds from [**1**, **9**]. The work in [**4**] does not explicitly address stability for the $\Sigma\Delta$ error diffusion algorithm, but instead relies on either a version of the additive noise model or an a priori assumption that the quantizer is scaled to avoid overflow.

6.2. A basic error bound. In this section, we prove mathematically rigorous error bounds for the generalized $\Sigma\Delta$ algorithm and the $\Sigma\Delta$ error diffusion algorithm. The error bound in Theorem 6.3 includes the settings of [**1**] and [**4**] and uses a generalized notion of the frame variation to quantify how the frame ordering affects the performance of the algorithms.

DEFINITION 6.2. Suppose that $F = \{f_n\}_{n=1}^N \subset \mathbb{R}^d$, and $c = \{c_n\}_{n=1}^N \subset \mathbb{R}$, and that p is a permutation of $\{1, 2, ..., N\}$. The *generalized frame variation of F with respect to c and p is defined by*

$$\sigma(F, p, c) = \sum_{n=1}^{N-1} \left\| f_{p(n)} - c_{n+1} f_{p(n+1)} \right\|. \tag{6.3}$$

THEOREM 6.3. *Fix $c = \{c_n\}_{n=1}^N \subseteq \mathbb{R}$ and consider the generalized $\Sigma\Delta$ algorithm (5.8) with quantization alphabet $\mathcal{A}_K^\delta$. Let $E = \{e_n\}_{n=1}^N$ be a frame for $\mathbb{R}^d$ and let $F = \{f_n\}_{n=1}^N$ be an associated dual frame. Let $M = \max_{1\leq n\leq N} |c_n|$ and suppose that $x \in \mathbb{R}^d$ satisfies*

$$\|x\| < \delta\left(K - \frac{M}{2}\right)\left(\max_{1\leq n\leq N} \|e_n\|\right)^{-1}. \tag{6.4}$$

Let p be a permutation of $\{1, 2, ..., N\}$, and suppose that $x_n = \langle x, e_{p(n)}\rangle$ is the input to the algorithm in Definition 5.3, and that $\{q_n\}_{n=1}^N$ is the quantized output. If one linearly reconstructs $\widetilde{x} \in \mathbb{R}^d$ by

$$\widetilde{x} = \sum_{n=1}^N q_n f_{p(n)}, \tag{6.5}$$

then

$$\|x - \widetilde{x}\| \leq \frac{\delta}{2}\left(\sigma(F, p, c) + \|f_{p(N)}\|\right). \tag{6.6}$$

PROOF. A computation using (5.8) and summation by parts gives

$$\begin{aligned}
x - \widetilde{x} &= \sum_{n=1}^{N} x_n f_{p(n)} - \sum_{n=1}^{N} q_n f_{p(n)} \\
&= \sum_{n=1}^{N} (x_n - q_n) f_{p(n)} \\
&= \sum_{n=1}^{N} (u_n - u_{n-1} c_n) f_{p(n)} \\
&= \sum_{n=1}^{N-1} u_n (f_{p(n)} - c_{n+1} f_{p(n+1)}) + u_N f_{p(N)} - u_0 c_1 f_{p(1)} \\
&= \sum_{n=1}^{N-1} u_n (f_{p(n)} - c_{n+1} f_{p(n+1)}) + u_N f_{p(N)}. \qquad (6.7)
\end{aligned}$$

It follows from (6.4) that

$$\forall\, 1 \leq n \leq N, \qquad |x_n| = |\langle x, e_{p(n)} \rangle| \leq \|x\| \|e_{p(n)}\| < \left(K - \frac{M}{2}\right)\delta.$$

Therefore, $|u_n| \leq \frac{\delta}{2}$ holds for all $1 \leq n \leq N$ by Proposition 6.1. This, together with (6.7), gives

$$\begin{aligned}
\|x - \widetilde{x}\| &\leq \sum_{n=1}^{N-1} |u_n| \; \|f_{p(n)} - c_{n+1} f_{p(n+1)}\| + |u_N| \; \|f_{p(N)}\| \\
&\leq \frac{\delta}{2} \left(\sigma(F, p, c) + \|f_{p(N)}\| \right).
\end{aligned}$$

□

In the case of standard first order $\Sigma\Delta$ quantization, Theorem 6.3 includes error bounds from [**1**]. Theorem 6.3 also includes an error bound for the $\Sigma\Delta$ error diffusion algorithm from [**4**], but does so without appealing to additive noise models and without a priori assumptions that the quantizer is scaled to avoid overflow. The condition (6.4) gives an explicit condition on the quantizer input sequence which, in view of Proposition 6.1, rigorously guarantees that the quantizer does not saturate.

The case when $\{c_n\}_{n=1}^N$ is given by $c_n = c_n^\star$ in (5.11) is important because it leads to small upper bounds in Theorem 6.3.

DEFINITION 6.4. Suppose $F = \{f_n\}_{n=1}^N \subset \mathbb{R}^d$ satisfies $\|f_n\| > 0$ for each $1 \leq n \leq N$. Let p be a permutation of $\{1, 2, ..., N\}$. The *projection frame variation* of the frame F with respect to p is defined as

$$\sigma^\star(F, p) = \sum_{n=1}^{N-1} \left\| f_{p(n)} - \left\langle f_{p(n)}, \frac{f_{p(n+1)}}{\|f_{p(n+1)}\|^2} \right\rangle f_{p(n+1)} \right\|. \qquad (6.8)$$

The next result shows that taking $c_n = c_n^\star$ minimizes the generalized frame variation, and hence gives best (smallest) control on the term $\sigma(F, p, c)$ in the error bound (6.6), cf. [**4**].

PROPOSITION 6.5. *Fix* $F = \{f_n\}_{n=1}^N \subset \mathbb{R}^d$ *that satisfies* $\|f_n\| > 0$ *for each* $1 \leq n \leq N$. *Fix a permutation* p *of* $\{1, 2, ..., N\}$. *If* $c = \{c_n\}_{n=1}^N \subset \mathbb{R}$, *then*

$$\sigma^\star(F, p) \leq \sigma(F, p, c). \tag{6.9}$$

Moreover, equality holds in (6.9) *if and only if* $c = \{c_n\}_{n=1}^N$ *is given by* $c_n = c_n^\star$ *as in* (5.11).

PROOF. For each $1 \leq n \leq N-1$, let

$$T_n(\lambda) = \|f_{p(n)} - \lambda f_{p(n+1)}\|^2 = \|f_{p(n)}\|^2 - 2\lambda\langle f_{p(n)}, f_{p(n+1)}\rangle + \lambda^2\|f_{p(n+1)}\|^2.$$

A direct computation shows the unique minimizer of $T_n(\lambda)$ is given by

$$\lambda^* = \frac{\langle f_{p(n)}, f_{p(n+1)}\rangle}{\|f_{p(n+1)}\|^2}.$$

In particular, it can be checked that $T_n'(\lambda^*) = 0$ and $T_n''(\lambda^*) > 0$.

□

EXAMPLE 6.6. For $N \geq 3$, let $E_N = \{e_n^N\}_{n=1}^N \subset \mathbb{R}^2$ be the Nth roots of unity frame for $\mathbb{R}^2$, and let $F_N = \{f_n\}_{n=1}^N$ be the canonical dual frame $f_n = \frac{2}{N}e_n$. In this example, we compare the frame variation terms in the upper bound (6.6) for the standard first order $\Sigma\Delta$ algorithm and the first order $\Sigma\Delta$ error diffusion algorithm, when the identity permutation $p(n) = n$ of $\{1, 2, \cdots, N\}$ is used.

The standard first order $\Sigma\Delta$ algorithm corresponds to using Definition 5.3 where $c = \{c_n\}_{n=1}^N$ satisfies $c_n = 1, 1 \leq n \leq N$. To compute the generalized frame variation, note that for each $1 \leq n \leq N-1$,

$$\|f_n^N - c_{n+1}f_{n+1}^N\| = \frac{2}{N}\|e_n^N - e_{n+1}^N\| = \frac{2}{N}\sqrt{2 - 2\cos(2\pi/N)}.$$

Thus,

$$\begin{aligned}\sigma(F_N, p, c) &= \sum_{n=1}^{N-1} \|f_n^N - f_{n+1}^N\| \\ &= \frac{2}{N}\sum_{n=1}^{N-1} \sqrt{2 - 2\cos(2\pi/N)} \\ &= \frac{2(N-1)}{N}\sqrt{2 - 2\cos(2\pi/N)}. \qquad (6.10)\end{aligned}$$

The $\Sigma\Delta$ error diffusion algorithm corresponds to using $\{c_n^\star\}_{n=1}^N$ defined by (5.11). Note that for $1 \leq n \leq N-1$, one has

$$c_{n+1}^\star = \left\langle f_n, \frac{f_{n+1}}{\|f_{n+1}\|^2}\right\rangle = \langle e_n^N, e_{n+1}^N\rangle = \cos(2\pi/N).$$

A computation shows that for each $1 \leq n \leq N-1$,

$$\|f_n^N - c_{n+1}^\star f_{n+1}^N\| = \frac{2}{N}\|e_n^N - c_{n+1}^\star e_{n+1}^N\| = \frac{2}{N}\sin(2\pi/N).$$

Thus,

$$\sigma^\star(F_N,p) = \sigma(F_N,p,c^\star) = \sum_{n=1}^{N-1} \|f_n^N - c_{n+1}^\star f_{n+1}^N\|$$
$$= \frac{2}{N}\sum_{n=1}^{N-1} \sin(2\pi/N)$$
$$= \frac{2(N-1)}{N}\sin(2\pi/N). \tag{6.11}$$

Since $0 < \sin\theta < \sqrt{2-2\cos\theta}$ holds for $0<\theta<\pi$, it follows that

$$\sigma^\star(F_N,p) < \sigma(F_N,p,c). \tag{6.12}$$

However, in (6.12), the improvement of (6.11) over (6.10) in this example is quite modest since both are of order $1/N$, and are asymptotically the same

$$\lim_{N\to\infty} N\ \sigma(F_N,p,c) = 4\pi = \lim_{N\to\infty} N\ \sigma^\star(F_N,p). \tag{6.13}$$

Although the generalized frame variation only gives (generally non-sharp) upper bounds on the error in (6.6), in this example the equivalence in (6.13) is consistent with the intuition for error diffusion since $c_n^\star = \cos(2\pi/N) \approx 1 = c_n$ when N is large. Example 6.7 shows a situation where this is not the case.

Finally, note that for all $N \geq 3$,

$$\sigma(F_N,p,c) = \frac{2(N-1)}{N}\sqrt{2-2\cos(2\pi/N)} \leq \frac{4\pi}{N}. \tag{6.14}$$

Using (6.6), (6.12) and (6.14), both the standard first order $\Sigma\Delta$ algorithm and the $\Sigma\Delta$ error diffusion algorithm satisfy the error bound

$$\|x-\widetilde{x}\| \leq \frac{\delta}{2}\left(\frac{4\pi}{N}+\frac{2}{N}\right) = \frac{\delta(2\pi+1)}{N}. \tag{6.15}$$

EXAMPLE 6.7. Here we repeat Example 6.6 with the following modifications: (i) we only consider even values of $N \geq 4$, and (ii) instead of the identity permutation, we use the permutation $p = p_N$ defined by

$$p(2n-1) = n,\ 1 \leq n \leq \frac{N}{2}, \quad \text{and} \quad p(2n) = \frac{N}{2}+n,\ 1\leq n\leq \frac{N}{2}.$$

For example, when $N=8$, this permutation corresponds to the frame ordering

$$e_1^8, e_5^8, e_2^8, e_6^8, e_3^8, e_7^8, e_4^8, e_8^8.$$

Moreover, it can be checked that the permutation p provides a greedy ordering of E_N and F_N.

The standard first order $\Sigma\Delta$ algorithm corresponds to using Definition 5.3 where $c = \{c_n\}_{n=1}^N$ satisfies $c_n = 1, 1 \leq n \leq N$. To compute the generalized frame variation, similar computations as in Example 6.6 show that for each $1 \leq n \leq N-1$,

(6.16)

$$\|f_{p(n)}^N - c_{n+1} f_{p(n+1)}^N\| = \frac{2}{N}\|e_{p(n)}^N - e_{p(n+1)}^N\| = \begin{cases} \frac{4}{N}, & \text{if } n \in 2\mathbb{Z}+1, \\ \frac{2}{N}\sqrt{2+2\cos(2\pi/N)}, & \text{if } n\in 2\mathbb{Z}. \end{cases}$$

Thus,

$$\begin{aligned}\sigma(F_N,p,c) &= \sum_{n=1}^{N-1} \|f^N_{p(n)} - f^N_{p(n+1)}\| \\ &= \left(\frac{N}{2}\right)\frac{4}{N} + \left(\frac{N}{2}-1\right)\frac{2}{N}\sqrt{2+2\cos(2\pi/N)} \\ &= 2 + \left(\frac{N-2}{N}\right)\sqrt{2+2\cos(2\pi/N)}. \qquad (6.17)\end{aligned}$$

The $\Sigma\Delta$ error diffusion algorithm corresponds to using $\{c^\star_n\}_{n=1}^N$ defined by (5.11). Similar computations as in Example 6.6 show that for $1 \leq n \leq N-1$, one has

$$c^\star_{n+1} = \left\langle f_{p(n)}, \frac{f_{p(n+1)}}{\|f_{p(n+1)}\|^2}\right\rangle = \langle e^N_{p(n)}, e^N_{p(n+1)}\rangle = \begin{cases} -1, & \text{if } n \in 2\mathbb{Z}+1, \\ -\cos(2\pi/N), & \text{if } n \in 2\mathbb{Z}. \end{cases}$$

A similar computation as in Example 6.6 shows that

$$\|f^N_{p(n)} - c^\star_{n+1} f^N_{p(n+1)}\| = \begin{cases} 0, & \text{if } n \in 2\mathbb{Z}+1, \\ \frac{2}{N}\sin(2\pi/N), & \text{if } n \in 2\mathbb{Z}. \end{cases}$$

Thus,

$$\begin{aligned}\sigma^\star(F_N,p) &= \sum_{n=1}^{N-1} \|f^N_{p(n)} - c^\star_{n+1} f^N_{p(n+1)}\| \\ &= \left(\frac{N}{2}-1\right)\frac{2}{N}\sin(2\pi/N) \\ &= \left(\frac{N-2}{N}\right)\sin(2\pi/N). \qquad (6.18)\end{aligned}$$

Note that for the $\Sigma\Delta$ error diffusion algorithm,

$$\lim_{N\to\infty} N\,\sigma^\star(F_N,p) = 2\pi, \qquad (6.19)$$

so that $\sigma^\star(F_N,p)$ is of the order $1/N$. Moreover, the constant 2π in (6.19) for the greedy ordering is smaller than the constant 4π in (6.13), for the natural ordering. However, for the standard $\Sigma\Delta$ algorithm, one has the weaker result

$$\lim_{N\to\infty} \sigma(F_N,p,c) = 4.$$

Finally, since $\sigma^\star(F_N,p) \leq 2\pi/N$, the error bound (6.6) for the $\Sigma\Delta$ error diffusion algorithm yields

$$\|x - \widetilde{x}\| \leq \frac{\delta}{2}\left(\frac{2\pi}{N} + \frac{2}{N}\right) = \frac{\delta(\pi+1)}{N}. \qquad (6.20)$$

Using (6.17),

$$\sigma(F_N,p,c) = 2 + \left(\frac{N-2}{N}\right)\sqrt{2+2\cos(2\pi/N)} \leq 4. \qquad (6.21)$$

Using (6.21), and the error bound (6.6) for the standard $\Sigma\Delta$ algorithm yields

$$\|x - \widetilde{x}\| \leq \frac{\delta}{2}\left(4 + \frac{2}{N}\right) = \delta\left(2 + \frac{1}{N}\right). \qquad (6.22)$$

Compared to (6.20), the bound (6.22) is quite weak since it does not vanish as $N \to \infty$.

Although (6.20) and (6.22) are simply upper bounds on the error, both are consistent with the actual performance that is observed in the numerical experiments of Example 6.11. In particular, the standard $\Sigma\Delta$ algorithm performs poorly with the greedy ordering, but the $\Sigma\Delta$ error diffusion algorithm performs well, see also Examples 6.8 and 6.9.

EXAMPLE 6.8. We draw $\{z_j\}_{j=1}^{10000} \subset \mathbb{R}^2$ independently at random according to the uniform distribution on $\{z \in \mathbb{R}^2 : \|z\| \leq 1\}$. Let $\{e_n^7\}_{n=1}^7 \subset \mathbb{R}^2$ be the 7th roots of unity frame for $\mathbb{R}^2$, taken in the natural ordering, so that each z_j has the representation

$$z_j = \frac{2}{7}\sum_{n=1}^{7} \langle z_j, e_n \rangle e_n.$$

For each $1 \leq j \leq 10000$, the frame coefficients $x_n^j = \langle z_j, e_n \rangle, 1 \leq n \leq N$, are given as input to the standard first order $\Sigma\Delta$ algorithm with quantization alphabet

$$\mathcal{A}_4^{1/4} = \left\{ \frac{-7}{8}, \frac{-5}{8}, \frac{-3}{8}, \frac{-1}{8}, \frac{1}{8}, \frac{3}{8}, \frac{5}{8}, \frac{7}{8} \right\},$$

to obtain the quantized output $\{q_n^j\}_{n=1}^N$. We then reconstruct linearly by

$$\widetilde{z}_j = \frac{2}{7}\sum_{n=1}^{7} q_n^j e_n^7.$$

Figure 10 shows a frequency chart of the computed errors $\|z_j - \widetilde{z}_j\|, 1 \leq j \leq 10000$.

Next, we repeat this experiment, but instead use the $\Sigma\Delta$ error diffusion algorithm. Figure 11 shows a frequency chart of the computed errors for the $\Sigma\Delta$ error diffusion algorithm.

Figures 10 and 11 show that the $\Sigma\Delta$ error diffusion algorithm outperforms the standard $\Sigma\Delta$ algorithm.

FIGURE 10. Approximation errors for the standard first order $\Sigma\Delta$ algorithm using the natural ordering

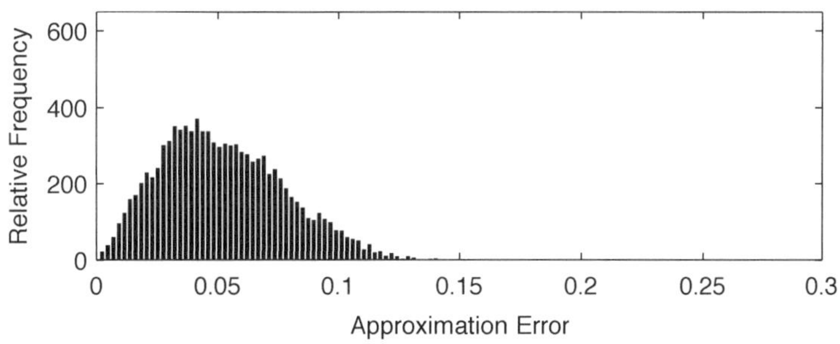

EXAMPLE 6.9. Here, we repeat Example 6.8, but instead use the greedy ordering given in Figure 6. Figure 12 shows a frequency chart of the computed errors for the standard $\Sigma\Delta$ algorithm with the greedy ordering. Figure 13 shows a frequency

FIGURE 11. Approximation errors for the $\Sigma\Delta$ error diffusion algorithm using the natural ordering

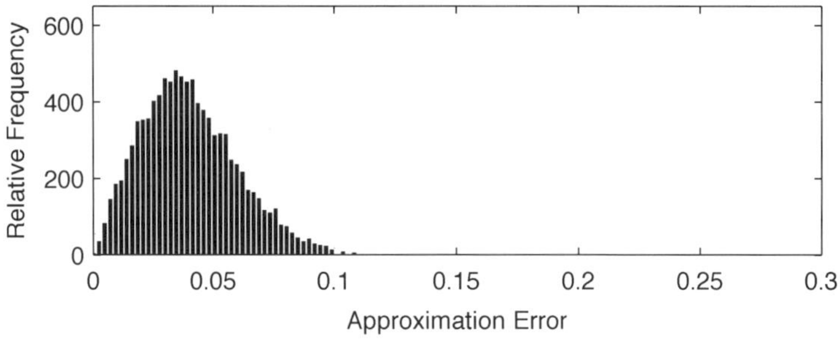

FIGURE 12. Approximation errors for the standard first order $\Sigma\Delta$ algorithm using the greedy ordering.

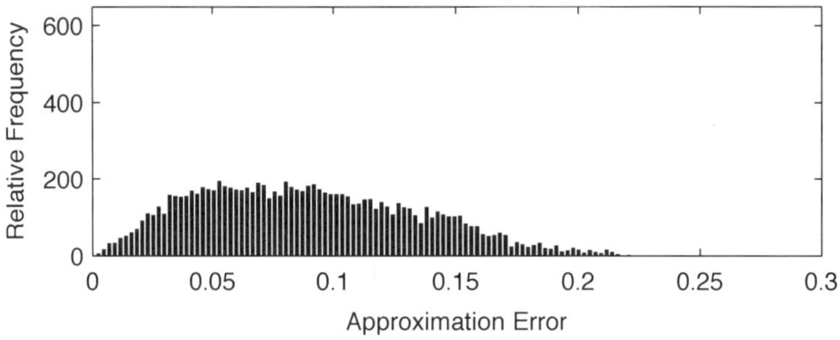

FIGURE 13. Approximation errors for the $\Sigma\Delta$ error diffusion algorithm using a greedy ordering.

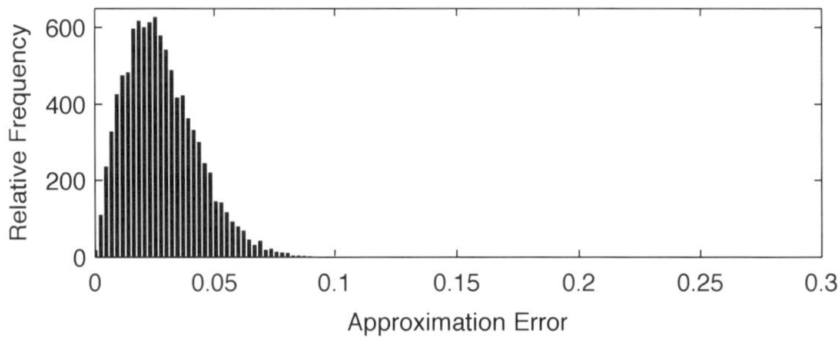

chart of the computed errors for the error diffusion algorithm with the greedy ordering. In this case, the standard $\Sigma\Delta$ algorithm performs especially poorly compared to the $\Sigma\Delta$ error diffusion algorithm. This is consistent with observations of Example 6.7 and Example 6.11.

EXAMPLE 6.10. Let $x = \begin{pmatrix} 1/\pi \\ \sqrt{3/17} \end{pmatrix}$. Let $\{e_n^N\}_{n=1}^N \subset \mathbb{R}^2$ be the Nth roots of unity frame for $\mathbb{R}^2$, taken in the natural ordering, so that

$$x = \frac{2}{N} \sum_{n=1}^{N} \langle x, e_n^N \rangle e_n^N.$$

For each N, we use the frame coefficients $x_n^N = \langle x, e_n^N \rangle, 1 \leq n \leq N$ as the input to the standard $\Sigma\Delta$ algorithm (5.1), with quantization alphabet $\mathcal{A}_2^1 = \{-1, 1\}$, to obtain the quantized output $\{q_n^N\}_{n=1}^N$. We then reconstruct linearly by

$$\widetilde{x}_N = \frac{2}{N} \sum_{n=1}^{N} q_n^N e_n^N.$$

Figure 14 shows a log-log plot of the error $\|x - \widetilde{x}_N\|$ against the frame size N. For details about the splitting present in Figure 14, and refined error bounds, see [**1**].

Next, we repeat this experiment, but instead use the $\Sigma\Delta$ error diffusion algorithm. Figure 15 shows a log-log plot of the error against the frame size.

FIGURE 14. Approximation errors (log-log scale) of the vector $\left(\frac{1}{\pi}, \sqrt{\frac{3}{17}}\right)$, for the standard $\Sigma\Delta$ algorithm (natural ordering) and various N^{th} roots of unity. See Example 6.10.

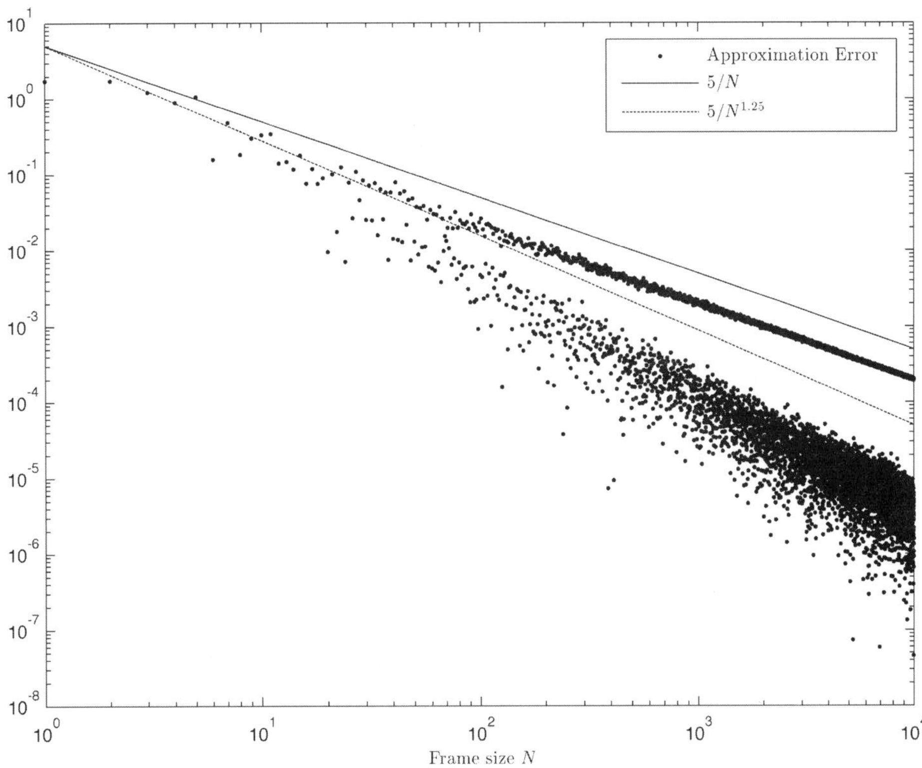

FIGURE 15. Approximation errors (log-log scale) of the vector $\left(\frac{1}{\pi}, \sqrt{\frac{3}{17}}\right)$, for the $\Sigma\Delta$ error diffusion algorithm (natural ordering) and various N^{th} roots of unity. See Example 6.10.

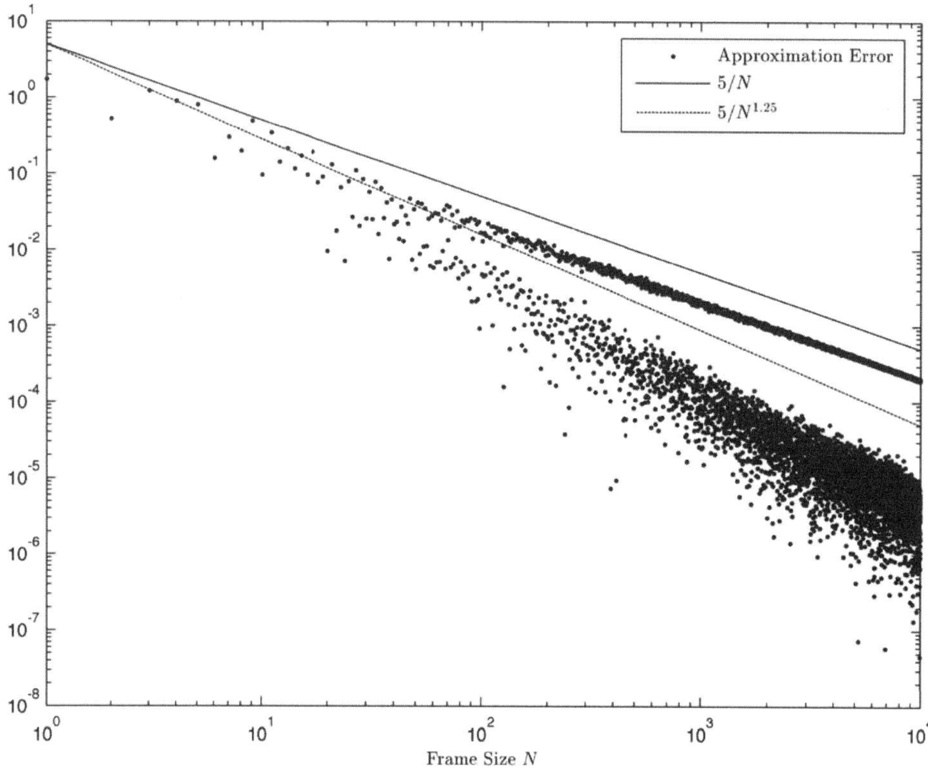

EXAMPLE 6.11. Here, we repeat Example 6.10 with the following changes: (1) only even values of the frame size N are considered, and (2) the greedy frame ordering from Example 6.7 is used. Figure 16 shows a log-log plot of the error against the frame size (only with N even) for the standard $\Sigma\Delta$ algorithm. Figure 17 shows a log-log plot of the error against the frame size for the $\Sigma\Delta$ error diffusion algorithm.

7. Acknowledgements

A. Powell was supported by NSF DMS Grant 1211687, and also gratefully acknowledges the hospitality and support of the Academia Sinica Institute of Mathematics (Taipei, Taiwan). Ö. Yılmaz was funded in part by a Natural Sciences and Engineering Research Council of Canada (NSERC) Discovery Grant (22R82411), an NSERC Accelerator Award (22R68054) and an NSERC Collaborative Research and Development Grant DNOISE II (22R07504).

FIGURE 16. Approximation errors (log-log scale) of the vector $\left(\frac{1}{\pi}, \sqrt{\frac{3}{17}}\right)$, for the regular $\Sigma\Delta$ algorithm (greedy ordering) and various even N^{th} roots of unity. See Example 6.11.

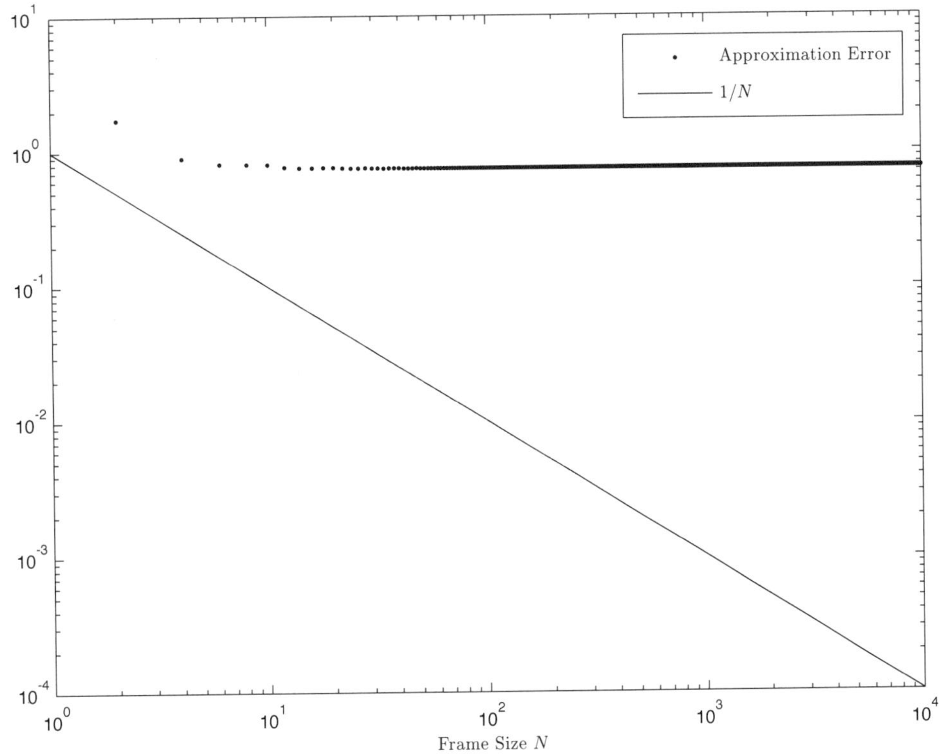

References

[1] J. J. Benedetto, A. M. Powell, and Ö. Yılmaz, *Sigma-Delta ($\Sigma\Delta$) quantization and finite frames*, IEEE Trans. Inform. Theory **52** (2006), no. 5, 1990–2005, DOI 10.1109/TIT.2006.872849. MR2234460 (2007a:94030)

[2] J. Blum, M. Lammers, A. M. Powell, and Ö. Yılmaz, *Sobolev duals in frame theory and sigma-delta quantization*, J. Fourier Anal. Appl. **16** (2010), no. 3, 365–381, DOI 10.1007/s00041-009-9105-x. MR2643587 (2011e:42062)

[3] B. G. Bodmann and V. I. Paulsen, *Frame paths and error bounds for sigma-delta quantization*, Appl. Comput. Harmon. Anal. **22** (2007), no. 2, 176–197, DOI 10.1016/j.acha.2006.05.010. MR2295294 (2007m:94108)

[4] P. T. Boufounos and A. V. Oppenheim, *Quantization noise shaping on arbitrary frame expansions*, EURASIP Journal on Applied Signal Processing, Special issue on Frames and Overcomplete Representations in Signal Processing, Communications, and Information Theory **2006** (2006), Article ID 53807, 12 pages.

[5] P. G. Casazza and J. Kovačević, *Equal-norm tight frames with erasures*, Adv. Comput. Math. **18** (2003), no. 2-4, 387–430, DOI 10.1023/A:1021349819855. Frames. MR1968127 (2004e:42046)

[6] *Finite frames*, Applied and Numerical Harmonic Analysis, Birkhäuser/Springer, New York, 2013. Theory and applications; Edited by Peter G. Casazza and Gitta Kutyniok. MR2964005

FIGURE 17. Approximation errors (log-log scale) of the vector $\left(\frac{1}{\pi}, \sqrt{\frac{3}{17}}\right)$, for the $\Sigma\Delta$ error diffusion algorithm (greedy ordering) and various even N^{th} roots of unity. See Example 6.11

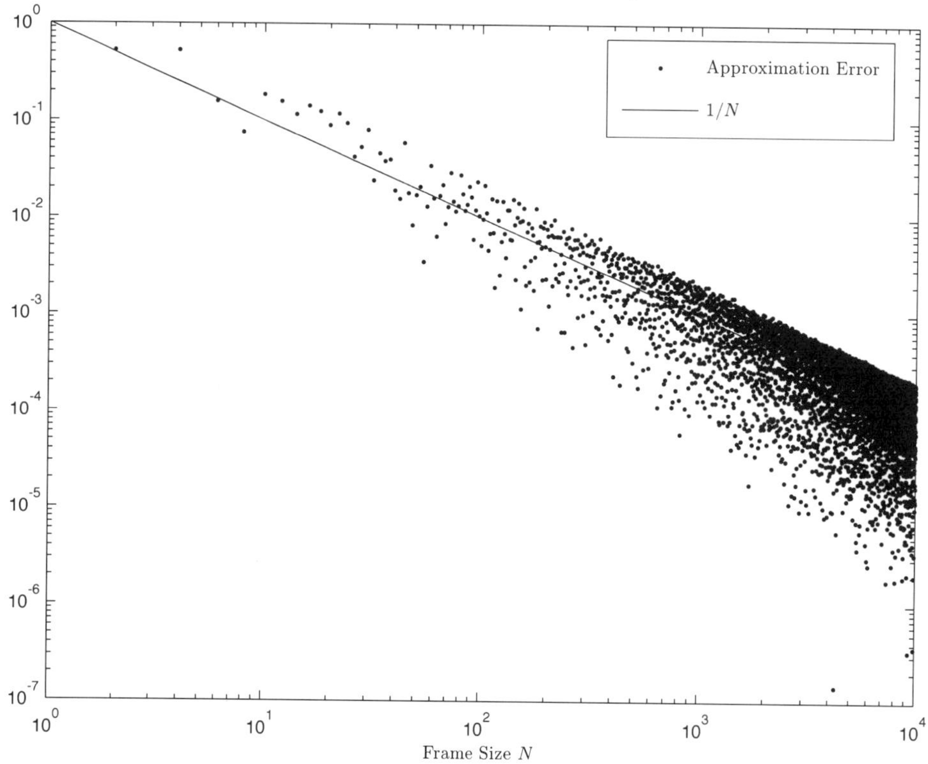

[7] E. Chou, C. Sinan Güntürk, F. Krahmer, R. Saab, and Ö. Yılmaz, *Noise-shaping quantization methods for frame-based and compressive sampling systems*, preprint (2015), arXiv:1502.05807.

[8] O. Christensen, *An introduction to frames and Riesz bases*, Applied and Numerical Harmonic Analysis, Birkhäuser Boston, Inc., Boston, MA, 2003. MR1946982 (2003k:42001)

[9] I. Daubechies and R. DeVore, *Approximating a bandlimited function using very coarsely quantized data: a family of stable sigma-delta modulators of arbitrary order*, Ann. of Math. (2) **158** (2003), no. 2, 679–710, DOI 10.4007/annals.2003.158.679. MR2018933 (2004i:42033)

[10] V. K. Goyal, M. Vetterli, and N. T. Thao, *Quantized overcomplete expansions in* $\mathbf{R}^N$*: analysis, synthesis, and algorithms*, IEEE Trans. Inform. Theory **44** (1998), no. 1, 16–31, DOI 10.1109/18.650985. MR1486646 (99a:94004)

[11] C. S. Güntürk, M. Lammers, A. M. Powell, R. Saab, and Ö. Yılmaz, *Sobolev duals for random frames and* $\Sigma\Delta$ *quantization of compressed sensing measurements*, Found. Comput. Math. **13** (2013), no. 1, 1–36, DOI 10.1007/s10208-012-9140-x. MR3009528

[12] C. S. Güntürk, *One-bit sigma-delta quantization with exponential accuracy*, Comm. Pure Appl. Math. **56** (2003), no. 11, 1608–1630, DOI 10.1002/cpa.3044. MR1995871 (2005f:94058)

[13] D. Jimenez, L. Wang, and Y. Wang, *White noise hypothesis for uniform quantization errors*, SIAM J. Math. Anal. **38** (2007), no. 6, 2042–2056 (electronic), DOI 10.1137/050636929. MR2299440 (2008a:94059)

[14] F. Krahmer, R. Saab, and Ö. Yilmaz, *Sigma-Delta quantization of sub-Gaussian frame expansions and its application to compressed sensing*, Inf. Inference **3** (2014), no. 1, 40–58, DOI 10.1093/imaiai/iat007. MR3311448

[15] M. Lammers, A. M. Powell, and Ö. Yılmaz, *Alternative dual frames for digital-to-analog conversion in sigma-delta quantization*, Adv. Comput. Math. **32** (2010), no. 1, 73–102, DOI 10.1007/s10444-008-9088-1. MR2574568 (2010m:42073)

[16] S.R. Norsworthy, R. Schreier, and G.C. Temes (eds.), *Delta-sigma-converters: Theory, design and simulation*, Wiley-IEEE, 1996.

[17] A. M. Powell, R. Saab, and Ö. Yılmaz, *Quantization and finite frames*, Finite frames, Appl. Numer. Harmon. Anal., Birkhäuser/Springer, New York, 2013, pp. 267–302, DOI 10.1007/978-0-8176-8373-3_8. MR2964013

[18] R. Saab, R. Wong, and Ö. Yılmaz, *Quantization of compressive samples with stable and robust recovery*, preprint (2015), arXiv:1504.00087.

[19] Ö. Yılmaz, *Coarse quantization of highly redundant time-frequency representations of square-integrable functions*, Appl. Comput. Harmon. Anal. **14** (2003), no. 2, 107–132, DOI 10.1016/S1063-5203(03)00020-4. MR1981204 (2004f:94029)

Department of Mathematics, Louisiana State University, Baton Rouge, Louisiana 70803
E-mail address: fdunke1@lsu.edu

Department of Mathematics, Vanderbilt University, Nashville, Tennessee 37240
E-mail address: alexander.m.powell@vanderbilt.edu

Department of Mathematics, Huntingdon College, Montgomery, Alabama 36016
E-mail address: aspaeth@hawks.huntingdon.edu

Department of Mathematics, University of British Columbia, Vancouver, BC V6T 1Z2, Canada
E-mail address: oyilmaz@math.ubc.ca

Proceedings of Symposia in Applied Mathematics
Volume **73**, 2016
http://dx.doi.org/10.1090/psapm/073/00632

Frames and Phaseless Reconstruction

Radu Balan

ABSTRACT. Frame design for phaseless reconstruction is now part of the broader problem of nonlinear reconstruction and is an emerging topic in harmonic analysis. The problem of phaseless reconstruction can be simply stated as follows. Given the magnitudes of the coefficients generated by a linear redundant system (frame), we want to reconstruct the unknown input. This problem first occurred in X-ray crystallography starting in the early 20th century. The same nonlinear reconstruction problem shows up in speech processing, particularly in speech recognition.

In this lecture we shall cover existing analysis results as well as stability bounds for signal recovery including: necessary and sufficient conditions for injectivity, Lipschitz bounds of the nonlinear map and its left inverses, stochastic performance bounds, and algorithms for signal recovery.

1. Introduction

These lecture notes concern the problem of finite dimensional vector reconstruction from magnitudes of frame coefficients.

Variants of this problem appear in several areas of engineering and science. In particular in X-ray crystallography one measures the magnitudes of the Fourier transform of the electron density from which one infers the atomic structure of the crystal [**Fin82**]. In speech processing, automatic speech recognition engines typically use cepstral coefficients, which are absolute values of linear combinations of the short time Fourier transform coefficients [**HLO80, Ba10**]. The phaseless reconstruction problem unifies these and other similar problems [**BCE06**].

While the problem can be stated in the more general context of infinite dimensional Hilbert spaces, in these lectures we focus exclusively on the finite dimensional case. In this case any spanning set is a frame (see [**Ca00**] for a complete definition and list of properties). Specifically let $H = \mathbb{C}^n$ denote the n dimensional complex Hilbert space and let $\mathcal{F} = \{f_1, \dots, f_m\}$ be a set of $m \geq n$ vectors that span H. Fix a real linear space V, that is also a subset of H, $V \subset H$. Our problem is to study when a vector $x \in V$ can be reconstructed from magnitudes of its frame coefficients $\{|\langle x, f_k\rangle|, 1 \leq k \leq m\}$ and how to do so efficiently. This setup covers both the real case and the complex case as studied before in literature: in the real case

2010 *Mathematics Subject Classification.* 15A29, 65H10, 90C26.

Key words and phrases. Frames, phase retrieval, nonlinear reconstruction, projective space, Lipschitz analysis, Fisher Information Matrix, Cramer-Rao Lower Bound.

Financial support from NSF Grants DMS-1109498 and DMS-1413249 is gratefully acknowledged.

$\mathcal{F} \subset V = \mathbb{R}^n$; in the complex case $V = H = \mathbb{C}^n$. Note we assume V is a real linear space which may not be closed under multiplication with complex scalars. While the analysis both of the real case and complex case is presented in a unified way, the reader should be aware that the two cases are not equally easy. In particular the geometric criterion (4) of Theorem 3.2 does not have a counterpart in the complex case.

Consider the following additional notation. Let

$$T : H \to \mathbb{C}^m \ , \ (T(x))_k = \langle x, f_k \rangle \ , \ 1 \le k \le m \tag{1.1}$$

denote the frame *analysis map*. Its adjoint is called the *synthesis map* and is defined by

$$T^* : \mathbb{C}^m \to H \ , \ T^*(c) = \sum_{k=1}^{m} c_k f_k. \tag{1.2}$$

We now define the main nonlinear function we discuss in this paper

$$x \mapsto (|\langle x, f_k \rangle|)_{1 \le k \le m}.$$

For two vectors $x, y \in H$, consider the equivalence relation $x \sim y$ if and only if there is a constant c of magnitude 1 so that $x = cy$. Thus $x \sim y$ if and only if $x = e^{i\varphi} y$ for some real φ. Let $\hat{H} = H/\sim$ denote the quotient space. Note the nonlinear map is well defined on $\hat{H}$ since $|\langle cx, f_k \rangle| = |\langle x, f_k \rangle|$ for all scalars c with $|c| = 1$. We let α denote the quotient map

$$\alpha : \hat{H} \to \mathbb{R}^m \ , \ (\alpha(x))_k = |\langle x, f_k \rangle| \ , \ 1 \le k \le m. \tag{1.3}$$

For purposes that will become clear later, let us also define the map

$$\beta : \hat{H} \to \mathbb{R}^m \ , \ (\beta(x))_k = |\langle x, f_k \rangle|^2 \ , \ 1 \le k \le m. \tag{1.4}$$

For the subspace V denote by $\hat{V}$ the set of equivalence classes $\hat{V} = \{\hat{x} \ , \ x \in V\}$.

Definition 1.1. The frame $\mathcal{F}$ is called a *phase retrievable frame with respect to a set* V if the restriction $\alpha|_{\hat{V}}$ is injective.

In these lecture notes we study the following problems:

(1) Find necessary and sufficient conditions for $\alpha|_{\hat{V}}$ to be a one-to-one (injective) map;
(2) Study Lipschitz properties of the maps α, β and their inverses;
(3) Study robustness guarantees (such as Cramer-Rao Lower Bounds) for any inversion algorithm;
(4) Recovery using convex algorithms (Linear Tensor recovery, and PhaseLift);
(5) Recovery using iterative algorithms (Gerchberg-Saxton, Wirtinger flow, regularized least-squares).

2. Geometry of $\hat{H}$ and $\mathcal{S}^{p,q}$ Spaces

2.1. $\hat{H}$. Recall $\hat{H} = \widehat{\mathbb{C}^n} = \mathbb{C}^n/\sim = \mathbb{C}^n/T^1$ where $T^1 = \{z \in \mathbb{C} \ , \ |z| = 1\}$. Algebraically $\widehat{\mathbb{C}^n}$ is a homogeneous space invariant to multiplications by positive real scalars. In particular any $x \in \widehat{\mathbb{C}^n} \setminus \{0\}$ has a unique decomposition $x = rp$, where $r = \|x\| > 0$ and $p \in \mathbb{CP}^{n-1}$ is in the projective space $\mathbb{CP}^{n-1} = \mathbb{P}(\mathbb{C}^n)$. Thus, topologically,

$$\widehat{\mathbb{C}^n} = \{0\} \cup \left((0, \infty) \times \mathbb{CP}^{n-1}\right).$$

The subset

$$\overset{\circ}{\widehat{\mathbb{C}^n}} = \widehat{\mathbb{C}^n} \setminus \{0\} = (0,\infty) \times \mathbb{CP}^{n-1}$$

is a real analytic manifold.

Now consider the set $\hat{V}$ of equivalence classes associated to vectors in V. Similar to $\hat{H}$, $\hat{V}$ admits the following decomposition

$$\hat{V} = \{0\} \cup ((0,\infty) \times \mathbb{P}(V)),$$

where $\mathbb{P}(V) = \{\,\{zx\ ,\ z \in \mathbb{C}\}\ ,\ x \in V, x \neq 0\}$ denotes the *projective space associated to* V. The interior subset

$$\overset{\circ}{\hat{V}} = \hat{V} \setminus \{0\} = (0,\infty) \times \mathbb{P}(V)$$

is a real analytic manifold of (real) dimension $1 + \dim_{\mathbb{R}} \mathbb{P}(V)$.

Two important cases are as follows:

- Real case. $V = \mathbb{R}^n$ embedded as $x \in \mathbb{R}^n \mapsto x + i0 \in \mathbb{C}^n = H$. Then two vectors $x, y \in V$ are $\sim$ equivalent if and only if $x = y$ or $x = -y$. Similarly, the projective space $\mathbb{P}(V)$ is diffeomorphically equivalent to the real projective space $\mathbb{RP}^{n-1}$ which is of (real) dimension $n-1$. Thus

$$\dim_{\mathbb{R}}(\overset{\circ}{\hat{V}}) = n.$$

- Complex case. $V = \mathbb{C}^n$ which has real dimension $2n$. Then the projective space $\mathbb{P}(V) = \mathbb{CP}^{n-1}$ has real dimension $2n-2$ (it is also a Khäler manifold) and thus

$$\dim_{\mathbb{R}}(\overset{\circ}{\hat{V}}) = 2n-1.$$

The significance of the real dimension of $\overset{\circ}{\hat{V}}$ is encoded in the following result:

THEOREM 2.1 ([**BCE06**]). *If $m \geq 1 + \dim_{\mathbb{R}}(\overset{\circ}{\hat{V}})$ then for a (Zariski) generic frame $\mathcal{F}$ of m elements, the set of vectors $x \in V$ such that $\alpha^{-1}(\alpha(\hat{x}))$ has one point in $\hat{V}$ has dense interior in V.*

The real case of this result is contained in Theorem 2.9, whereas the complex case is contained in Theorem 3.4. Both can be found in [**BCE06**].

2.2. $\mathcal{S}^{p,q}$**.** Consider now $\mathrm{Sym}(H) = \{T : \mathbb{C}^n \to \mathbb{C}^n\ ,\ T = T^*\}$, the real vector space of self-adjoint operators over $H = \mathbb{C}^n$ endowed with the Hilbert-Schmidt scalar product $\langle T, S\rangle_{HS} = \mathrm{trace}(TS)$. We also use the notation $\mathrm{Sym}(W)$ for the real vector space of symmetric operators over a (real or complex) vector space W. In both cases self-adjoint means the operator T satisfies $\langle Tx, y\rangle = \langle x, Ty\rangle$ for every x, y in the underlying vector space W. T^* means the adjoint operator of T, and therefore the transpose conjugate of T, when T is a matrix. When T is a an operator acting on a real vector space, T^T denotes its adjoint. For two vectors $x, y \in \mathbb{C}^n$ we denote

$$[\![x, y]\!] = \frac{1}{2}(xy^* + yx^*) \in \mathrm{Sym}(\mathbb{C}^n), \tag{2.1}$$

their symmetric outer product. On $\mathrm{Sym}(H)$ and $B(H) = \mathbb{C}^{n\times n}$ we consider the class of p-norms defined by the p-norm of the vector of singular values:

$$\|T\|_p = \begin{cases} \max_{1\leq k\leq n} \sigma_k(T) & for \quad p = \infty \\ \left(\sum_{k=1}^n \sigma_k^p\right)^{1/p} & for \quad 1 \leq p < \infty \end{cases}, \tag{2.2}$$

where $\sigma_k = \sqrt{\lambda_k(T^*T)}$, $1 \leq k \leq n$, are the singular values of T, with $\lambda_k(S)$, $1 \leq k \leq n$, denoting the eigenvalues of S.

Fix two integers $p, q \geq 0$ and set

$$\begin{aligned} \mathcal{S}^{p,q}(H) &= \{T \in \mathrm{Sym}(H)\ ,\ T \text{ has at most } p \text{ strictly positive eigenvalues} \\ &\quad \text{and at most } q \text{ strictly negative eigenvalues}\}, \end{aligned} \tag{2.3}$$

$$\begin{aligned} \mathring{\mathcal{S}}^{p,q}(H) &= \{T \in \mathrm{Sym}(H)\ ,\ T \text{ has exactly } p \text{ strictly positive eigenvalues} \\ &\quad \text{and exactly } q \text{ strictly negative eigenvalues}\}. \end{aligned} \tag{2.4}$$

For instance $\mathring{\mathcal{S}}^{0,0}(H) = \mathcal{S}^{0,0}(H) = \{0\}$ and $\mathring{\mathcal{S}}^{1,0}(H)$ is the set of all non-negative operators of rank exactly one. When there is no confusion we shall drop the underlying vector space $H = \mathbb{C}^n$ from notation.

The following basic properties can be found in [**Ba13**], Lemma 3.6; in fact, the last statement is a special instance of the Witt's decomposition theorem.

LEMMA 2.2.

(1) *For any $p_1 \leq p_2$ and $q_1 \leq q_2$, $\mathcal{S}^{p_1,q_1} \subset \mathcal{S}^{p_2,q_2}$;*

(2) *For any nonnegative integers p, q the following disjoint decomposition holds true*

$$\mathcal{S}^{p,q} = \cup_{r=0}^{p} \cup_{s=0}^{q} \mathring{\mathcal{S}}^{r,s}, \tag{2.5}$$

where by convention $\mathring{\mathcal{S}}^{p,q} = \emptyset$ for $p + q > n$.

(3) *For any $p, q \geq 0$,*

$$-\mathcal{S}^{p,q} = \mathcal{S}^{q,p}. \tag{2.6}$$

(4) *For any linear operator $T : H \to H$ (symmetric or not, invertible or not) and nonnegative integers p, q,*

$$T^*\mathcal{S}^{p,q}T \subset \mathcal{S}^{p,q}. \tag{2.7}$$

However if T is invertible then $T^*\mathcal{S}^{p,q}T = \mathcal{S}^{p,q}$.

(5) *For any nonnegative integers p, q, r, s,*

$$\mathcal{S}^{p,q} + \mathcal{S}^{r,s} = \mathcal{S}^{p,q} - \mathcal{S}^{s,r} = \mathcal{S}^{p+r,q+s}. \tag{2.8}$$

The spaces $\mathcal{S}^{1,0}$ and $\mathcal{S}^{1,1}$ play a special role in the following section. Next we summarize their properties (see Lemmas 3.7 and 3.9 in [**Ba13**], and the comment after Lemma 9 in [**BCMN13**]).

LEMMA 2.3 (Space $\mathcal{S}^{1,0}$). *The following statements hold true:*

(1) $\mathring{\mathcal{S}}^{1,0} = \{xx^*\ ,\ x \in H, x \neq 0\}$;

(2) $\mathcal{S}^{1,0} = \{xx^*\ ,\ x \in H\} = \{0\} \cup \{xx^*\ ,\ x \in H, x \neq 0\}$;

(3) *The set $\mathring{\mathcal{S}}^{1,0}$ is a real analytic manifold in $\mathrm{Sym}(\mathbb{C}^n)$ of real dimension $2n-1$. As a real manifold, its tangent space at $X = xx^*$ is given by*

$$T_X\mathring{\mathcal{S}}^{1,0} = \left\{ [\![x, y]\!] = \frac{1}{2}(xy^* + yx^*)\ ,\ y \in \mathbb{C}^n \right\}. \tag{2.9}$$

The $\mathbb{R}$-linear embedding $\mathbb{C}^n \mapsto T_X\mathring{\mathcal{S}}^{1,0}$ given by $y \mapsto [\![x, y]\!]$ has null space $\{iax\ ,\ a \in \mathbb{R}\}$.

LEMMA 2.4 (Space $\mathcal{S}^{1,1}$). *The following statements hold true:*

(1) $\mathcal{S}^{1,1} = \mathcal{S}^{1,0} - \mathcal{S}^{1,0} = \mathcal{S}^{1,0} + \mathcal{S}^{0,1} = \{[\![x, y]\!]\ ,\ x, y \in H\}$;

(2) *For any vectors* $x, y, u, v \in H$,

$$xx^* - yy^* = [\![x+y, x-y]\!] = [\![x-y, x+y]\!], \tag{2.10}$$

$$[\![u,v]\!] = \frac{1}{4}(u+v)(u+v)^* - \frac{1}{4}(u-v)(u-v)^*. \tag{2.11}$$

Additionally, for any $T \in \mathcal{S}^{1,1}$ *let* $T = a_1 e_1 e_1^* - a_2 e_2 e_2^*$ *be its spectral factorization with* $a_1, a_2 \geq 0$ *and* $\langle e_i, e_j \rangle = \delta_{i,j}$*. Then*

$$T = [\![\sqrt{a_1}e_1 + \sqrt{a_2}e_2, \sqrt{a_1}e_1 - \sqrt{a_2}e_2]\!].$$

(3) *The set* $\mathring{\mathcal{S}}^{1,1}$ *is a real analytic manifold in* $\mathrm{Sym}(\mathbb{C}^n)$ *of real dimension* $4n-4$*. Its tangent space at* $X = [\![x,y]\!]$ *is given by*

$$T_X\mathring{\mathcal{S}}^{1,1} = \{[\![x,u]\!] + [\![y,v]\!] = \frac{1}{2}(xu^* + ux^* + yv^* + vy^*)\ ,\ u,v \in \mathbb{C}^n\}. \tag{2.12}$$

The $\mathbb{R}$*-linear embedding* $\mathbb{C}^n \times \mathbb{C}^n \mapsto T_X\mathring{\mathcal{S}}^{1,1}$ *given by* $(u,v) \mapsto [\![x,u]\!] + [\![y,v]\!]$ *has null space* $\{a(ix,0) + b(0,iy) + c(y,-x) + d(iy,ix)\ ,\ a,b,c,d \in \mathbb{R}\}$.

(4) *Let* $T = [\![u,v]\!] \in \mathcal{S}^{1,1}$*. Then its eigenvalues and p-norms are:*

$$a_+ = \frac{1}{2}\left(\mathrm{real}(\langle u,v\rangle) + \sqrt{\|u\|^2\|v\|^2 - (\mathrm{imag}(\langle u,v\rangle))^2}\right) \geq 0, \tag{2.13}$$

$$a_- = \frac{1}{2}\left(\mathrm{real}(\langle u,v\rangle) - \sqrt{\|u\|^2\|v\|^2 - (\mathrm{imag}(\langle u,v\rangle))^2}\right) \leq 0, \tag{2.14}$$

$$\|T\|_1 = \sqrt{\|u\|^2\|v\|^2 - (\mathrm{imag}(\langle u,v\rangle))^2}, \tag{2.15}$$

$$\|T\|_2 = \sqrt{\frac{1}{2}\left(\|u\|^2\|v\|^2 + (\mathrm{real}(\langle u,v\rangle))^2 - (\mathrm{imag}(\langle u,v\rangle))^2\right)}, \tag{2.16}$$

$$\|T\|_\infty = \frac{1}{2}\left(|\mathrm{real}(\langle u,v\rangle)| + \sqrt{\|u\|^2\|v\|^2 - (\mathrm{imag}(\langle u,v\rangle))^2}\right). \tag{2.17}$$

(5) *Let* $T = xx^* - yy^* \in \mathcal{S}^{1,1}$*. Then its eigenvalues and p-norms are:*

$$a_+ = \frac{1}{2}\left(\|x\|^2 - \|y\|^2 + \sqrt{(\|x\|^2 + \|y\|^2)^2 - 4|\langle x,y\rangle|^2}\right), \tag{2.18}$$

$$a_- = \frac{1}{2}\left(\|x\|^2 - \|y\|^2 - \sqrt{(\|x\|^2 + \|y\|^2)^2 - 4|\langle x,y\rangle|^2}\right), \tag{2.19}$$

$$\|T\|_1 = \sqrt{(\|x\|^2 + \|y\|^2)^2 - 4|\langle x,y\rangle|^2}, \tag{2.20}$$

$$\|T\|_2 = \sqrt{\|x\|^4 + \|y\|^4 - 2|\langle x,y\rangle|^2}, \tag{2.21}$$

$$\|T\|_\infty = \frac{1}{2}\left(|\|x\|^2 - \|y\|^2| + \sqrt{(\|x\|^2 + \|y\|^2)^2 - 4|\langle x,y\rangle|^2}\right). \tag{2.22}$$

Note the above results hold true for the case of symmetric operators over the real subspace V. In particular the factorization at Lemma 2.4(1) implies that

$$\mathcal{S}^{1,1}(V) = \mathcal{S}^{1,0}(V) - \mathcal{S}^{1,0}(V) = \mathcal{S}^{1,0}(V) + \mathcal{S}^{0,1}(V) = \{[\![u,v]\!]\ ,\ u,v \in V\}. \tag{2.23}$$

More generally this result holds for subsets $V \subset H$ that are closed under addition and subtraction (such as modules over $\mathbb{Z}$).

2.3. Metrics. The space $\hat{H} = \widehat{\mathbb{C}^n}$ admits two classes of distances (metrics). The first class is the "natural metric" induced by the quotient space structure. The second metric is a matrix norm-induced distance.

Fix $1 \leq p \leq \infty$.

The *natural metric* denoted by $D_p : \hat{H} \times \hat{H} \to \mathbb{R}$ is defined by

$$D_p(\hat{x}, \hat{y}) = \min_{\varphi \in [0,2\pi)} \|x - e^{i\varphi} y\|_p, \tag{2.24}$$

where $x \in \hat{x}$ and $y \in \hat{y}$. In the case $p = 2$ the distance becomes

$$D_2(\hat{x}, \hat{y}) = \sqrt{\|x\|^2 + \|y\|^2 - 2|\langle x, y\rangle|}.$$

By abuse of notation we use also $D_p(x, y) = D_p(\hat{x}, \hat{y})$ since the distance does not depend on the choice of representatives.

The *matrix norm-induced distance* denoted by $d_p : \hat{H} \times \hat{H} \to \mathbb{R}$ is defined by

$$d_p(\hat{x}, \hat{y}) = \|xx^* - yy^*\|_p, \tag{2.25}$$

where again $x \in \hat{x}$ and $y \in \hat{y}$. In the case $p = 2$ we obtain

$$d_2(x, y) = \sqrt{\|x\|^4 + \|y\|^4 - 2|\langle x, y\rangle|^2}.$$

By abuse of notation we use also $d_p(x, y) = d_p(\hat{x}, \hat{y})$ since again the distance does not depend on the choice of representatives.

As analyzed in [**BZ14**], Proposition 2.4, D_p is not Lipschitz equivalent to d_p, however D_p is an equivalent distance to D_q and similarily, d_p is equivalent to d_q, for any $1 \leq p, q \leq q$ (see also [**BZ15b**] for the last claim below):

LEMMA 2.5.

(1) *For each $1 \leq p \leq \infty$, D_p and d_p are distances (metrics) on $\hat{H}$;*

(2) *$(D_p)_{1\leq p\leq\infty}$ are equivalent distances; that is, each D_p induces the same topology on $\hat{H}$ and, for every $1 \leq p, q \leq \infty$, the identity map $i : (\hat{H}, D_p) \to (\hat{H}, D_q)$, $i(x) = x$, is Lipschitz continuous with Lipschitz constant*

$$\mathrm{Lip}^D_{p,q,n} = \max(1, n^{\frac{1}{q}-\frac{1}{p}}).$$

(3) *$(d_p)_{1\leq p\leq\infty}$ are equivalent distances, that is, each d_p induces the same topology on $\hat{H}$ and, for every $1 \leq p, q \leq \infty$, the identity map $i : (\hat{H}, d_p) \to (\hat{H}, d_q)$, $i(x) = x$, is Lipschitz continuous with Lipschitz constant*

$$\mathrm{Lip}^d_{p,q,n} = \max(1, 2^{\frac{1}{q}-\frac{1}{p}}).$$

(4) *The identity map $i : (\hat{H}, D_p) \to (\hat{H}, d_p)$, $i(x) = x$ is continuous, but it is not Lipschitz continuous. The identity map $i : (\hat{H}, d_p) \to (\hat{H}, D_p)$, $i(x) = x$ is continuous but it is not Lipschitz continuous. Hence the induced topologies on $(\hat{H}, D_p)$ and $(\hat{H}, d_p)$ are the same, but the corresponding distances are not Lipschitz equivalent.*

(5) *The metric space $(\hat{H}, d_p)$ is isometrically isomorphic to $\mathcal{S}^{1,0}$ endowed with the p-norm. The isomorphism is given by the map*

$$\kappa_\beta : \hat{H} \to \mathcal{S}^{1,0} \ , \ x \mapsto [\![x, x]\!] = xx^*.$$

(6) *The metric space* $(\hat{H}, D_2)$ *is Lipschitz isomorphic (not isometric) with* $\mathcal{S}^{1,0}$ *endowed with the* 2*-norm. The bi-Lipschitz map*

$$\kappa_\alpha : \hat{H} \to \mathcal{S}^{1,0} \ , \ x \mapsto \kappa_\alpha(x) = \begin{cases} \frac{1}{\|x\|} xx^* & if \quad x \neq 0 \\ 0 & otherwise \end{cases}$$

has lower Lipschitz constant 1 *and upper Lipschitz constant* $\sqrt{2}$.

Note the Lipschitz constant $\mathrm{Lip}^D_{p,q,n}$ is equal to the operator norm of the identity map between $(\mathbb{C}^n, \|\cdot\|_p)$ and $(\mathbb{C}^n, \|\cdot\|_q)$: $\mathrm{Lip}^D_{p,q,n} = \|I\|_{l^p(\mathbb{C}^n)\to l^q(\mathbb{C}^n)}$. Note also the equality $\mathrm{Lip}^d_{p,q,n} = \mathrm{Lip}^D_{p,q,2}$. A consequence of the last two claims in the above result is that while the identity map between $(\hat{H}, D_p)$ and $(\hat{H}, d_q)$ is not bi-Lipschitz, the map $x \mapsto \frac{1}{\sqrt{\|x\|}} x$ is bi-Lipschitz.

3. The Injectivity Problem

In this section we summarize existing results on the injectivity of the maps α and β. Our plan is to present the real and the complex case in a unified way.

Recall that V is a real vector space which is also a subset of $H = \mathbb{C}^n$. The special two cases are $V = \mathbb{R}^n$ (the real case) and $V = \mathbb{C}^n$ (the complex case).

First we describe the *realification procedure* of H and V. Consider the $\mathbb{R}$-linear map $\jmath : \mathbb{C}^n \to \mathbb{R}^{2n}$ defined by

$$\jmath(x) = \begin{bmatrix} \mathrm{real}(x) \\ \mathrm{imag}(x) \end{bmatrix}.$$

Let $\mathcal{V} = \jmath(V)$ be the embedding of V into $\mathbb{R}^{2n}$, and let Π denote the orthogonal projection (with respect to the real scalar product on $\mathbb{R}^{2n}$) onto $\mathcal{V}$. Let J denote the folowing orthogonal antisymmetric $2n \times 2n$ matrix

$$J = \begin{bmatrix} 0 & -I_n \\ I_n & 0 \end{bmatrix}, \tag{3.1}$$

where I_n denotes the $n \times n$ identity matrix. Note that $J^T = -J$, $J^2 = -I_{2n}$ and $J^{-1} = -J$.

Each vector f_k of the frame set $\mathcal{F} = \{f_1, \dots, f_m\}$ is mapped by $\jmath$ onto a vector in $\mathbb{R}^{2n}$ denoted by φ_k, and a symmetric operator in $\mathcal{S}^{2,0}(\mathbb{R}^{2n})$ denoted by Φ_k:

$$\varphi_k = \jmath(f_k) = \begin{bmatrix} \mathrm{real}(f_k) \\ \mathrm{imag}(f_k) \end{bmatrix}, \ \Phi_k = \varphi_k\varphi_k^T + J\varphi_k\varphi_k^T J^T. \tag{3.2}$$

Note that when $f_k \neq 0$ the symmetric form Φ_k has rank 2 and belongs to $\mathring{\mathcal{S}}^{2,0}$. Its spectrum has two distinct eigenvalues: $\|\varphi_k\|^2 = \|f_k\|^2$ with multiplicity 2, and 0 with multiplicity $2n-2$. Furthermore, $\frac{1}{\|\varphi_k\|^2}\Phi_k$ is a rank 2 projection.

Let $\xi = \jmath(x)$ and $\eta = \jmath(y)$ denote the realifications of vectors $x, y \in \mathbb{C}^n$. Then a bit of algebra shows that

$$\begin{aligned} (3.3) \qquad\qquad \langle x, f_k\rangle &= \langle \xi, \varphi_k\rangle + i\langle \xi, J\varphi_k\rangle, \\ \langle F_k, xx^*\rangle_{HS} = \mathrm{trace}\,(F_k xx^*) = |\langle x, f_k\rangle|^2 = \langle \Phi_k\xi, \xi\rangle &= \langle \Phi_k, \xi\xi^T\rangle_{HS}, \\ \langle F_k, [\![x,y]\!]\rangle_{HS} = \mathrm{trace}\,(F_k[\![x,y]\!]) = \mathrm{real}(\langle x, f_k\rangle\langle f_k, y\rangle) &= \langle \Phi_k\xi, \eta\rangle = \mathrm{trace}(\Phi_k[\![\xi,\eta]\!]) \\ &= \langle \Phi_k, [\![\xi,\eta]\!]\rangle_{HS}, \end{aligned}$$

where $F_k = [\![f_k, f_k]\!] = f_k f_k^* \in \mathcal{S}^{1,0}(H)$.

The following objects play an important role in the subsequent theory:

$$R:\mathbb{C}^n\to \mathrm{Sym}(\mathbb{C}^n)\ ,\ R(x)=\sum_{k=1}^m |\langle x,f_k\rangle|^2 f_k f_k^*\ ,\ x\in\mathbb{C}^n, \tag{3.4}$$

$$\mathcal{R}:\mathbb{R}^{2n}\to \mathrm{Sym}(\mathbb{R}^{2n})\ ,\ \mathcal{R}(\xi)=\sum_{k=1}^m \Phi_k\xi\xi^T\Phi_k\ ,\ \xi\in\mathbb{R}^{2n}, \tag{3.5}$$

$$\mathcal{S}:\mathbb{R}^{2n}\to \mathrm{Sym}(\mathbb{R}^{2n})\ ,\ \mathcal{S}(\xi)=\sum_{k:\Phi_k\xi\neq 0}\frac{1}{\langle\Phi_k\xi,\xi\rangle}\Phi_k\xi\xi^T\Phi_k\ ,\ \xi\in\mathbb{R}^{2n}, \tag{3.6}$$

$$\mathcal{Z}:\mathbb{R}^{2n}\to \mathbb{R}^{2n\times m}\ ,\ \mathcal{Z}(\xi)=\left[\ \Phi_1\xi\ \mid\ \dots\ \mid\ \Phi_m\xi\ \right]\ ,\ \xi\in\mathbb{R}^{2n}. \tag{3.7}$$

Note $\mathcal{R}=\mathcal{Z}\mathcal{Z}^T$.

Following [**BBCE07**] we note that $|\langle x,f_k\rangle|^2$ is the Hilbert-Schmidt scalar product between two rank 1 symmetric forms:

$$|\langle x,f_k\rangle|^2=\mathrm{trace}\,(F_kX)=\langle F_k,X\rangle_{HS},$$

where $X=xx^*$. Thus the nonlinear map β induces a linear map on the real vector space $\mathrm{Sym}(\mathbb{C}^n)$ of symmetric forms over $\mathbb{C}^n$:

$$\mathbb{A}:\mathrm{Sym}(\mathbb{C}^n)\to\mathbb{R}^m\ ,\ (\mathbb{A}(T))_k=\langle T,F_k\rangle_{HS}=\langle Tf_k,f_k\rangle\ ,\ 1\leq k\leq m \tag{3.8}$$

Similarly it induces a linear map on $\mathrm{Sym}(\mathbb{R}^{2n})$, the space of symmetric forms over $\mathbb{R}^{2n}=\jmath(\mathbb{C}^n)$, that is denoted by $\mathcal{A}$:

$$\begin{aligned}\mathcal{A}:\mathrm{Sym}(\mathbb{R}^{2n})\to\mathbb{R}^m\ ,\ (\mathcal{A}(T))_k&=\langle T,\Phi_k\rangle_{HS}\\&=\langle T\varphi_k,\varphi_k\rangle+\langle TJ\varphi_k,J\varphi_k\rangle\ ,\ 1\leq k\leq m.\end{aligned} \tag{3.9}$$

Now we are ready to state a necessary and sufficient condition for injectivity that works in both the real and the complex case:

THEOREM 3.1 ([**HMW11**, **BCMN13**, **Ba13**]). *Let $H=\mathbb{C}^n$ and let V be a real vector space that is also a subset of H, $V\subset H$. Denote by $\mathcal{V}=\jmath(V)$ the realification of V. Assume $\mathcal{F}$ is a frame for V. The following statements are equivalent:*

1. *The frame $\mathcal{F}$ is phase retrievable with respect to V;*
2. $\ker\mathbb{A}\cap\left(\mathcal{S}^{1,0}(V)-\mathcal{S}^{1,0}(V)\right)=\{0\}$;
3. $\ker\mathbb{A}\cap\mathcal{S}^{1,1}(V)=\{0\}$;
4. $\ker\mathbb{A}\cap(\mathcal{S}^{2,0}(V)\cup\mathcal{S}^{1,1}(V)\cup\mathcal{S}^{0,2})=\{0\}$;
5. *There do not exist vectors $u,v\in V$ with $[\![u,v]\!]\neq 0$ so that*
$$\mathrm{real}\,(\langle u,f_k\rangle\langle f_k,v\rangle)=0\ ,\ \forall 1\leq k\leq m;$$
6. $\ker\mathcal{A}\cap\left(\mathcal{S}^{1,0}(\mathcal{V})-\mathcal{S}^{1,0}(\mathcal{V})\right)=\{0\}$;
7. $\ker\mathcal{A}\cap\mathcal{S}^{1,1}(\mathcal{V})=\{0\}$;
8. *There do not exist vectors $\xi,\eta\in\mathcal{V}$, with $[\![\xi,\eta]\!]\neq 0$ so that*
$$\langle\Phi_k\xi,\eta\rangle=0\ ,\ \forall 1\leq k\leq m.$$

Proof.

(1) $\Leftrightarrow$ (2) It is immediate once we notice that any element in the null space of $\mathbb{A}$ of the form xx^*-yy^* implies $\mathbb{A}(xx^*)=\mathbb{A}(yy^*)$ for some $x,y\in V$ with $\hat{x}\neq\hat{y}$.

(2) $\Leftrightarrow$ (3) and (3) $\Leftrightarrow$ (5) are consequences of (2.23).

For (4) first note that $\ker\mathbb{A}\cap\mathcal{S}^{2,0}(V)=\{0\}=\ker\mathbb{A}\cap\mathcal{S}^{0,2}(V)$ since $\mathcal{F}$ is a frame for V. Thus (3) $\Leftrightarrow$ (4).

(6),(7) and (8) are simply restatements of (2),(3) and (4) using the realification procedure. □

In the case (4) above, note that $\mathcal{S}^{2,0}(V) \cup \mathcal{S}^{1,1}(V) \cup \mathcal{S}^{0,2}(V)$ is the set of all rank less than or equal to 2 symmetric operators in $\mathrm{Sym}(V)$ (This statement has been proposed in [**BCMN13**]).

The above general injectivity result is next made more explicit in the cases $V = \mathbb{C}^n$ and $V = \mathbb{R}^n$.

THEOREM 3.2 ([**BCE06, Ba12**]). *(The real case) Assume $\mathcal{F} \subset \mathbb{R}^n$. The following are equivalent:*

(1) *$\mathcal{F}$ is phase retrievable for $V = \mathbb{R}^n$;*
(2) *$R(x)$ is invertible for every $x \in \mathbb{R}^n$, $x \neq 0$;*
(3) *There do not exist vectors $u, v \in \mathbb{R}^n$ with $u \neq 0$ and $v \neq 0$ so that*

$$\langle u, f_k\rangle\langle f_k, v\rangle = 0 \ , \ \forall 1 \leq k \leq m;$$

(4) *For any disjoint partition of the frame set $\mathcal{F} = \mathcal{F}_1 \cup \mathcal{F}_2$, either $\mathcal{F}_1$ spans $\mathbb{R}^n$ or $\mathcal{F}_2$ spans $\mathbb{R}^n$.*

Recall a set $\mathcal{F} \subset \mathbb{C}^n$ is called *full spark* if any subset of n vectors is linearly independent. Then an immediate corollary of the above result is the following

COROLLARY 3.3 ([**BCE06**]). *Assume $\mathcal{F} \subset \mathbb{R}^n$. Then*

(1) *If $\mathcal{F}$ is phase retrievable for $\mathbb{R}^n$ then $m \geq 2n - 1$;*
(2) *If $m = 2n - 1$, then $\mathcal{F}$ is phase retrievable if and only if $\mathcal{F}$ is full spark;*

Proof.
Indeed, the first claim follows from Theorem 3.2(4): If $m \leq 2n - 2$ then there is a partition of $\mathcal{F}$ into two subsets each of cardinal less than or equal to $n - 1$. Thus neither set can span $\mathbb{R}^n$. Contradiction.

The second claim is immediate from the same statement as above. □

A more careful analysis of Theorem 3.2(4) gives a recipe of constructing two non-similar vectors $x, y \in \mathbb{R}^n$ so that $\alpha(x) = \alpha(y)$. Indeed, if $\mathcal{F} = \mathcal{F}_1 \cup \mathcal{F}_2$ so that $\dim \mathrm{span}(\mathcal{F}_1) < n$ and $\dim \mathrm{span}(\mathcal{F}_2) < n$ then there are non-zero vectors $u, v \in \mathbb{R}^n$ with $\langle u, f_k\rangle = 0$ for all $k \in I$, and $\langle v, f_k\rangle = 0$ for all $k \in I^c$. Here I is the index set of frame vectors in $\mathcal{F}_1$ and I^c denotes its complement in $\{1, \ldots, m\}$. Set $x = u + v$ and $y = u - v$. Then $|\langle x, f_k\rangle| = |\langle v, f_k\rangle| = |\langle y, f_k\rangle|$ for all $k \in I$, and $|\langle x, f_k\rangle| = |\langle u, f_k\rangle| = |\langle y, f_k\rangle|$ for all $k \in I^c$. Thus $\alpha(x) = \alpha(y)$, but $x \neq y$ and $x \neq -y$.

THEOREM 3.4 ([**BCMN13, Ba13**]). *(The complex case) The following are equivalent:*

(1) *$\mathcal{F}$ is phase retrievable for $H = \mathbb{C}^n$;*
(2) $\mathrm{rank}(\mathcal{Z}(\xi)) = 2n - 1$ *for all $\xi \in \mathbb{R}^{2n}$, $\xi \neq 0$;*
(3) $\dim \ker \mathcal{R}(\xi) = 1$ *for all $\xi \in \mathbb{R}^{2n}$, $\xi \neq 0$;*
(4) *There do not exist $\xi, \eta \in \mathbb{R}^{2n}$, $\xi \neq 0$ and $\eta \neq 0$ so that $\langle J\xi, \eta\rangle = 0$ and*

$$\langle \Phi_k \xi, \eta\rangle = 0 \ , \ \forall 1 \leq k \leq m. \tag{3.10}$$

In terms of cardinality, here is what we know:

THEOREM 3.5 ([**Mi67, HMW11, BH13, Ba15, MV13, CEHV13, Viz15**]).

(1) [**HMW11**] *If $\mathcal{F}$ is a phase retrievable frame for $\mathbb{C}^n$ then*

$$m \geq 4n - 2 - 2b + \begin{cases} 2 & \text{if } n \text{ odd and } b = 3 \bmod 4 \\ 1 & \text{if } n \text{ odd and } b = 2 \bmod 4 \\ 0 & \text{otherwise} \end{cases} , \tag{3.11}$$

where $b = b(n)$ denotes the number of 1's in the binary expansion of $n-1$.

(2) [**BH13**] *For any positive integer n there is a frame with $m = 4n-4$ vectors so that $\mathcal{F}$ is phase retrievable for $\mathbb{C}^n$;*

(3) [**CEHV13**] *If $m \geq 4n-4$ then a (Zariski) generic frame is phase retrievable for $\mathbb{C}^n$;*

(4) [**Ba15**] *The set of phase retrievable frames is open in $\mathbb{C}^n \times \cdots \times \mathbb{C}^n$. In particular phase retrievable property is stable under small perturbations.*

(5) [**CEHV13**] *If $n = 2^k + 1$ and $m \leq 4n - 5$ then $\mathcal{F}$ cannot be phase retrievable for $\mathbb{C}^n$.*

(6) [**Viz15**] *For $n = 4$ there is a frame with $m = 11 < 4n - 4 = 12$ vectors that is phase retrievable for $\mathbb{C}^n$.*

4. Robustness of Reconstruction

In this section we analyze stability bounds for reconstruction. Specifically we analyze two types of margins:

- Deterministic, worst-case type bounds: These bounds are given by lower Lipschitz constants of the forward nonliner analysis maps;
- Stochastic, average type bounds: Cramer-Rao Lower Bounds (CRLB).

4.1. Bi-Lipschitzianity of the Nonlinear Analysis Maps. In Section 2 we introduced two distances on $\hat{H}$. As the following theorem shows, the nonlinear maps α and β are bi-Lipschitz with respect to the corresponding distance:

THEOREM 4.1. [**Ba12, EM12, BCMN13, Ba13, BW13, BZ14, BZ15a**], [**BZ15b**] *Let $\mathcal{F}$ be a phase retrievable frame for V, a real linear space, subset of $H = \mathbb{C}^n$. Then:*

(1) *The nonlinear map $\alpha : (\hat{V}, D_2) \to (\mathbb{R}^m, \|\|_2)$ is bi-Lipschitz. Specifically there are positive constants $0 < A_0 \leq B_0 < \infty$ so that*

$$\sqrt{A_0} D_2(x,y) \leq \|\alpha(x) - \alpha(y)\|_2 \leq \sqrt{B_0} D_2(x,y) \ , \ \forall x, y \in V. \tag{4.1}$$

(2) *The nonlinear map $\beta : (\hat{V}, d_1) \to (\mathbb{R}^m, \|\|_2)$ is bi-Lipschitz. Specifically there are positive constants $0 < a_0 \leq b_0 < \infty$ so that*

$$\sqrt{a_0} d_1(x,y) \leq \|\beta(x) - \beta(y)\|_2 \leq \sqrt{b_0} d_1(x,y) \ , \ \forall x, y \in V. \tag{4.2}$$

The converse is also true: If either (4.1) or (4.2) holds true for all $x, y \in V$ then $\mathcal{F}$ is phase retrievable for V.

The choice of distance D_2 and d_1 in the statement of this theorem is only for reasons of convenience since these specific constants will appear later in the text. Any other distance D_p instead of D_2, and d_q instead of d_1 would work. The Lipschitz constants would be different, of course. This result was first obtained for the real case in [**EM12**] for the map α and in [**Ba12**] for the map β. The complex case for map β was shown independently in [**BCMN13**] and [**Ba13**]. The complex case for the more challenging map α was proved in [**BZ15b**]. The paper [**BW13**] computes the optimal bound A_0 in the real case. The statement presented here (Theorem 4.1) unifies these two cases.

On the other hand the condition that $\mathcal{F}$ is phase retrievable for V is equivalent to the existence of a lower bound for a family of quadratic forms. We state this condition now:

THEOREM 4.2. *Let $\mathcal{F} \subset H = \mathbb{C}^n$ and let V be a real vector space, subset of H. Denote by $\mathcal{V} = \jmath(V) \subset \mathbb{R}^{2n}$ the realification of V, and let Π denote the projection onto $\mathcal{V}$. Then the following statements are equivalent:*

(1) *$\mathcal{F}$ is phase retrievable for V;*

(2) *There is a constant $a_0 > 0$ so that*

$$\Pi\mathcal{R}(\xi)\Pi \geq a_0 \Pi P_{J\xi}^{\perp}\Pi \ , \ \forall \xi \in \mathcal{V}, \|\xi\| = 1, \tag{4.3}$$

where $P_{J\xi}^{\perp} = I_{2n} - P_{J\xi} = I_{2n} - J\xi\xi^T J^T$ is the orthogonal projection onto the orthogonal complement to $J\xi$;

(3) *There is $a_0 > 0$ so that for all $\xi, \eta \in \mathbb{R}^{2n}$,*

$$\sum_{k=1}^{m} |\langle \Pi\Phi_k \Pi\xi, \eta\rangle|^2 \geq a_0 \left(\|\Pi\xi\|^2 \|\Pi\eta\|^2 - |\langle J\Pi\xi, \Pi\eta\rangle|^2 \right). \tag{4.4}$$

Note the same constant a_0 can be chosen in (4.2) and (4.3) and (4.4). This result was shown separately for the real and complex case. Here we state these conditions in a unified way.

Proof.

(1) $\Leftrightarrow$ (2) If $\mathcal{F}$ is a phase retrievable frame for V then, by Theorem 3.1(8), for all vectors $\xi, \eta \in \mathcal{V}$, with $[\![\xi, \eta]\!] \neq 0$ we have $\langle \Phi_k \xi, \eta \rangle \neq 0$, for some $1 \leq k \leq m$. Take $\mu \in \mathbb{R}^{2n}$ and set $\eta = \Pi\mu$. Normalize ξ to $\|\xi\| = 1$. Then

$$\sum_{k=1}^{m} |\langle \Phi_k \xi, \eta \rangle|^2 = \langle \mathcal{R}(\xi)\Pi\mu, \Pi\mu \rangle,$$

and by (2.15),

$$\|[\![\xi, \eta]\!]\|_1^2 = \|\xi\|^2 \|\eta\|^2 - |\langle \xi, J\eta \rangle|^2 = \|\Pi\mu\|^2 - |\langle J\xi, \Pi\mu \rangle|^2 = \langle (I_{2n} - J\xi\xi^T J^T)\Pi\mu, \Pi\mu \rangle.$$

Thus if μ satisfies $[\![\xi, \Pi\mu]\!] = 0$ then it must also satisfy $\Pi\mu = tJ\xi$ for some real t. In this case $\Pi\mu$ lies in the null space of $\mathcal{R}(\xi)$. In particular this proves that the following quotient of quadratic forms

$$\frac{\langle \Pi\mathcal{R}(\xi)\Pi\mu, \mu \rangle}{\langle \Pi(I_{2n} - J\xi\xi^T J^T)\Pi\mu, \mu \rangle}$$

is bounded above and below away from zero. This proves that (4.3) must hold for some $a_0 > 0$. Conversely, if (4.3) holds true, then for every $\xi, \eta \in \mathcal{V}$ with $[\![\xi, \eta]\!] \neq 0$, $\langle P_{J\xi}^{\perp}\eta, \eta \rangle \neq 0$ and thus $\langle \Phi_k \xi, \eta \rangle \neq 0$ for some k. This shows that $\mathcal{F}$ is a phase retrievable frame for V.

(2) $\Leftrightarrow$ (3) This follows by writing out (4.3) explicitly.

□

REMARK 4.3. Condition (2) of this theorem expressed by Equation (4.3) can be used to check if a given frame is phase retrievable as we explain next.

In the real case, $\Pi = I_n \oplus 0$, and this condition reduces to

$$R(x) = \sum_{k=1}^{m} |\langle x, f_k\rangle|^2 f_k f_k^T \geq a_0 \|x\|^2 I_H \ , \ \forall x \in H = \mathbb{R}^n.$$

In turn this is equivalent to any of the conditions of Theorem 3.2.

In the complex case the condition (4.3) turns into

$$\lambda_{2n-1}(\mathcal{R}(\xi)) \geq a_0 \ , \ \forall \xi \in \mathbb{R}^{2n}, \|\xi\| = 1, \tag{4.5}$$

where $\lambda_{2n-1}(\mathcal{R}(\xi))$ denotes the next to the smallest eigenvalue of $\mathcal{R}(\xi)$. The algorithm requires an upper bound for $b_0 = \max_{\|\xi\|=1} \lambda_1(\mathcal{R}(\xi))$. For instance $b_0 \leq B \max_k \|f_k\|^2$, where B is the frame upper bound [**Ba15**]. The condition (4.5) can be checked using an ε-net of the unit sphere in $\mathbb{R}^{2n}$. Specifically let $\{\xi_j^\varepsilon\}$ be such an ε-net, that is $\|\xi_j^\varepsilon\| = 1$ and $\|\xi_j^\varepsilon - \xi_k^\varepsilon\| < \varepsilon$ for all $j \neq k$. Set $a_0 = \frac{1}{2}\min_j \lambda_{2n-1}(\mathcal{R}(\xi_j^\varepsilon))$. If $2b_0\varepsilon \leq a_0$ then stop, otherwise set $\varepsilon = \frac{1}{2}\varepsilon$ and construct a new ε-net.

The condition $2b_0\varepsilon \leq a_0$ guarantees that for every $\xi \in \mathbb{R}^{2n}$ with $\|\xi\| = 1$, $\lambda_{2n-1}(\mathcal{R}(\xi)) \geq a_0$ since (see also [**Ba15**] for a similar derivation)

$$\|\mathcal{R}(\xi) - \mathcal{R}(\xi_j^\varepsilon)\| \leq \sqrt{b_0^2 \|\xi - \xi_j^\varepsilon\| \|\xi + \xi_j^\varepsilon\|} \leq 2b_0 \|\xi - \xi_j^\varepsilon\| \leq 2b_0\varepsilon$$

and by Weyl's perturbation theorem (see III.2.6 in [**Bh97**])

$$\lambda_{2n-1}(\mathcal{R}(x)) \geq \lambda_{2n-1}(\mathcal{R}(\xi_j^\varepsilon)) - \|\mathcal{R}(\xi) - \mathcal{R}(\xi_j^\varepsilon)\| \geq 2a_0 - 2b\varepsilon \geq a_0.$$

Unfortunately such an approach has at least an NP computational cost since the cardinality of an ε-net is of the order $\left(\frac{1}{\varepsilon}\right)^n$.

The computations of lower bounds is fairly subtle. In fact there is a distinction between local bounds and global bounds. Specifically for every $z \in V$ we define the following bounds:

The *type I local lower Lipschitz bounds* are defined by:

$$A(z) = \lim_{r\to 0} \inf_{x,y\in V, D_2(x,z)<r, D_2(y,z)<r} \frac{\|\alpha(x) - \alpha(y)\|_2^2}{D_2(x,y)^2}, \tag{4.6}$$

$$a(z) = \lim_{r\to 0} \inf_{x,y\in V, d_1(x,z)<r, d_1(y,z)<r} \frac{\|\beta(x) - \beta(y)\|_2^2}{d_1(x,y)^2}. \tag{4.7}$$

The *type II local lower Lipschitz bounds* are defined by:

$$\tilde{A}(z) = \lim_{r\to 0} \inf_{y\in V, D_2(y,z)<r} \frac{\|\alpha(z) - \alpha(y)\|_2^2}{D_2(z,y)^2}, \tag{4.8}$$

$$\tilde{a}(z) = \lim_{r\to 0} \inf_{y\in V, d_1(y,z)<r} \frac{\|\beta(z) - \beta(y)\|_2^2}{d_1(z,y)^2}. \tag{4.9}$$

Similarly the *type I local upper Lipschitz bounds* are defined by:

$$B(z) = \lim_{r\to 0} \sup_{x,y\in V, D_2(x,z)<r, D_2(y,z)<r} \frac{\|\alpha(x) - \alpha(y)\|_2^2}{D_2(x,y)^2}, \tag{4.10}$$

$$b(z) = \lim_{r\to 0} \sup_{x,y\in V, d_1(x,z)<r, d_1(y,z)<r} \frac{\|\beta(x) - \beta(y)\|_2^2}{d_1(x,y)^2} \tag{4.11}$$

and the *type II local upper Lipschitz bounds* are defined by:

$$\tilde{B}(z) = \lim_{r\to 0} \sup_{y\in V, D_2(y,z)<r} \frac{\|\alpha(z)-\alpha(y)\|_2^2}{D_2(z,y)^2}, \tag{4.12}$$

$$\tilde{b}(z) = \lim_{r\to 0} \sup_{y\in V, d_1(y,z)<r} \frac{\|\beta(z)-\beta(y)\|_2^2}{d_1(z,y)^2}. \tag{4.13}$$

The *global lower bounds* are defined by

$$A_0 = \inf_{x,y\in V, D_2(x,y)>0} \frac{\|\alpha(x)-\alpha(y)\|_2^2}{D_2(x,y)^2}, \tag{4.14}$$

$$a_0 = \inf_{x,y\in V, d_1(x,y)>0} \frac{\|\beta(x)-\beta(y)\|_2^2}{d_1(x,y)^2}, \tag{4.15}$$

whereas the *global upper bounds* are defined by

$$B_0 = \sup_{x,y\in V, D_2(x,y)>0} \frac{\|\alpha(x)-\alpha(y)\|_2^2}{D_2(x,y)^2}, \tag{4.16}$$

$$b_0 = \sup_{x,y\in V, d_1(x,y)>0} \frac{\|\beta(x)-\beta(y)\|_2^2}{d_1(x,y)^2} \tag{4.17}$$

and represent the square of the corresponding Lipschitz constants .

Due to homogeneity $A_0 = A(0)$, $B_0 = B(0)$, $a_0 = a(0)$, and $b_0 = b(0)$. On the other hand, for $z \neq 0$, $A(z) = A(\frac{z}{\|z\|})$, $B(z) = B(\frac{z}{\|z\|})$, $a(z) = a(\frac{z}{\|z\|})$, and $b(z) = b(\frac{z}{\|z\|})$. Note that $A(z)$ stands for the local lower Lipschitz bound of type I at z, whereas A denotes the optimal lower frame bound of $\mathcal{F}$.

The exact expressions of these bounds are summarized by the following results. For any $I \subset \{1, 2, \ldots, m\}$ let $\mathcal{F}[I] = \{f_k\ ,\ k \in I\}$ denote the subset indexed by I. Also let $\sigma_1^2[I]$ and $\sigma_n^2[I]$ denote the upper and the lower frame bound of set $\mathcal{F}[I]$, respectively. Thus:

$$\sigma_1^2[I] = \lambda_{\max}\left(\sum_{k\in I} f_k f_k^*\right) \ ,\ \sigma_n^2[I] = \lambda_{\min}\left(\sum_{k\in I} f_k f_k^*\right).$$

As usual, I^c denotes the complement of the index set I, that is $I^c = \{1, \ldots, m\} \setminus I$.

THEOREM 4.4 ([**BW13, BCMN13**]). *(The real case) Assume $\mathcal{F} \subset \mathbb{R}^n$ is a phase retrievable frame for $\mathbb{R}^n$. Let A and B denote its optimal lower and upper frame bound, respectively. Then:*

(1) *For every $0 \neq x \in \mathbb{R}^n$, $A(x) = \sigma_n^2[\mathrm{supp}(\alpha(x))]$, where $\mathrm{supp}(\alpha(x)) = \{k\ ,\ \langle x, f_k\rangle \neq 0\}$;*
(2) *For every $x \in \mathbb{R}^n$, $\tilde{A}(x) = A$;*
(3) *$A_0 = A(0) = \min_I(\sigma_n^2[I] + \sigma_n^2[I^c])$;*
(4) *For every $x \in \mathbb{R}^n$, $B(x) = \tilde{B}(x) = B$;*
(5) *$B_0 = B(0) = \tilde{B}(0) = B$, the optimal upper frame bound;*
(6) *For every $0 \neq x \in \mathbb{R}^n$, $a(x) = \tilde{a}(x) = \lambda_{\min}(R(x))/\|x\|^2$;*
(7) *$a_0 = a(0) = \tilde{a}(0) = \min_{\|x\|=1} \lambda_{\min}(R(x))$;*
(8) *For every $0 \neq x \in \mathbb{R}^n$, $b(x) = \tilde{b}(x) = \lambda_{\max}(R(x))/\|x\|^2$;*
(9) *$b_0 = b(0) = \tilde{b}(0) = \max_{\|x\|=1} \lambda_{\max}(R(x))$;*

(10) a_0 *is the largest constant so that*

$$R(x) \geq a_0 \|x\|^2 I_n \ , \ \forall x \in \mathbb{R}^n,$$

or, equivalently,

$$\sum_{k=1}^{m} |\langle x, f_k\rangle|^2 |\langle y, f_k\rangle|^2 \geq a_0 \|x\|^2 \|y\|^2 \ , \ \forall x, y \in \mathbb{R}^n;$$

(11) b_0 *is the* 4th *power of the frame analysis operator norm* $T : (\mathbb{R}^n, \|\cdot\|_2) \to (\mathbb{R}^m, \|\cdot\|_4)$,

$$b_0 = \|T\|^4_{B(l^2,l^4)} = \max_{\|x\|_2=1} \sum_{k=1}^{m} |\langle x, f_k\rangle|^4.$$

The complex case is subtler. The following result presents some of the local and global Lipschitz bounds.

THEOREM 4.5 ([**BZ15b**]). *(The complex case) Assume* $\mathcal{F}$ *is phase retrievable for* $H = \mathbb{C}^n$ *and* A, B *are its optimal frame bounds. Then:*

(1) *For every* $0 \neq z \in \mathbb{C}^n$, $A(z) = \lambda_{2n-1}(\mathcal{S}(\jmath(z)))$ *(the next to the smallest eigenvalue);*
(2) $A_0 = A(0) > 0$;
(3) *For every* $z \in \mathbb{C}^n$, $\tilde{A}(z) = \lambda_{2n-1}\left(\mathcal{S}(\jmath(z)) + \sum_{k:\langle z, f_k\rangle=0} \Phi_k\right)$ *(the next to the smallest eigenvalue);*
(4) $\tilde{A}(0) = A$, *the optimal lower frame bound;*
(5) *For every* $z \in \mathbb{C}^n$, $B(z) = \tilde{B}(z) = \lambda_1\left(\mathcal{S}(\jmath(z)) + \sum_{k:\langle z, f_k\rangle=0} \Phi_k\right)$ *(the largest eigenvalue);*
(6) $B_0 = B(0) = \tilde{B}(0) = B$, *the optimal upper frame bound;*
(7) *For every* $0 \neq z \in \mathbb{C}^n$, $a(z) = \tilde{a}(z) = \lambda_{2n-1}(\mathcal{R}(\jmath(z)))/\|z\|^2$ *(the next to the smallest eigenvalue);*
(8) *For every* $0 \neq z \in \mathbb{C}^n$, $b(z) = \tilde{b}(z) = \lambda_1(\mathcal{R}(\jmath(z)))/\|z\|^2$ *(the largest eigenvalue);*
(9) a_0 *is the largest constant so that*

$$\mathcal{R}(\xi) \geq a_0(I - J\xi\xi^T J^T) \ , \ \forall \xi \in \mathbb{R}^{2n}, \|\xi\| = 1,$$

or, equivalently,

$$\sum_{k=1}^{m} |\langle \Phi_k \xi, \eta\rangle|^2 \geq a_0 \left(\|\xi\|^2 \|\eta\|^2 - |\langle J\xi, \eta\rangle|^2\right) \ , \ \forall \xi, \eta \in \mathbb{R}^{2n};$$

(10) $b(0) = \tilde{b}(0) = b_0$ *is the* 4th *power of the frame analysis operator norm* $T : (\mathbb{C}^n, \|\cdot\|_2) \to (\mathbb{R}^m, \|\cdot\|_4)$,

$$b_0 = \|T\|^4_{B(l^2,l^4)} = \max_{\|x\|_2=1} \sum_{k=1}^{m} |\langle x, f_k\rangle|^4;$$

(11) $\tilde{a}(0)$ *is given by*

$$\tilde{a}(0) = \min_{\|z\|=1} \sum_{k=1}^{m} |\langle z, f_k\rangle|^4.$$

The results presented so far show that both α and β admit left inverses that are Lipschitz continuous on their domains of definition. One remaining problem is to know whether these left inverses can be extended to Lipschitz maps over the entire $\mathbb{R}^m$. The following two results provide a positive answer (see [**BZ14, BZ15b**] for details).

THEOREM 4.6 ([**BZ15b**]). *Assume $\mathcal{F} \subset H = \mathbb{C}^n$ is a phase retrievable frame for $\mathbb{C}^n$. Let $\sqrt{A_0}$ be the lower Lipschitz constant of the map $\alpha : (\hat{H}, D_2) \to (\mathbb{R}^m, \|\cdot\|_2)$. Then there is a Lipschitz map $\omega : (\mathbb{R}^m, \|\cdot\|_2) \to (\hat{H}, D_2)$ so that: (i) $\omega(\alpha(x)) = x$ for all $x \in \hat{H}$, and (ii) its Lipschitz constant is $\mathrm{Lip}(\omega) \leq \frac{4+3\sqrt{2}}{\sqrt{A_0}}$.*

THEOREM 4.7 ([**BZ14, BZ15a**]). *Assume $\mathcal{F} \subset H = \mathbb{C}^n$ is a phase retrievable frame for $\mathbb{C}^n$. Let $\sqrt{a_0}$ be the lower Lipschitz constant of the map $\beta : (\hat{H}, d_1) \to (\mathbb{R}^m, \|\cdot\|_2)$. Then there is a Lipschitz map $\psi : (\mathbb{R}^m, \|\cdot\|_2) \to (\hat{H}, d_1)$ so that: (i) $\psi(\beta(x)) = x$ for all $x \in \hat{H}$, and (ii) its Lipschitz constant is $\mathrm{Lip}(\psi) \leq \frac{4+3\sqrt{2}}{\sqrt{a_0}}$.*

Sketch of Proof

Proofs of both results follow a similar strategy. First both metric spaces $(\hat{H}, d_1)$ and $(\hat{H}, D_2)$ are bi-Lipschitz isomorphic with $\mathcal{S}^{1,0}$ via Lemma 2.5. Then one uses Kirszbraun's Theorem (see, e.g., [**BL00, HG13, WW75**]) to obtain an isometric Lipschitz extension of the left inverse of α (or β) from its range to the entire $(\mathbb{R}^m, \|\cdot\|_2)$ into $(\mathrm{Sym}(H), \|\cdot\|_2)$. The final step is to construct a Lipschitz map $\pi : \mathrm{Sym}(H) \to \mathcal{S}^{1,0}(H)$ so that $\pi(x^*) = xx^*$ for every $x \in H$. This map is realized as $\pi(A) = (\lambda_1 - \lambda_2)P_1$, where $\lambda_1 \geq \lambda_2$ are the two largest eigenvalues of A, and P_1 is the principal eigenprojector. Using the integration contour from [**ZB06**] and Weyl's inequalities (see III.2 in [**Bh97**]) the authors of [**BZ15b**] obtained that π is Lipschitz with $\mathrm{Lip}(\pi) \leq 3 + 2\sqrt{2}$ for $\pi : (\mathrm{Sym}(H), \|\cdot\|_2) \to (\mathcal{S}^{1,0}, \|\cdot\|_2)$.

4.2. Fisher Information Matrices and Cramer-Rao Lower Bounds. Throughout this section assume $\mathcal{F} = \{f_1, \ldots, f_m\} \subset H = \mathbb{C}^n$ is a phase retrievable frame for V, where $V \subseteq H$ is a real linear space, and $x \in V$.

Consider two measurement processes. The first is the Additive White Gaussian Noise (AWGN) model

$$y_k = |\langle x, f_k \rangle|^2 + \nu_k \ , \ 1 \leq k \leq m, \tag{4.18}$$

where $(\nu_k)_{1 \leq k \leq m}$ are independent and identically distributed (i.i.d.) realizations of a normal random variable of zero mean and variance σ^2. The second process is a non-Additive White Gaussian Noise (nonAWGN) model where the noise is added prior to taking the absolute value:

$$y_k = |\langle x, f_k \rangle + \mu_k|^2 \ , \ 1 \leq k \leq m, \tag{4.19}$$

where $(\mu_k)_{1 \leq k \leq m}$ are i.i.d. realizations of a Gaussian complex process with zero mean and variance ρ^2.

First we present the *Fisher Information matrices* $\mathbb{I}$ for these two processes. The general definition of the Fisher Information Matrix is (see [**Ky10**])

$$\mathbb{I}(x) = \mathbb{E}[(\nabla_x \log p(y; x))(\nabla_x \log p(y; x))^T].$$

Following [**BCMN13**] and [**Ba13**] for the AWGN model (4.18) we obtain:

$$\mathbb{I}^{\mathrm{AWGN}}(x) = \frac{4}{\sigma^2}\mathcal{R}(\xi) = \frac{4}{\sigma^2}\sum_{k=1}^{m} \Phi_k \xi \xi^T \Phi_k, \tag{4.20}$$

where $\xi = \jmath(x) \in \mathbb{R}^{2n}$. In general $\mathbb{I}(x)$ has rank at most $2n-1$ because $J\xi$ is always in its kernel.

In [**Bal15**] the Fisher information matrix for the nonAWGN model (4.19) is shown to have the following form:

$$\begin{aligned} \mathbb{I}^{\text{nonAWGN}}(x) &= \frac{4}{\rho^4}\sum_{k=1}^{m}\left(G_1\left(\frac{\langle \Phi_k\xi,\xi\rangle}{\rho^2}\right)-1\right)\Phi_k\xi\xi^*\Phi_k \\ &= \frac{4}{\rho^2}\sum_{k=1}^{m} G_2\left(\frac{\langle \Phi_k\xi,\xi\rangle}{\rho^2}\right)\frac{1}{\langle \Phi_k\xi,\xi\rangle}\Phi_k\xi\xi^*\Phi_k, \end{aligned} \tag{4.21}$$

where the two universal scalar functions $G_1, G_2 : \mathbb{R}^+ \to \mathbb{R}^+$ are given by

$$\begin{aligned} G_1(a) &= \frac{e^{-a}}{a}\int_0^\infty \frac{I_1^2(2\sqrt{at})}{I_0(2\sqrt{at})}te^{-t}dt = \frac{e^{-a}}{8a^3}\int_0^\infty \frac{I_1^2(t)}{I_0(t)}t^3e^{-\frac{t^2}{4a}}dt \\ G_2(a) &= a(G_1(a)-1), \end{aligned} \tag{4.22}$$

where I_0 and I_1 are the modifed Bessel functions of the first kind and order 0 and 1, respectively. Both Fisher information matrices have the same null space spanned by $J\xi$.

Next we present a lower bound on the variance of any unbiased estimator for x. Let $z_0 \in V$ be a fixed vector. Define

$$V_{z_0} = \{x \in V \ , \ \langle x, z_0\rangle > 0, \} \tag{4.23}$$

where $\langle\cdot,\cdot\rangle$ is the complex scalar product in H. Set $E_{z_0} = \text{span}_{\mathbb{R}}(V_{z_0})$ the real vector space spanned by V_{z_0}. Note $E_{z_0} = \{x \in V \ , \ \text{imag}(\langle x, z_0\rangle) = 0\}$.

To make (4.18) identifiable we select the representative $x \in V_{z_0}$ of the class $\hat{x}$. This is a mild condition since it only asks for the class $\hat{x}$ not to be orthogonal to z_0 with respect to the scalar product of H. An estimator $\omega : \mathbb{R}^m \to E_{z_0}$ is unbiased if $\mathbb{E}[\omega(\beta(x)+\nu)] = x$ for all $x \in V_{z_0}$. Here the expectation is taken with respect to the noise random variable.

A careful analysis (see [**Ba13**]) of the estimation process shows that the *Cramer-Rao Lower Bound (CRLB)* for either measurement process (4.18) and (4.19) is given by $(\Pi_{z_0}\mathbb{I}(x)\Pi_{z_0})^\dagger$, where Π_{z_0} is the orthogonal projection onto $\mathcal{V}_{z_0} = \jmath(E_{z_0})$ in $\mathbb{R}^{2n}$, and upper script $\dagger$ denotes the Moore-Penrose pseudo-inverse. Here $\mathbb{I}(x)$ stands for the Fisher information matrix $\mathbb{I}^{\text{AWGN}}(x)$ or $\mathbb{I}^{\text{nonAWGN}}(x)$. Then the covariance of any unbiased estimator $\omega : \mathbb{R}^m \to E_{z_0}$ is bounded as follows:

$$\text{Cov}[\omega] \geq (\Pi_{z_0}\mathbb{I}(x)\Pi_{z_0})^\dagger . \tag{4.24}$$

In particular, if one choses $z_0 = x$ then Π_{z_0} becomes the orthogonal projection onto the range of $\mathbb{I}(x)$ and $(\Pi_{z_0}\mathbb{I}(x)\Pi_{z_0})^\dagger = \mathbb{I}(x)^\dagger$ (see (4.25) below).

In the real case, $\mathcal{F} \subset V = \mathbb{R}^n \subset \mathbb{C}^n$, the Fisher information matrices of the AWGN model (4.18) and of the nonAWGN model (4.19) take the form

$$\mathbb{I}^{\text{AWGN}}(x) = \frac{4}{\sigma^2}\begin{bmatrix} R(x) & 0 \\ 0 & 0 \end{bmatrix} \ , \ \mathbb{I}^{\text{nonAWGN}}(x) = \frac{4}{\sigma^2}\sum_{k=1}^{m} G_2\left(\frac{|\langle x, f_k\rangle|^2}{\rho^2}\right)\begin{bmatrix} f_kf_k^T & 0 \\ 0 & 0 \end{bmatrix}.$$

Restricting to the real component of the estimator, the CRLB for the AWGN model (4.18) becomes

$$\text{Cov}[\omega^{\text{AWGN}}] \geq \frac{\sigma^2}{4}R(x)^{-1}$$

whereas for the nonAWGN model (4.19), the CRLB becomes

$$\mathrm{Cov}[\omega^{\mathrm{nonAWGN}}] \geq \frac{\rho^2}{4}\left(\sum_{k=1}^{m} G_2(\frac{|\langle x, f_k\rangle|^2}{\rho^2}) f_k f_k^T\right)^{-1}.$$

In the complex case $\mathcal{F} \subset V = H = \mathbb{C}^n$, $\Pi_{z_0} = I_{2n} - J\psi_0\psi_0^T J^T$ with $\psi_0 = \jmath(z_0)$ and the CRLB for AWGN becomes

$$\mathrm{Cov}[\omega] \geq \frac{\sigma^2}{4}((I_{2n} - J\psi_0\psi_0^T J^T)\mathcal{R}(\xi)(I_{2n} - J\psi_0\psi_0^T J^T))^\dagger.$$

Since $\mathcal{F}$ is phase retrievable for H, by Theorem 4.5(9) $\mathcal{R}(\xi)$ satisfies the lower bound $\mathcal{R}(\xi) \geq a_0(\|\xi\|^2 I_{2n} - J\xi\xi^T J^T)$. A little bit of algebra shows

$$\begin{aligned} a_0|\langle x, z_0\rangle|^2\Pi_{z_0} &= a_0|\langle \xi, \psi_0\rangle|^2\Pi_{z_0} \\ &\leq a_0|\langle \xi, \psi_0\rangle|^2(I_{2n} - J\psi_0\psi_0^T J^T) + a_0\left[\|\xi - \langle \xi, \psi_0\rangle\psi_0\|^2 I_{2n}\right. \\ &\quad \left. -J(\xi - \langle \xi, \psi_0\rangle\psi_0)(\xi - \langle \xi, \psi_0\rangle\psi_0)^T J^T\right] \\ &= (I_{2n} - J\psi_0\psi_0^T J^T)\mathcal{R}(\xi)(I_{2n} - J\psi_0\psi_0^T J^T). \end{aligned}$$

In particular this inequality shows that, if an unbiased estimator $\omega^0 : \mathbb{R}^m \to E_{z_0}$ for the AWGN model (4.18) achieves the CRLB then its covariance matrix is upper bounded by

$$\mathrm{Cov}[\omega^0] \leq \frac{\sigma^2}{4a_0|\langle x, z_0\rangle|^2}\Pi_{z_0}.$$

This result was derived in [**Ba13**].

Finally, if the global phase is provided by an oracle by correlating the estimated signal with the original signal x, then we can choose $z_0 = x$ and $\Pi_x = I_{2n} - J\xi\xi^T J^T$. But then

$$\Pi_x \mathbb{I}^{\mathrm{AWGN}}(x)\Pi_x = \mathbb{I}^{\mathrm{AWGN}}(x)\ ,\ \Pi_x \mathbb{I}^{\mathrm{nonAWGN}}(x)\Pi_x = \mathbb{I}^{\mathrm{nonAWGN}}(x)$$

which implies the CRLBs:

$$\mathrm{Cov}[\omega^{\mathrm{AWGN}}] \geq \left(\mathbb{I}^{\mathrm{AWGN}}(x)\right)^\dagger\ ,\ \mathrm{Cov}[\omega^{\mathrm{nonAWGN}}] \geq \left(\mathbb{I}^{\mathrm{nonAWGN}}(x)\right)^\dagger. \tag{4.25}$$

5. Reconstruction Algorithms

We present two types of reconstruction algorithms:

- Rank 1 tensor recovery: Linear Reconstruction, PhaseLift;
- Iterative algorithms: Gerchberg-Saxton, Mean-Squares Optimization: Wirtinger flow and IRLS.

The literature contains more algorithms than those presented here, see e.g. [**WAM12, ABFM12, Ba10, FMNW13, Fin82**].

Throughout this section we assume $\mathcal{F}$ is a phase retrievable frame for $H = \mathbb{C}^n$. We let $y = (y_k)_{1 \leq k \leq m}$ denote the vector of measurements. We analyze two cases: The noiseless case, when $y = \beta(x)$, and the additive noise case, when $y = \beta(x) + \nu$, where $\nu \in \mathbb{R}^m$ denotes the noise.

5.1. Rank 1 Tensor Recovery. The matrix recovery algorithms attempt to estimate the rank 1 matrix $X = xx^*$ from the measurements $y = (y_k)_{1 \le k \le m}$. We present two such algorithms: the linear reconstruction algorithm using lifting , and PhaseLift. An extension of the linear reconstruction algorithm from a matrix to a higher order tensor setting is also included.

5.1.1. *Linear Reconstruction Using Lifting.*

(i) Order 2 Tensor Embedding.

Linear reconstruction works well when the frame has high redundancy. Specifically if $m \ge \dim_{\mathbb{R}}(\mathrm{Sym}(H)) = n^2$ then, generically, the set of rank 1 operators $\{F_k = f_k f_k^* \ , \ 1 \le k \le m\}$ is a frame for $\mathrm{Sym}(H)$. In this case the measurements are linear on the space of matrices:

$$y_k = \langle F_k, X\rangle_{HS} + \nu_k \ , \ 1 \le k \le m.$$

Let $\{\tilde{F}_k \ , \ 1 \le k \le m\}$ denote the canonical dual frame to $\{F_k \ , \ 1 \le k \le m\}$. Then the minimum Frobenius norm estimate of X is given by the linear formula $X_{\text{est}} = \sum_{k=1}^m y_k \tilde{F}_k$. The class $\hat{x}$ is recovered using the spectral decomposition of X_{est},

$$X_{\text{est}} = \sum_{j=1}^d \lambda_{r(j)} P_j,$$

where d denotes the number of distinct eigenvalues of X_{est}, $\lambda_1 \ge \lambda_2 \ge \cdots \ge \lambda_n$ are the eigenvalues of X_{est}, P_j is the orthogonal projection onto the eigenspace associated to eigenvalue $\lambda_{r(j)}$, and $r(j+1) = r(j) + \mathrm{rank}(P_j)$, with $r(1) = 1$. The least squares estimator x^{LS} of $\hat{x}$ from X_{est} minimizes $\|X_{\text{est}} - xx^*\|^2$. The solution is unique when the top eigenvalue of X_{est} is simple, $\lambda_1 > \lambda_2$. In this case let $P = ee^*$ for some unit norm vector e. The least squares estimator x^{LS} is given explicitly by

$$x^{LS} = \begin{cases} \sqrt{\lambda_1} e & if \quad \lambda_1 \ge 0 \\ 0 & otherwise \end{cases} . \tag{5.1}$$

In the case $\lambda_1 = \lambda_2 > 0$ there are infinitely many possible top eigenvectors. Unfortunately any such choice produces an estimator that is discontinuous as a function of y. On the other hand the following estimator

$$x^{\text{Lip}} = \begin{cases} \sqrt{\lambda_1 - \lambda_2} e & if \quad \lambda_1 > \lambda_2 \\ 0 & otherwise \end{cases} \tag{5.2}$$

is an exact reconstruction scheme in the absence of noise and it is a Lipschitz continuous map with respect to the measurement vector y. Its Lipschitz constant and its performance with respect to additive noise are described in [**BZ15b**].

(ii) Higher Order Tensor Embeddings.

The idea of lifting into the space $\mathcal{S}^{1,0}$ of rank 1 matrices can be extended to spaces of higher order tensors (see [**Ba09**]). Fix an integer $d \ge 1$ and denote by $O_{n,d} = \{\gamma \in \mathbb{N}^d \ , \ 1 \le \gamma(1) \le \cdots \le \gamma(d) \le n\}$ the set of ordered d-tuples of positive integers up to n. Let $\mathcal{P}_{d,d}(Z_1, \ldots, Z_n)$ denote the real linear space of symmetric homogeneous polynomials in n variables $Z_1, \ldots, Z_n$ of degree (d, d), meaning that each monomial has degree d in variables $Z_1, \ldots, Z_n$ and degree d in conjugate variables $\overline{Z_1}, \ldots, \overline{Z_n}$,

$$P = \sum_{\gamma,\delta \in O_{n,d}} c_{\gamma,\delta} Z_{\gamma(1)} \cdots Z_{\gamma(d)} \overline{Z_{\delta(1)} \cdots Z_{\delta(d)}} \ , \ c_{\gamma,\delta} = \overline{c_{\delta,\gamma}} \in \mathbb{C}.$$

In the case $d = 1$, $\mathcal{P}_{1,1}(Z_1, \ldots, Z_n)$ is a linear $\mathbb{R}$-space isomorphic to $\mathrm{Sym}(\mathbb{C}^n)$. In general, $\mathcal{P}_{d,d}(Z_1, \ldots, Z_n)$ is isomorphic to the $\mathbb{R}$-linear space of (d,d)-sesquilinear functionals over $\mathbb{C}^n$ (denoted by $\Lambda_{d,d}(\mathbb{C}^n)$ in [**Ba09**]). For a given ordered d-tuple $\gamma \in O_{n,d}$ we denote by $\Pi(\gamma)$ the collection of all permutations of d elements that produce distinct d-tuples when applied to γ. Let $d_1, d_2, \ldots, d_n$ denote, respectively, the number of repetitions of $1, 2, \ldots, n$ in γ. Then the cardinal of $\Pi(\gamma)$ is given by the multinomial formula $\mathrm{Card}(\Pi(\gamma)) = \frac{d!}{d_1! \cdots d_n!}$. On $\mathcal{P}_{d,d}(Z_1, \ldots, Z_n)$ consider the sesquilinear scalar product $\langle\langle \cdot, \cdot \rangle\rangle$ so that

$$\{Z^{(\gamma,\delta)} := (\mathrm{Card}(\Pi(\gamma))\,\mathrm{Card}(\Pi(\delta)))^{1/2}\, Z_{\gamma(1)} \cdots Z_{\gamma(d)} \overline{Z_{\delta(1)} \cdots Z_{\delta(d)}} \ ; \ \gamma, \delta \in O_{n,d}\}$$

is an orthonormal basis. Let

$$\kappa_{d,d} : \widehat{\mathbb{C}^n} \times \cdots \times \widehat{\mathbb{C}^n} \to \mathcal{P}_{d,d}(Z_1, \ldots, Z_n) \ , \ \kappa_{d,d}(x^1, \ldots, x^d) = \prod_{k=1}^{d} |x_1^k Z_1 + \cdots + x_n^k Z_n|^2.$$

Note $x \mapsto \kappa_{d,d}(x, x, \ldots, x)$ is an embedding of $\widehat{\mathbb{C}^n}$ into $\mathcal{P}_{d,d}(Z_1, \ldots, Z_n)$. Let $P = \kappa_{d,d}(x, x, \ldots, x)$ and $Q_{k_1,\ldots,k_d} = \kappa_{d,d}(f_{k_1}, \ldots, f_{k_d})$. Then a little algebra shows that

$$\langle\langle P, Q_{k_1,\ldots,k_d} \rangle\rangle = |\langle x, f_{k_1}\rangle|^2 \cdots |\langle x, f_{k_d}\rangle|^2. \tag{5.3}$$

Now the phase retrieval problem can be restated as the problem of finding a homogeneous polynomial P of rank 1 (that is, of the form $\kappa_{d,d}(x, \ldots, x)$) so that

$$\langle\langle P, Q_{k_1,\ldots,k_d} \rangle\rangle = y_{k_1} \cdots y_{k_d} \ , \ \forall (k_1, \ldots, k_d) \in O_{m,d}. \tag{5.4}$$

The number of equations in (5.4) is

$$M_{m,d} = \mathrm{Card}(O_{m,d}) = \binom{m+d-1}{d},$$

whereas the dimension of the real linear space $\mathcal{P}_{d,d}(Z_1, \ldots, Z_n)$ is

$$N_{n,d} = \dim_{\mathbb{R}} \mathbb{P}_{d,d}(Z_1, \ldots, Z_n) = (\mathrm{Card}(O_{n,d}))^2 = \binom{n+d-1}{d}^2.$$

If d is so that $M_{m,d} \geq N_{n,d}$ and the set of polynomials $\mathcal{Q} = \{Q_\varepsilon \ ; \ \varepsilon \in O_{m,d}\}$ forms a frame for $\mathcal{P}_{d,d}(Z_1, \ldots, Z_n)$ then P can be obtained by solving a linear system (albeit of dimension growing exponentially with n and m). In particular if the set $\mathcal{Q}$ forms a d-design the reconstruction is particularly simple. The case $d = 2$ has been also explored in [**BE15**]. In the absence of noise $P \in \kappa_{d,d}(\widehat{\mathbb{C}^n})$ and thus $\hat{x}$ is found by solving a factorization problem. In the presence of noise, P is no longer of rank 1 and a different estimation procedure should be used. For instance one can find the "closest" rank 1 homogeneous polynomial in $\mathcal{P}_{d,d}(Z_1, \ldots, Z_n)$ and invert $\kappa_{d,d}$ to estimate x.

5.1.2. *PhaseLift.* Consider the noiseless case $y = \beta(x)$. The main idea is embodied in the following feasibility problem:

$$\text{find}_{\text{subject to:}\mathbb{A}(X)=y, X=X^*\geq 0, \text{rank}(X)=1} X.$$

Except for the condition $\text{rank}(X) = 1$, the optimization problem would be convex. However the rank constraint destroys the convexity property. Once a solution X is found, the vector x can be obtained by solving the factorization problem $X = xx^*$.

The feasibility problem admits at most a unique solution and so does the following optimization problem:

$$(5.5) \qquad \min_{\mathbb{A}(X)=y, X=X^*\geq 0} \text{rank}(X),$$

which is still non-convex. The insight provided by matrix completion theory and exploited in [**CSV12, CESV12**] is to replace $\text{rank}(X)$ by $\text{trace}(X)$ which is convex. Thus one obtains

$$(5.6) \qquad (\text{PhaseLift}) \quad \min_{\mathbb{A}(X)=y, X=X^*\geq 0} \text{trace}(X),$$

which is a convex optimization problem (a semi-definite program). In [**CL12**] the authors proved that for random frames, with high probability, the problem (5.6) has the same solution as the problem (5.5):

THEOREM 5.1. *Assume each vector f_k is drawn independently from $\mathcal{N}(0, I_n/2) + i\mathcal{N}(0, I_n/2)$, or each vector is drawn independently from the uniform distribution on the complex sphere of radius $\sqrt{n}$. Then there are universal constants $c_0, c_1, \gamma > 0$ so that for $m \geq c_0 n$, for every $x \in \mathbb{C}^n$ the problem (5.6) has the same solution as (5.5) with probability at least $1 - c_1 e^{-\gamma n}$.*

As explained in [**DH14**] and [**CL12**], the minimization of trace is not necessary; in the absence of noise it reduces to a feasibility problem.

The PhaseLift algorithm is also robust to noise. Consider the measurement process

$$y = \beta(x) + \nu,$$

for some $\nu \in \mathbb{R}^m$ noise vector. Consider the following modified optimization problem:

$$(5.7) \qquad \min_{X=X^*\geq 0} \|\mathbb{A}(X) - y\|_1.$$

In [**CL12**] the following result has been shown:

THEOREM 5.2. *Consider the same stochastic process for a random frame $\mathcal{F}$. There is a universal constant $C_0 > 0$ so that for all $x \in C^n$ the solution to (5.7) obeys*

$$\|X - xx^*\|_2 \leq C_0 \frac{\|\nu\|_1}{m}.$$

For the Gaussian model this holds with the same probability as in the noiseless case, whereas the probability of failure is exponentially small in n in the uniform model. The principal eigenvector x^0 of X (normalized by the square root of the principal eigenvalue) obeys

$$D_2(x^0, x) \leq C_0 \min(\|x\|_2, \frac{\|\nu\|_1}{m\|x\|_2}).$$

5.2. Iterative Algorithms. We present two classes of iterative algorithms: the Gerchberg-Saxton algorithm and mean-squares minimization algorithms.

5.2.1. *The Gerchberg-Saxton Algorithm.* Let $c = (c_k)_{1 \leq k \leq m} \in \mathbb{C}^m$ denote a sequence of the frame coefficients $c_k = \langle x, f_k \rangle$. Let $E = \{(\langle x, f_k \rangle)_{1 \leq k \leq m} \ , \ x \in \mathbb{C}^n\}$ denote the range of frame coefficients. Assume the measurements are all nonnegative, $y_k \geq 0$ for all k (otherwise rectify at 0). Denote by $\{\tilde{f}_k \ , \ 1 \leq k \leq m\}$ the canonical dual frame of $\{f_1, \ldots, f_m\}$. The Gerchberg-Saxton algorithm first introduced in [**GS72**] iterates between two sets of constraints: $|c_k| = \sqrt{y_k}$, and $c \in E$. Let $x^0 \in H$ be an initialization and set $t = 0$. The algorithm repeats the following steps:

(1) Linear Analysis: $c_k = \langle x^t, f_k \rangle$, $1 \leq k \leq m$;
(2) Magnitude Adjustment: $d_k = \sqrt{y_k} c_k / |c_k|$, $1 \leq k \leq m$;
(3) Linear Synthesis: $x^{t+1} = \sum_{k=1}^m d_k \tilde{f}_k$;
(4) Increment $t = t + 1$;

until a stopping criterion is achieved. The main advantage of this algorithm is its simplicity. It can easily incorporate additional constraints on x or its transform c. Unfortunately it suffers of a couple of disadvantages. Namely, the convergence is not guaranteed, and furthermore, when it converges it only converges to a local minimum. However, despite these shortcomings, the algorithm performs relatively well when x is highly constrained, for instance when all of its entries are nonnegative (see e.g. [**Fin82**]).

5.2.2. *Mean-Squares Minimization Algorithms.* Consider again the measurement process

$$y_k = |\langle x, f_k \rangle|^2 + \nu_k \ , \ 1 \leq k \leq m.$$

The Least-Squares criterion

$$\min_{x \in \mathbb{C}^n} \sum_{k=1}^m ||\langle x, f_k \rangle|^2 - y_k|^2$$

can be understood as the Maximum Likelihood Estimator (MLE) when the additive noise vector $\nu \in \mathbb{R}^m$ is normal distributed with zero mean and covariance $\sigma^2 I_m$. However the optimization problem is not convex and has many local minima.

We present two algorithms that minimize the mean-squares error: the Wirtinger flow and the Iterative Regularized Least-Squares.

(i) Gradient Descent Using the Wirtinger Flow

This algorithm has been introduced in [**CLS14**]. The idea is to follow the gradient descent for the criterion

$$f(x) = \sum_{k=1}^m |y_k - |\langle x, f_k \rangle|^2|^2. \tag{5.8}$$

The initialization is performed using the spectral method. Specifically:

Step 1. Initialization. Compute the principal eigenvector of $R_y = \sum_{k=1}^m y_k f_k f_k^*$ using, e.g., the power method. Let (e_1, a_1) be the eigen-pair with $e_1 \in \mathbb{C}^n$, $\|e_1\| = 1$, and $a_1 \in \mathbb{R}$. Initialize:

$$x^0 = \sqrt{n \frac{\sum_{k=1}^m y_k}{\sum_{k=1}^m \|f_k\|^2}} e_1 \ , \ t = 0. \tag{5.9}$$

Step 2. Iteration. Repeat:

2.1 Gradient descent:

$$x^{t+1} = x^t - \frac{\mu_{t+1}}{\|x^0\|^2}\left(\frac{1}{m}\sum_{k=1}^{m}(|\langle x^t, f_k\rangle|^2 - y_k)\langle x^t, f_k\rangle f_k\right). \tag{5.10}$$

2.2 Update

$$\mu^{t+1} = \min(\mu_{\max}, 1 - e^{-\tau/\tau_0}). \tag{5.11}$$

Step 3. Stopping. Stop after a fixed number of iterations or an error criterion is achieved.

The authors of [**CLS14**] showed that this algorithm converges with high probability to the exact solution in the absence of noise:

THEOREM 5.3 ([**CLS14**]). *Let $x \in \mathbb{C}^n$ and $y = \beta(x)$ with $m \geq c_0 n \log n$, where c_0 is a sufficiently large constant. Then the Wirtinger flow initial estimate x^0, normalized to have squared Euclidean norm equal to $\frac{1}{m}\sum_{k=1}^{m} y_k$, obeys $D_2(x^0, x) \leq \frac{1}{8}\|x\|$ with probability at least $1 - 10e^{-\gamma n} - \frac{8}{n^2}$, where γ is a fixed positive numerical constant. Further, take a constant learning parameter sequence, $\mu^t = \mu$ for all $t \geq 1$ and assume $\mu \leq \frac{c_1}{n}$ for some fixed numerical constant c_1. Then there is an event of probability at least $1 - 13e^{-\gamma n} - me^{-1.5m} - \frac{8}{n^2}$, such that on this event, starting from any initial solution x^0 obeying $D_2(x^0, x) \leq \frac{1}{8}\|x\|$, we have*

$$D_2(x^t, x) \leq \frac{1}{8}\left(1 - \frac{\mu}{4}\right)^{t/2}\|x\|. \tag{5.12}$$

(ii) The Iterative Regularized Least-Squares Algorithm

The iterative algorithm described next tries to find the global minimum using a regularization term. Consider the following optimization criterion:

$$\begin{aligned} J(u, v; \lambda, \mu) &= \sum_{k=1}^{m}\left|\frac{1}{2}(\langle u, f_k\rangle\langle f_k, v\rangle + \langle v, f_k\rangle\langle f_k, u\rangle) - y_k\right|^2 \\ &\quad + \lambda\|u\|_2^2 + \mu\|u - v\|_2^2 + \lambda\|v\|_2^2. \end{aligned} \tag{5.13}$$

The *Iterative Regularized Least-Squares (IRLS)* algorithm presented in [**Ba13**] works as follows.

Fix a stopping criterion, such as a tolerance ε, a desired level of signal-to-noise-ratio snr, or a maximum number of steps T. Fix an initialization parameter $\rho \in (0, 1)$, a learning rate $\gamma \in (0, 1)$, and a saturation parameter $\mu_{\min} > 0$.

Step 1. Initialization. Compute the principal eigenvector of $R_y = \sum_{k=1}^{m} y_k f_k f_k^*$ using e.g. the power method. Let (e_1, a_1) be the eigen-pair with $e_1 \in \mathbb{C}^n$, $\|e_1\| = 1$, and $a_1 \in \mathbb{R}$. If $a_1 \leq 0$ then set $x = 0$ and exit. Otherwise initialize:

$$x^0 = \sqrt{\frac{(1-\rho)a_1}{\sum_{k=1}^{m}|\langle e_1, f_k\rangle|^4}}\, e_1 \;,\; \lambda_0 = \rho a_1 \;,\; \mu_0 = \rho a_1 \;,\; t = 0. \tag{5.14}$$

Step 2. Iteration. Perform:

2.1 Solve the least-squares problem:

$$x^{t+1} = \operatorname{argmin}_u J(u, x^t; \lambda_t, \mu_t)$$

using the conjugate gradient method.

2.2 Update:

$$\lambda_{t+1} = \gamma\lambda_t \;,\; \mu_t = \max(\gamma\mu_t, \mu_{\min}) \;,\; t = t + 1.$$

Step 3. Stopping. Repeat Step 2 until

- The error criterion is achieved: $J(x^t, x^t; 0, 0) < \varepsilon$;
- The desired signal-to-noise-ratio is reached: $\frac{\|x^t\|^2}{J(x^t,x^t;0,0)} > \text{snr}$; or
- The maximum number of iterations is reached: $t > T$.

The final estimate can be x^T or the best estimate obtained in the iteration path: $x^{\text{est}} = x^{t_0}$ where $t_0 = \text{argmin}_t J(x^t, x^t; 0, 0)$.

The initialization (5.14) is performed for the following reason. Consider the modified criterion:

$$\begin{aligned} H(x;\lambda) &= J(x,x;\lambda,0) = \|\beta(x)-y\|_2^2 + 2\lambda\|x\|_2^2 \\ &= \sum_{k=1}^{m} |\langle x, f_k\rangle|^4 + 2\langle(\lambda I_n - R_y)x, x\rangle + \|y\|_2^2. \end{aligned}$$

In general this function is not convex in x, except for large values of λ. Specifically for $\lambda > a_1$, the largest eigenvalue of R_y, $x \mapsto H(x;\lambda)$ is convex and has a unique global minimum at $x = 0$. For $a_1 - \varepsilon < \lambda < a_1$ the criterion is no longer convex, but the global minimum stays in a neighborhood of the origin. Neglecting the 4th order terms, the critical points are given by the eigenvectors of R_y. Choosing $\lambda = \rho a_1$ and $x = s e_1$, the optimal value of s for $s \mapsto H(se_1; \rho a_1)$ is given in (5.14).

The path of iterates $(x^t)_{t\geq 0}$ can be thought of as trying to approximate the measured vector y with $\mathbb{A}([\![x^{t-1}, x^t]\!])$, where $\mathbb{A}$ is defined in (3.8). The parameter μ penalizes the unique negative eigenvalue of $[\![x^{t-1}, x^t]\!]$; the larger the value of μ_t the smaller the iteration step $\|x^{t+1} - x^t\|$ and the smaller the deviation of the matrix $[\![x^{t+1}, x^t]\!]$ from a rank 1 matrix; the smaller the parameter μ_t the larger in magnitude the negative eigenvalue of $[\![x^{t+1}, x^t]\!]$. This fact explains why in the noisy case the iterates first decrease the matching error $\|\mathbb{A}(x^t(x^t)^* - y\|_2$ to some t_0, and then they may start to increase this error; instead the rank 2 self-adjoint operator $T = [\![x^{t+1}, x^t]\!]$ always decreases the matching error $\|\mathbb{A}(T) - y\|_2$.

At any point on the path, if the value of criterion J is smaller than the value reached at the target vector x, then the algorithm is guaranteed to converge near x. Specifically in [**Ba13**] the following result has been proved:

THEOREM 5.4 ([**Ba13**], Theorem 5.6). *Fix $0 \neq z_0 \in \mathbb{C}^n$. Assume the frame $\mathcal{F}$ is so that $\ker \mathbb{A} \cap \mathcal{S}^{2,1} = \{0\}$. Then there is a constant $A_3 > 0$ that depends on $\mathcal{F}$ so that for every $x \in \mathbb{C}^n$ with $\langle x, z_0\rangle > 0$ and $\nu \in \mathbb{C}^n$ that produce $y = \beta(x) + \nu$ if there are $u, v \in \mathbb{C}^n$ so that $J(u,v;\lambda,\mu) < J(x,x;\lambda,\mu)$ then*

$$\|[\![u,v]\!] - xx^*\|_1 \leq \frac{4\lambda}{A_3} + \frac{2\|\nu\|_2}{\sqrt{A_3}}. \tag{5.15}$$

Moreover, let $[\![u,v]\!] = a_1 e_1 e_1^* + a_2 e_2 e_2^*$ be its spectral factorization with $a_1 \geq 0 \geq a_2$ and $\|e_1\| = \|e_2\| = 1$. Set $\tilde{x} = \sqrt{a_1} e_1$. Then

$$D_2(x,\tilde{x})^2 \leq \frac{4\lambda}{A_3} + \frac{2\|\nu\|_2}{\sqrt{A_3}} + \frac{\|\nu\|_2^2}{4\mu} + \frac{\lambda\|x\|_2^2}{2\mu}. \tag{5.16}$$

The kernel requirement on $\mathbb{A}$ is satisfied for generic frames when $m \geq 6n$. In particular this condition requires the frame $\mathcal{F}$ is phase retrievable for $\mathbb{C}^n$.

References

[ABFM12] B. Alexeev, A. S. Bandeira, M. Fickus, D. G. Mixon, *Phase Retrieval with Polarization*, SIAM J. Imaging Sci., **7** (1) (2014), 35–66.

[BE15] C. Bachoc and M. Ehler, *Signal reconstruction from the magnitude of subspace components*, IEEE Trans. Inform. Theory **61** (2015), no. 7, 4015–4027, DOI 10.1109/TIT.2015.2429634. MR3367817

[Ba09] R. Balan, *A Nonlinear Reconstruction Algorithm from Absolute Value of Frame Coefficients for Low Redundancy Frames*, Proceedings of SampTA Conference, Marseille, France May 2009.

[Ba10] R. Balan, *On Signal Reconstruction from Its Spectrogram*, Proceedings of the CISS Conference, Princeton NJ, May 2010.

[Ba12] R. Balan, *Reconstruction of Signals from Magnitudes of Redundant Representations*, available online arXiv:1207.1134v1 [math.FA] 4 July 2012.

[Ba13] R. Balan, *Reconstruction of Signals from Magnitudes of Redundant Representations: The Complex Case*, available online arXiv:1304.1839v1 [math.FA] 6 April 2013, to appear in Foundations of Computational Mathematics (2015).

[Ba15] R. Balan, *Stability of frames which give phase retrieval*, to appear in the Houston Journal of Mathematics 2015.

[Bal15] R. Balan, *The Fisher Information Matrix and the CRLB in a Non-AWGN Model for the Phase Retrieval Problem*, proceedings of SampTA 2015.

[BCE06] R. Balan, P. Casazza, and D. Edidin, *On signal reconstruction without phase*, Appl. Comput. Harmon. Anal. **20** (2006), no. 3, 345–356, DOI 10.1016/j.acha.2005.07.001. MR2224902 (2007b:94054)

[BBCE07] R. Balan, B. G. Bodmann, P. G. Casazza, and D. Edidin, *Painless reconstruction from magnitudes of frame coefficients*, J. Fourier Anal. Appl. **15** (2009), no. 4, 488–501, DOI 10.1007/s00041-009-9065-1. MR2549940 (2010m:42066)

[BW13] R. Balan and Y. Wang, *Invertibility and robustness of phaseless reconstruction*, Appl. Comput. Harmon. Anal. **38** (2015), no. 3, 469–488, DOI 10.1016/j.acha.2014.07.003. MR3323113

[BZ14] R. Balan and D. Zou, *Phase Retrieval using Lipschitz Continuous Maps*, available online arXiv:1403.2304v1.

[BZ15a] R. Balan and D. Zou, *On Lipschitz Inversion of Nonlinear Redundant Representations*, Contemporary Mathematics 650, "Trends in Harmonic Analysis and Its Applications", 15–22, 2015.

[BZ15b] R. Balan and D. Zou, *On Lipschitz Analysis and Lipschitz Synthesis for the Phase Retrieval Problem*, available online arXiv:1506.02092

[BCMN13] A. S. Bandeira, J. Cahill, D. G. Mixon, and A. A. Nelson, *Saving phase: injectivity and stability for phase retrieval*, Appl. Comput. Harmon. Anal. **37** (2014), no. 1, 106–125, DOI 10.1016/j.acha.2013.10.002. MR3202304

[BL00] Y. Benyamini and J. Lindenstrauss, *Geometric nonlinear functional analysis. Vol. 1*, American Mathematical Society Colloquium Publications, vol. 48, American Mathematical Society, Providence, RI, 2000. MR1727673 (2001b:46001)

[Bh97] R. Bhatia, *Matrix analysis*, Graduate Texts in Mathematics, vol. 169, Springer-Verlag, New York, 1997. MR1477662 (98i:15003)

[BH13] B. G. Bodmann and N. Hammen, *Stable phase retrieval with low-redundancy frames*, Adv. Comput. Math. **41** (2015), no. 2, 317–331, DOI 10.1007/s10444-014-9359-y. MR3337494

[CCPW13] J. Cahill, P.G. Casazza, J. Peterson, L. Woodland, *Phase retrieval by projections*, available online arXiv: 1305.6226v3

[CSV12] E. J. Candès, T. Strohmer, and V. Voroninski, *PhaseLift: exact and stable signal recovery from magnitude measurements via convex programming*, Comm. Pure Appl. Math. **66** (2013), no. 8, 1241–1274, DOI 10.1002/cpa.21432. MR3069958

[CESV12] E. J. Candès, Y. C. Eldar, T. Strohmer, and V. Voroninski, *Phase retrieval via matrix completion*, SIAM J. Imaging Sci. **6** (2013), no. 1, 199–225, DOI 10.1137/110848074. MR3032952

[CL12] E. J. Candès and X. Li, *Solving quadratic equations via PhaseLift when there are about as many equations as unknowns*, Found. Comput. Math. **14** (2014), no. 5, 1017–1026, DOI 10.1007/s10208-013-9162-z. MR3260258

[CLS14] E. J. Candès, X. Li, and M. Soltanolkotabi, *Phase retrieval via Wirtinger flow: theory and algorithms*, IEEE Trans. Inform. Theory **61** (2015), no. 4, 1985–2007, DOI 10.1109/TIT.2015.2399924. MR3332993

[Ca00] P. G. Casazza, *The art of frame theory*, Taiwanese J. Math. **4** (2000), no. 2, 129–201. MR1757401 (2001f:42046)

[CEHV13] C. Davis and W. M. Kahan, *Some new bounds on perturbation of subspaces*, Bull. Amer. Math. Soc. **75** (1969), 863–868. MR0246155 (39 #7460)

[DH14] L. Demanet and P. Hand, *Stable optimizationless recovery from phaseless linear measurements*, J. Fourier Anal. Appl. **20** (2014), no. 1, 199–221, DOI 10.1007/s00041-013-9305-2. MR3180894

[EM12] Y. C. Eldar and S. Mendelson, *Phase retrieval: stability and recovery guarantees*, Appl. Comput. Harmon. Anal. **36** (2014), no. 3, 473–494, DOI 10.1016/j.acha.2013.08.003. MR3175089

[FMNW13] M. Fickus, D. G. Mixon, A. A. Nelson, and Y. Wang, *Phase retrieval from very few measurements*, Linear Algebra Appl. **449** (2014), 475–499, DOI 10.1016/j.laa.2014.02.011. MR3191879

[Fin82] J.R. Fienup. Phase retrieval algorithms: A comparison. *Applied Optics*, 21(15):2758–2768, 1982.

[GS72] R. W. Gerchberg and W. O. Saxton, *A practical algorithm for the determination of the phase from image and diffraction plane pictures*, Optik 35, 237 (1972).

[HLO80] M. H. Hayes, J. S. Lim, and A. V. Oppenheim, *Signal reconstruction from phase or magnitude*, IEEE Trans. Acoust. Speech Signal Process. **28** (1980), no. 6, 672–680, DOI 10.1109/TASSP.1980.1163463. MR604824 (82a:94010)

[HMW11] T. Heinosaari, L. Mazzarella, and M. M. Wolf, *Quantum tomography under prior information*, Comm. Math. Phys. **318** (2013), no. 2, 355–374, DOI 10.1007/s00220-013-1671-8. MR3020161

[HG13] M. J. Hirn and E. Y. Le Gruyer, *A general theorem of existence of quasi absolutely minimal Lipschitz extensions*, Math. Ann. **359** (2014), no. 3-4, 595–628, DOI 10.1007/s00208-013-1003-5. MR3231008

[Ky10] S. M. Kay, **Fundamentals of Statistical Signal Processing. I. Estimation Theory**, Prentice Hall PTR, 18th Printing, 2010.

[MV13] D. Mondragon and V. Voroninski, *Determination of all pure quantum states from a minimal number of observables*, online arXiv:1306.1214v1 [math-ph] 5 June 2013.

[Mi67] R. J. Milgram, *Immersing projective spaces*, Ann. of Math. (2) **85** (1967), 473–482. MR0211412 (35 #2293)

[NQL82] H. Nawab, T. F. Quatieri, and J. S. Lim, *Signal Reconstruction from the Short-Time Fourier Transform Magnitude*, in Proceedings of ICASSP 1984.

[Viz15] C. Vinzant, *A small frame and a certificate of its injectivity*, available online arXiv:1502.0465v1 [math.FA] 16 Feb. 2015.

[WAM12] I. Waldspurger, A. d'Aspremont, S. Mallat, *Phase recovery, MaxCut and complex semidefinite programming*, Available online: arXiv:1206.0102

[WW75] J. H. Wells and L. R. Williams, *Embeddings and extensions in analysis*, Springer-Verlag, New York-Heidelberg, 1975. Ergebnisse der Mathematik und ihrer Grenzgebiete, Band 84. MR0461107 (57 #1092)

[ZB06] L. Zwald and G. Blanchard, *On the convergence of eigenspaces in kernel Principal Component Analysis*, Proc. NIPS 05, vol. 18, 1649-1656, MIT Press, 2006.

Department of Mathematics, Center for Scientific Computation and Mathematical Modeling, Norbert Wiener Center, University of Maryland, College Park, Maryland 20742

E-mail address: rvbalan@math.umd.edu

Proceedings of Symposia in Applied Mathematics
Volume **73**, 2016
http://dx.doi.org/10.1090/psapm/073/00633

Compressed sensing and dictionary learning

Guangliang Chen and Deanna Needell

Abstract. Compressed sensing is a new field that arose as a response to inefficient traditional signal acquisition schemes. Under the assumption that the signal of interest is *sparse*, one wishes to take a small number of linear samples and later utilize a reconstruction algorithm to accurately recover the compressed signal. Typically, one assumes the signal is sparse itself or with respect to some fixed orthonormal basis. However, in applications one instead more often encounters signals sparse with respect to a tight frame which may be far from orthonormal. In the first part of these notes, we will introduce the compressed sensing problem as well as recent results extending the theory to the case of sparsity in tight frames.

The second part of the notes focuses on dictionary learning which is also a new field and closely related to compressive sensing. Briefly speaking, a dictionary is a redundant system consisting of prototype signals that are used to express other signals. Due to the redundancy, for any given signal, there are many ways to represent it, but normally the sparsest representation is preferred for simplicity and easy interpretability. A good analog is the English language where the dictionary is the collection of all words (prototype signals) and sentences (signals) are short and concise combinations of words. Here we will introduce the problem of dictionary learning, its applications, and existing solutions.

Contents

1. Introduction
2. Background to Compressed signal processing
3. Compressed sensing with tight frames
4. Dictionary Learning: An introduction
5. Dictionary Learning: Algorithms
Acknowledgment
References

1. Introduction

Many signals of interest contain far less information than their ambient dimension suggests, making them amenable to compression. However, traditional

2010 *Mathematics Subject Classification.* 94A12, 41A45, 42A10.

Key words and phrases. ℓ_1-analysis, ℓ_1-synthesis, tight frames, dictionary sparsity, compressed sensing, dictionary learning, image denoising, K-SVD, geometric multiresolution analysis, online dictionary learning.

D. Needell was supported by NSF CAREER #1348721 and the Alfred P. Sloan Fellowship.

signal acquisition schemes sample the entire signal only to discard most of that information during the compression process. This wasteful and costly acquisition methodology leads one to ask whether there is an acquisition scheme in which the compressed samples are obtained directly, without the need for time and resources to observe the entire signal. Surprisingly, the answer is often *yes*.

Compressive signal processing (CSP) or compressed sensing (CS) is a new and fast growing field which seeks to resolve this dilemma [**Can06**, **Don06**, **BS07**, **DSP**]. Work in CSP demonstrates that for a certain class of signals, very few *compressive samples* are necessary to accurately represent the signal. In fact, the number of samples required is proportional to the amount of information one wishes to acquire from the signal, and only weakly dependent on the signal's ambient dimension. These samples can be acquired directly from the signal via a linear mapping, and thus the costly process of observing the entire signal is completely eliminated.

Once a signal is acquired via this CSP technology, one needs an efficient algorithm to recover the signal from the compressed samples. Fortunately, CSP has also provided us with methods for recovery which guarantee tractable and robust signal reconstruction. The CSP methodology continues to impact areas ranging from imaging [**WLD+06**, **LDP07**, **PPM**], analog-to-information conversion [**TWD+06**, **KLW+06**, **ME11**] and radar [**BS07**, **HS09**] to geophysical data analysis [**LH07**, **TSHM09**] and computational biology [**DSMB09**, **MSW+10**].

An important aspect of the application of CSP to real-world scenarios is that the sparsifying basis must be known. Often, they are carefully designed based on a mathematical model of the expected kind of signal, with corresponding requirement that they possess some desired theoretical property, such as the Restricted Isometry Property [**CT05**, **CT06**]. Typical choices are random matrices with subgaussian entries or random sign entries.

The dictionary learning problem is closely related to the CSP but arises in a different context, where the main goal is to find compact and meaningful signal representations and correspondingly use them in signal and image processing tasks, such as compression [**BE08**], denoising [**BCM05**, **EA06**, **MSE**], deblurring [**HX13**], and super-revolution [**PETM09**].

Specifically, given signal data $\mathbf{x}_1, \ldots, \mathbf{x}_n \in \mathbb{R}^L$, we train a *dictionary* $\mathbf{D} = [\mathbf{d}_1, \ldots, \mathbf{d}_m] \in \mathbb{R}^{m \times L}$, which can be thought of as an overcomplete basis consisting of elementary signals (called *atoms*). We then use the learned dictionary to represent a signal $\mathbf{x} \in \mathbb{R}^L$ by finding the coefficient vector γ that satisfies the equation $\mathbf{x} = \mathbf{D}\gamma$. When the dictionary forms a basis, there is exactly one solution and thus every signal is uniquely represented as a linear combination of the dictionary atoms. While mathematically this is very simple to operate, such a unique representation has very limited expressiveness.

When $\mathbf{D}$ is an overcomplete system, the problem has more than one solution. This gives us greater flexibility in choosing which coefficient to use for the signal, and allows us to seek the most informative representation, often measured by some cost function $C(\gamma)$:

$$\gamma^* = \arg\min C(\gamma) \qquad \text{subject to} \quad \mathbf{x} = \mathbf{D}\gamma.$$

For example, if one chooses $C(\gamma) = \|\gamma\|_0$, which counts the number of nonzero entries, the above program effectively searches for the sparsest solution, a problem commonly referred to as *sparse coding* [**MZ93**, **OF96**, **CDS98**, **BDE09**].

However, similarly to CSP, the choice of the dictionary $\mathbf{D}$ is crucial but it often requires extensive effort to build it. Traditionally, the signal processing community heavily depended on the Fourier and wavelet dictionaries, which perform quite well for 1-dimensional signals. However, these dictionaries are not adequate for representing more complex natural signal data, especially in higher dimensions, so better dictionary structures were sought.

A variety of dictionaries have been developed in response to the rising need. These dictionaries emerge from one of two sources – either a mathematical model of the data [**AR77, Mal89, Dau92, Bas80, Jan81, CD04, CDDY00**], or a set of realizations of the data [**AEB05, AEB06, ZCP+09, MBPS09, MBPS10, CM10, CM11b, ACM12**]. Dictionaries of the first type are often referred to as *analytic* dictionaries, because they are characterized by an analytic formulation and often equipped with a fast implicit implementation. In contrast, dictionaries of the second type deliver increased flexibility and possess the ability to adapt to specific signal data, and for this reason they are called *data-dependent* dictionaries. In this manuscript we focus on data-dependent dictionaries.

Organization. The rest of the lecture notes consists of four sections, the first two of which are devoted to compressive sensing and the last two to dictionary learning. In each part, we carefully present the background material, the problem being considered, existing solutions and theory, and its connection to other fields (including frames).

2. Background to Compressed signal processing

2.1. The CSP Model. In the model of CSP, the signal $\boldsymbol{f}$ in general is an element of $\mathbb{C}^d$. Linear measurements are taken of the form

$$y_i = \langle \boldsymbol{\phi}_i, \boldsymbol{f} \rangle \quad \text{for } i = 1, 2, \dots m,$$

where $m \ll d$. The vectors $\boldsymbol{\phi}_i$ can be viewed as columns from an $m \times d$ matrix $\boldsymbol{\Phi}$, which we call the *sampling operator*, and the measurement vector $\boldsymbol{y}$ is of the form $\boldsymbol{y} = \boldsymbol{\Phi}\boldsymbol{f}$. With $m \ll d$, $\boldsymbol{\Phi}$ clearly has a nontrivial nullspace and thus the problem of reconstructing $\boldsymbol{f}$ from $\boldsymbol{y}$ is ill-posed without further assumptions. The additional assumption in CSP is that the signals of interest contain far less information than the dimension d suggests. A means for quantifying this notion is called *sparsity*. We say that a signal $\boldsymbol{f} \in \mathbb{C}^d$ is s- *sparse* when it has at most s non-zero components:

$$\|\boldsymbol{f}\|_0 \stackrel{\text{def}}{=} |\operatorname{supp}(\boldsymbol{f})| \leq s \ll d, \tag{2.1}$$

where $\|\cdot\|_0$ denotes the ℓ_0 quasi-norm. For $1 \leq p < \infty$, $\|\cdot\|_p$ denotes the usual p-norm,

$$\|\boldsymbol{f}\|_p := \left(\sum_{i=1}^{d} |\boldsymbol{f}_i|^p \right)^{1/p},$$

and $\|\boldsymbol{f}\|_\infty = \max |f_i|$. In practice, signals are often encountered that are not exactly sparse, but whose coefficients decay rapidly. *Compressible* signals are those satisfying a power law decay:

$$|\boldsymbol{f}_k^*| \leq R k^{(-1/q)}, \tag{2.2}$$

where $\boldsymbol{f}^*$ is a non-increasing rearrangement of $\boldsymbol{f}$, R is some positive constant, and $0 < q < 1$. Note that in particular, sparse signals are compressible and for small values of q compressibility becomes essentially the same as sparsity. In any case, compressible signals are well approximated by sparse signals since the majority of the energy in the signal is captured by a few components. If we denote by $\boldsymbol{f}_s$ the vector consisting of the s largest coefficients in magnitude of $\boldsymbol{f}$, then we see that for compressible signals $\boldsymbol{f}$ and $\boldsymbol{f}_s$ are close,

$$\|\boldsymbol{f} - \boldsymbol{f}_s\|_2 \leq Rs^{1/2-1/q} \quad \text{and} \quad \|\boldsymbol{f} - \boldsymbol{f}_s\|_1 \leq Rs^{1-1/q}.$$

Therefore when working with compressible signals we may capture the majority of the information in the signal by taking advantage of sparsity.

One observes however that this definition (2.1) of sparsity requires that the signal itself contain few non-zeros. This notion can be generalized by asking instead that the signal $\boldsymbol{f}$ of interest be sparse with respect to some *sparsifying basis*. We fix some orthonormal basis, written as the columns of the matrix $\boldsymbol{D}$. Then formally, we will again call a signal $\boldsymbol{f}$ s-sparse when

$$\boldsymbol{f} = \boldsymbol{D}\boldsymbol{x} \quad \text{with} \quad \|\boldsymbol{x}\|_0 \leq s \ll d. \tag{2.3}$$

We call $\boldsymbol{x}$ the *coefficient vector*. We will say that $\boldsymbol{f}$ is compressible when its coefficient vector $\boldsymbol{x}$ satisfies the power law decay as in (2.2). Many signals in practice are compressible in this sense. Natural signals such as images are often compressible with respect to the identity or wavelet sparsifying basis [**CT05**, **CRT06b**, **CW08**]. Likewise, manmade signals such as those in radar, medical imaging, and communications applications are compressible with respect to the Fourier basis and other sparsifying bases [**BS07**, **Rom08**]. Since $\boldsymbol{D}$ is an orthonormal system, we may think of absorbing $\boldsymbol{D}$ into the sampling operator and attempting to estimate $\boldsymbol{f}$ by estimating $\boldsymbol{x}$. From this viewpoint we can assume $\boldsymbol{f}$ is sparse with respect to the coordinate basis and acknowledge that results for this class of signals apply also to the broader class which are sparse with respect to some fixed orthonormal basis.

2.2. Sampling Mechanisms. The sampling operator $\boldsymbol{\Phi}$ is a linear map from $\mathbb{C}^d$ to some lower dimensional space $\mathbb{C}^m$. It is clear that to recover a sparse signal $\boldsymbol{f}$ from its measurements $\boldsymbol{y} = \boldsymbol{\Phi}\boldsymbol{f}$ one at least needs that $\boldsymbol{\Phi}$ is one-to-one on all sparse vectors. Indeed, if this is the case then to recover $\boldsymbol{f}$ from $\boldsymbol{y}$ one simply solves the minimization problem

$$\hat{\boldsymbol{f}} = \operatorname*{argmin}_{g \in \mathbb{C}^d} \|\boldsymbol{g}\|_0 \quad \text{subject to } \boldsymbol{\Phi}\boldsymbol{g} = \boldsymbol{y}. \tag{2.4}$$

Then since $\boldsymbol{\Phi}$ does not map any two sparse vectors to the same image, it must be that we recover our signal, $\hat{\boldsymbol{f}} = \boldsymbol{f}$. This minimization problem however, is intractable and NP-Hard in general [**Mut05**, Sec. 9.2.2]. We thus consider slightly stronger requirements on $\boldsymbol{\Phi}$. The first assumption one can make on the sampling operator is that its columns are *incoherent*. For a matrix $\boldsymbol{\Phi}$ with unit norm columns $\{\boldsymbol{\phi}_i\}$, we define its coherence μ to be the largest correlation among the columns,

$$\mu = \max_{i \neq j} \langle \boldsymbol{\phi}_i, \boldsymbol{\phi}_j \rangle .$$

A sampling operator is *incoherent* when its coherence μ is sufficiently small. Incoherent operators are thus those which are approximately orthonormal on sparse vectors.

An alternative property which captures this idea was developed by Candès and Tao and is called the Restricted Isometry Property (RIP) [**CT05**, **CT06**]. The RIP implies incoherence and that the operator $\boldsymbol{\Phi}$ approximately preserves the geometry of all sparse vectors. Formally, they define the restricted isometry constant δ_s to be the smallest constant such that

$$(2.5)\qquad (1-\delta_s)\|\boldsymbol{f}\|_2^2 \le \|\boldsymbol{\Phi f}\|_2^2 \le (1+\delta_s)\|\boldsymbol{f}\|_2^2 \quad \text{for all } s\text{-sparse vectors } \boldsymbol{f}.$$

We say that the sampling operator $\boldsymbol{\Phi}$ has the RIP of order s when δ_s is sufficiently small, say $\delta_s \le 0.1$. The important question is of course what type of sampling operators have this property, and how large does the number m of samples have to be. Fortunately, the literature in CSP has shown that many classes of matrices possess this property when the number of measurements m is *nearly linear in the sparsity* s. Two of the most important examples are the following.

Subgaussian matrices. A random variable X is subgaussian if $\mathbb{P}(|X| > t) \le Ce^{-ct^2}$ for all $t > 0$ and some positive constants C, c. Thus subgaussian random variables have tail distributions that are dominated by that of the standard Gaussian random variable. Choosing $C = c = 1$, we trivially have that standard Gaussian matrices (those whose entries are distributed as standard normal random variables) are subgaussian. Choosing $C = \frac{1}{e}$ and $c = 1$, we see that Bernoulli matrices (those whose entries are uniform ± 1) are also subgaussian. More generally, any bounded random variable is subgaussian. It has been shown that if $\boldsymbol{\Phi}$ is an $m \times d$ subgaussian matrix then with high probability $\frac{1}{\sqrt{m}}\boldsymbol{\Phi}$ satisfies the RIP of order s when m is on the order of $s \log d$ [**MPTJ08, RV06**].

Partial bounded orthogonal matrices. Let $\boldsymbol{\Psi}$ be an orthogonal $d \times d$ matrix whose entries are bounded by $C/\sqrt{d}$ for some constant C. A $m \times d$ partial bounded orthogonal matrix is a matrix $\boldsymbol{\Phi}$ formed by choosing m rows of such a matrix $\boldsymbol{\Psi}$ uniformly at random. Since the $d \times d$ discrete Fourier transform matrix is orthogonal with entries bounded by $1/\sqrt{d}$, the $m \times d$ random partial Fourier matrix is a partial bounded orthogonal matrix. Rudelson and Vershynin showed that such matrices satisfy the RIP with high probability when the number of measurements m is on the order of $s \log^4 d$ [**RV08**].

Work continues to be done to demonstrate other types of random matrices which satisfy the RIP so that this assumption is quite viable in many practical applications (see e.g. [**HN07, KW11, PRT11**]). Matrices with structure such as the partial Fourier are particularly important in applications since they can utilize a fast-multiply.

2.3. Current Approaches to CSP.

Since the problem (2.4) is computationally infeasible, alternative methods are needed. In addition, the signal $\boldsymbol{f}$ is often compressible rather than sparse, and the samples are often corrupted by noise so that the measurement vector is actually $\boldsymbol{y} = \boldsymbol{\Phi f} + \boldsymbol{e}$ for some error vector $\boldsymbol{e}$. An ideal recovery method would thus possess the following properties.

Nonadaptive samples: The method should utilize sampling operators $\boldsymbol{\Phi}$ which do not depend on the signal $\boldsymbol{f}$. Note that the operators satisfying the RIP above possess the nonadaptivity property.

Optimal number of samples: The number m of samples required for reconstruction should be minimal.

Uniform Guarantees: One sampling operator should suffice for recovery of *all* signals.

Robust Recovery: The method should be stable and robust to noise and provide optimal error guarantees.

Computational Complexity: The algorithm should be computationally efficient.

There are currently two main approaches in CSP which provide methods for sparse reconstruction with these ideal properties in mind. The first solves an optimization problem and the second utilizes greedy algorithms to recover the signal.

2.3.1. *Optimization based methods.* Initial work in CSP [**DH01**, **CT05**, **Don06**, **CRT06b**, **Tro06**] considered the convex relaxation of the NP-Hard problem (2.4). The closest convex norm to the ℓ_0 quasi-norm is the ℓ_1-norm, and the geometry of the ℓ_1-ball promotes sparsity. We therefore estimate a compressible signal $\boldsymbol{f}$ by the minimizer $\hat{\boldsymbol{f}}$ to the following problem

$$\hat{\boldsymbol{f}} = \underset{\boldsymbol{g}\in\mathbb{C}^d}{\operatorname{argmin}} \|\boldsymbol{g}\|_1 \quad \text{subject to } \|\boldsymbol{\Phi g} - \boldsymbol{y}\|_2 \le \varepsilon, \tag{2.6}$$

where ε bounds the norm of the noise: $\|\boldsymbol{e}\|_2 \le \varepsilon$. This problem can be formulated as a linear program and so standard methods in Linear Programming can be used to solve it. Candès, Romberg and Tao showed that this ℓ_1-minimization problem provides the following error guarantee.

THEOREM 2.1 (Candès-Romberg-Tao [**CRT06b**]). *Let $\boldsymbol{\Phi}$ be a sampling operator which satisfies the RIP. Then for any signal $\boldsymbol{f}$ and noisy measurements $\boldsymbol{y} = \boldsymbol{\Phi f} + \boldsymbol{e}$ with $\|\boldsymbol{e}\|_2 \le \varepsilon$, the solution $\hat{\boldsymbol{f}}$ to* (2.6) *satisfies*

$$\|\hat{\boldsymbol{f}} - \boldsymbol{f}\|_2 \le C\left[\varepsilon + \frac{\|\boldsymbol{f} - \boldsymbol{f}_s\|_1}{\sqrt{s}}\right],$$

where $\boldsymbol{f}_s$ again denotes the vector of the s largest coefficients in magnitude of $\boldsymbol{f}$.

This result says that the recovery error is at most proportional to the norm of the noise in the samples and the tail of the signal. The error bound is optimal up to the precise value of the constant C [**CDD09**]. Note that when the signal $\boldsymbol{f}$ is exactly sparse and there is no noise in the samples that this result confirms that the signal $\boldsymbol{f}$ is reconstructed exactly [**CT05**]. For a compressible signal as in (2.2), this bound guarantees that

$$\|\hat{\boldsymbol{f}} - \boldsymbol{f}\|_2 \le C[\varepsilon + Rs^{1/2-1/q}].$$

Therefore, using Gaussian or Fourier samples, by solving a linear program we can achieve an optimal error bound of this form with number of samples m approximately $s\log d$. This result thus provides uniform guarantees with optimal error bounds using few nonadaptive samples. Although linear programming methods are becoming more and more efficient, for some applications the computational cost of this approach may still be burdensome. For that reason, greedy algorithms have been proposed and may provide some advantages.

2.3.2. *Greedy methods.* Orthogonal Matching Pursuit (OMP) is an early greedy algorithm for sparse reconstruction analyzed by Gilbert and Tropp [**TG07**]. Given an exactly s-sparse signal $\boldsymbol{f}$ with noiseless samples $\boldsymbol{y} = \boldsymbol{\Phi f}$, OMP iteratively identifies elements of the support of $\boldsymbol{f}$. Once the support T is located, the signal is reconstructed by $\hat{\boldsymbol{f}} = \boldsymbol{\Phi}_T^\dagger \boldsymbol{y}$ where $\boldsymbol{\Phi}_T$ denotes the restriction of $\boldsymbol{\Phi}$ to the columns indexed by T and $\boldsymbol{\Phi}_T^\dagger$ denotes its pseudo-inverse. The critical observation which allows OMP to succeed is that when $\boldsymbol{\Phi}$ is Gaussian or more generally is incoherent, $\boldsymbol{\Phi}^*\boldsymbol{\Phi}$ is close to the identity. Thus $\boldsymbol{u} := \boldsymbol{\Phi}^*\boldsymbol{y} = \boldsymbol{\Phi}^*\boldsymbol{\Phi f}$ is in a loose sense close to $\boldsymbol{f}$, and so OMP estimates that the largest coefficient of $\boldsymbol{u}$ is in the true support of

$\boldsymbol{f}$. The contribution from that column is subtracted from the samples, and OMP repeats this process. Gilbert and Tropp showed that OMP correctly recovers a *fixed* sparse signal with high probability. Indeed, in [**TG07**] they proved the following.

THEOREM 2.2 (OMP Signal Recovery [**TG07**]). *Let $\boldsymbol{\Phi}$ be an $m \times d$ (sub)Gaussian measurement matrix with $m \geq Cs \log d$ and let $\boldsymbol{f}$ be an s-sparse signal in $\mathcal{R}^d$. Then with high probability, OMP correctly reconstructs the signal $\boldsymbol{f}$ from its measurements $\boldsymbol{y} = \boldsymbol{\Phi f}$.*

Without modifications, OMP is not known to be robust to noise. Also, the result provides non-uniform guarantees; for a given sampling operator and fixed signal, OMP recovers the signal with high probability. In fact, a uniform guarantee from OMP has been proved to be impossible [**Rau08**]. However, the strong advantage of OMP over previous methods is its extremely low computational cost. Using an efficient implementation reduces the overall cost of OMP to $O(smd)$ in general, and even faster when the sampling operator has a fast-multiply.

CoSaMP. Motivated by a breakthrough greedy algorithm analyzed with the RIP [**NV07b**, **NV07a**], Needell and Tropp developed the greedy method Compressive Sampling Matching Pursuit (CoSaMP) [**NT08b**, **NT08a**], which is similar in spirit to OMP. This algorithm again has a similar structure to OMP. In each iteration, multiple components are selected to be in the support of the estimation. Then the signal is estimated using this support and then pruned to maintain sparsity. A critical difference between CoSaMP and the other two matching pursuits is that in CoSaMP elements of the support which are incorrectly identified may be removed from the estimation in future iterations. Formally, the CoSaMP template is described by the pseudocode below. Note that parameters within the algorithm can of course be tuned for optimal performance.

COMPRESSIVE SAMPLING MATCHING PURSUIT (COSAMP) [**NT08b**]

INPUT: Sampling operator $\boldsymbol{\Phi}$, sample vector $\boldsymbol{y} = \boldsymbol{\Phi f}$, sparsity level s
OUTPUT: s-sparse reconstructed vector $\hat{\boldsymbol{f}} = \boldsymbol{a}$
PROCEDURE:

Initialize: Set $\boldsymbol{a}^0 = \mathbf{0}$, $\boldsymbol{v} = \boldsymbol{y}$, $k = 0$. Repeat the following steps and increment k until the halting criterion is true.
Signal Proxy: Set $\boldsymbol{u} = \boldsymbol{\Phi}^*\boldsymbol{v}$, $\Omega = \operatorname{supp} \boldsymbol{u}_{2s}$ and merge the supports: $T = \Omega \cup \operatorname{supp} \boldsymbol{a}^{k-1}$.
Signal Estimation: Using least-squares, set $\boldsymbol{b}|_T = \boldsymbol{\Phi}_T^\dagger \boldsymbol{y}$ and $\boldsymbol{b}|_{T^c} = \mathbf{0}$.
Prune: To obtain the next approximation, set $\boldsymbol{a}^k = \boldsymbol{b}_s$.
Sample Update: Update the current samples: $\boldsymbol{v} = \boldsymbol{y} - \Phi \boldsymbol{a}^k$.

Several halting criteria are offered in [**NT08b**], the simplest of which is to halt after $6s$ iterations. The authors prove the following guarantee for the CoSaMP algorithm.

THEOREM 2.3 (CoSaMP [**NT08b**]). *Suppose that $\boldsymbol{\Phi}$ is an $m \times d$ sampling matrix satisfying the RIP. Let $\boldsymbol{y} = \boldsymbol{\Phi f} + \boldsymbol{e}$ be a vector of samples of an arbitrary* *signal, contaminated with noise. The algorithm CoSaMP produces an s-sparse approximation $\hat{\boldsymbol{f}}$ that satisfies*

$$\|\hat{\boldsymbol{f}} - \boldsymbol{f}\|_2 \leq C\left[\varepsilon + \frac{\|\boldsymbol{f} - \boldsymbol{f}_s\|_1}{\sqrt{s}}\right],$$

where $\boldsymbol{f}_s$ is a best (s)-sparse approximation to $\boldsymbol{f}$. The running time is $\mathrm{O}(\mathcal{L} \cdot \log(\|\boldsymbol{f}\|_2))$, where $\mathcal{L}$ bounds the cost of a matrix–vector multiply with $\boldsymbol{\Phi}$ or $\boldsymbol{\Phi}^*$. Working storage is $\mathrm{O}(d)$.

CoSaMP utilizes minimal nonadaptive samples and provides uniform guarantees with optimal error bounds. The computational cost is proportional to the cost of applying the sampling operator and CoSaMP is therefore the first algorithm to provide optimality at every critical aspect. In addition, under a Gaussian noise model, CoSaMP and other greedy methods have guaranteed recovery error similar to the best possible obtained when the support of the signal is known [**GE12**].

Other greedy methods like the iterative hard thresholding algorithm (IHT) can also provide analogous guarantees [**BD09**]. IHT can be described by the simple recursive iteration

$$\boldsymbol{x}_k = H_s(\boldsymbol{x}_{k-1} + \boldsymbol{\Phi}(\boldsymbol{y} - \boldsymbol{\Phi}\boldsymbol{x}_{k-1})),$$

where H_s is the thresholding operator which sets all but the largest (in magnitude) s entries to zero, and $\boldsymbol{x}_0$ can be chosen as an arbitrary starting estimate. We focus mainly on the CoSaMP greedy method and its adaptation to tight frames in these notes, but see e.g. [**BD09, Blu11, GNE$^+$12**] for similar adaptations of methods like IHT.

2.3.3. *Total variation methods.* In numerous CSP applications, the signals of interest are images. Natural images tend to be compressible with respect to some orthonormal basis such as the wavelet basis. With this notion of sparsity, one can use ℓ_1-minimization or a greedy method to recover the image from a small number of measurements. The standard results in CSP then guarantee the reconstruction error will be small, relative to the noise level and the compressibility of the signal. However, the errors using this approach arise as artifacts from high frequency oscillations, and their structure often appears displeasing to the eye, and makes image analysis challenging. An alternative is to consider the sparsity of the signal with respect to the image *gradient*, rather than some orthonormal basis. Minimizing the ℓ_1-norm of the gradient leads to the well-known *total variation* program.

The key to the total variation problem is that because of the structure of natural images, their gradients tend to be sparse. In other words, the matrix whose entries are the distances between neighboring pixels of a natural image is a compressible matrix. Concretely, we define the *total variation* (TV) of an image $\boldsymbol{X}$ as

$$\|\boldsymbol{X}\|_{TV} \stackrel{\text{def}}{=} \sum_{j,k}\sqrt{(\boldsymbol{X}_{j+1,k} - \boldsymbol{X}_{j,k})^2 + (\boldsymbol{X}_{j,k+1} - \boldsymbol{X}_{j,k})^2} = \sum_{j,k}|(\nabla\boldsymbol{X})_{j,k}|,$$

where $\nabla\boldsymbol{X}$ denotes the (discrete) gradient of the image. The gradient can then be defined by writing

$$\boldsymbol{X}_x : \mathbb{C}^{N\times N} \to \mathbb{C}^{(N-1)\times N}, \qquad (\boldsymbol{X}_x)_{j,k} = \boldsymbol{X}_{j+1,k} - \boldsymbol{X}_{j,k} \tag{2.7}$$

$$\boldsymbol{X}_y : \mathbb{C}^{N\times N} \to \mathbb{C}^{N\times(N-1)}, \qquad (\boldsymbol{X}_y)_{j,k} = \boldsymbol{X}_{j,k+1} - \boldsymbol{X}_{j,k}, \tag{2.8}$$

and then setting

$$(2.9)\qquad \big[\boldsymbol{\nabla X}\big]_{j,k} \stackrel{\text{def}}{=} \begin{cases} \big((\boldsymbol{X}_x)_{j,k}, (\boldsymbol{X}_y)_{j,k}\big), & 1 \le j \le N-1, \quad 1 \le k \le N-1 \\ \big(0, (\boldsymbol{X}_y)_{j,k}\big), & j = N, \quad 1 \le k \le N-1 \\ \big((\boldsymbol{X}_x)_{j,k}, 0\big), & k = N, \quad 1 \le j \le N-1 \\ \big(0, 0\big), & j = k = N. \end{cases}$$

Since images have a compressible gradient, we may consider minimizing with respect to the TV-norm:

$$\hat{\boldsymbol{X}} = \operatorname*{argmin}_{\boldsymbol{M}} \|\boldsymbol{M}\|_{TV} \quad \text{subject to} \quad \|\boldsymbol{y} - \mathcal{A}(\boldsymbol{M})\|_2 \le \varepsilon,$$

where $\boldsymbol{y} = \mathcal{A}(\boldsymbol{X}) + \boldsymbol{e}$ are noisy measurements with bounded noise $\|\boldsymbol{e}\|_2 \le \varepsilon$.

Instead of searching for a sparse image in the wavelet basis, the total variation problem searches for an image with a sparse gradient. This reduces the high frequency oscillatory artifacts from the recovered image, as seen in Figure 1.

(a) (b) (c)

Figure 1 *Images from* [**NW13**]. *(a) Original cameraman image, (b) its reconstruction from 20% random Fourier coefficients using total-variation minimization and (c) ℓ_1-minimization of its Haar wavelet coefficients.*

The benefits of using total variation norm minimization have been observed extensively and the method is widely used in practice (see e.g. [**CRT06b**, **CRT06a**, **CR05**, **OSV03**, **CSZ06**, **LDP07**, **LDSP08**, **LW11**, **NTLC08**, **MYZC08**], [**KTMJ08**, **Kee03**]). Despite this, theoretical results showing robust recovery via TV have only been obtained very recently. In [**NW12**, **NW13**], Needell and Ward prove the first robust theoretical result for total variation minimization:

Theorem 2.4 (Needell and Ward [**NW12**, **NW13**]). *From* $\mathrm{O}(s\log(N))$ *linear RIP measurements with noise level* ε, *for any* $\boldsymbol{X} \in \mathbb{C}^{N\times N}$, *the solution* $\hat{\boldsymbol{f}}$ *to the TV minimization problem satisfies*

$$\|\boldsymbol{X} - \hat{\boldsymbol{X}}\|_2 \lesssim \log(N) \cdot \left[\frac{\|\nabla[\boldsymbol{X}] - \nabla[\boldsymbol{X}]_s\|_1}{\sqrt{s}} + \varepsilon\right]$$

Analogous to the bounds of ℓ_1-optimization, this result guarantees that the recovery error is at most proportional to the noise in the samples and the "tail" of the compressible gradient. The proof technique relies on the development of an improved Sobolev inequality, and the error guarantees obtained are optimal up to the logarithmic factor. The linear measurements can be obtained from a RIP sampling operator, and have also been extended to higher dimensional arrays, see [**NW12**, **NW13**] for details.

2.4. Matrix recovery by CSP. In many applications, the signals of interest are better represented by matrices than by vectors. Such a signal may still be sparse in the sense described previously (2.1), in which case the theory above extends naturally. Alternatively, a data matrix may possess some low-rank structure. Then the question becomes, given measurements of such a low-rank data matrix, can one recover the matrix? This problem gained popularity by the now famous NetFlix problem in collaborative filtering [**RS05, Sre04**]. In this problem, the data matrix consists of user ratings for movies. Since not every user rates every movie and not every movie is rated by every user, only partial information about the true rating data matrix is known. From these partial measurements one wishes to obtain the true matrix containing the missing entries so that preferences can be inferred and movie recommendations can be made to the users.

A similar problem arises in the triangulation from partial data. Here, one is given some information about distances between objects in a network and wishes to recover the (low-dimensional) geometry of the network [**LLR95, SY07, Sch86, Sin08**]. This type of problem of course appears in many applications including remote sensing, wireless communications, and global positioning.

Formally speaking, in all of these problems we are given measurements $\boldsymbol{y} = \mathcal{A}(\boldsymbol{X})$ and wish to recover the low-rank data matrix $\boldsymbol{X}$. In general the measurement operator is of the form $\mathcal{A} : \mathbb{R}^{n \times n} \to \mathbb{R}^m$ and acts on a matrix $\boldsymbol{X}$ by

$$(\mathcal{A}(\boldsymbol{X}))_i = \langle \boldsymbol{A}_i, \boldsymbol{X} \rangle \tag{2.10}$$

where $\boldsymbol{A}_i$ are $n \times n$ matrices and $\langle \cdot, \cdot \rangle$ denotes the usual matrix inner product:

$$\langle \boldsymbol{A}, \boldsymbol{B} \rangle \stackrel{\text{def}}{=} \operatorname{trace}(\boldsymbol{A}^* \boldsymbol{B}).$$

Analogous to the program (2.4), one considers solving

$$\hat{\boldsymbol{X}} = \operatorname*{argmin}_{\boldsymbol{M}} \operatorname{rank}(\boldsymbol{M}) \quad \text{such that} \quad \mathcal{A}(\boldsymbol{M}) = \boldsymbol{y}.$$

However, as in the case of (L_0), the problem (2.4) is not computationally feasible in general. We thus consider instead its relaxation, which minimizes the ℓ_1-norm of its singular values.

$$\hat{\boldsymbol{X}} = \operatorname*{argmin}_{\boldsymbol{M}} \|\boldsymbol{M}\|_* \quad \text{such that} \quad \mathcal{A}(\boldsymbol{M}) = \boldsymbol{y}. \tag{2.11}$$

Here $\|\cdot\|_*$ denotes the *nuclear norm* which is defined by

$$\|\boldsymbol{X}\|_* = \operatorname{trace}(\sqrt{\boldsymbol{X}^* \boldsymbol{X}}) = \|\sigma(\boldsymbol{X})\|_1,$$

where $\sigma(\boldsymbol{X})$ is the vector of singular values of $\boldsymbol{X}$.

This program (2.11) can be cast as a semidefinite program and is thus numerically feasible. Work in CSP has shown [**NRWY10, RFP07, OH10, CP09**] that $m \geq Cnr$ measurements suffice to recover any $n \times n$ rank-r matrix via (2.11).

2.4.1. *Matrix Decomposition.* In addition to the recovery of a low-rank structure, one may also simultaneously wish to recover a sparse component of a data matrix. That is, given a data matrix $\boldsymbol{X}$, one seeks to identify a low-rank component $\boldsymbol{L}$ and a sparse component $\boldsymbol{S}$ such that $\boldsymbol{X} = \boldsymbol{L} + \boldsymbol{S}$. For example, in a surveillance video one may wish to decompose the foreground from the background to detect moving targets. This problem has received much attention recently and can be interpreted as a robust version of principal component analysis [**CSPW11, EJCW09**]. The applications of this problem are numerous and include video surveillance, facial recognition, collaborative filtering and many more.

A particular challenge of this application is in the well-posedness of the decomposition problem. For example, if the sparse component also has some low-rank structure or the low-rank component is sparse, the problem does not have a unique solution. Thus some assumptions are placed on the structure of both components. Current results assume that the low-rank component $\boldsymbol{L}$ satisfies an incoherence condition (see Section 1.3 of [**EJCW09**]) which guarantees that its singular vectors are sufficiently spread and that the sparsity pattern in the sparse component $\boldsymbol{S}$ is selected uniformly at random.

The proposed method for solving this decomposition problem is *Principal Component Pursuit* [**EJCW09, ZLW⁺10**] which solves the following convex optimization problem

$$(\hat{\boldsymbol{L}}, \hat{\boldsymbol{S}}) = \underset{\boldsymbol{L},\boldsymbol{S}}{\operatorname{argmin}} \|\boldsymbol{L}\|_* + \lambda\|\boldsymbol{S}\|_1 \quad \text{subject to} \quad \boldsymbol{L} + \boldsymbol{S} = \boldsymbol{X}. \tag{2.12}$$

Under the assumption that the low-rank component $\boldsymbol{L}$ has spread singular vectors and that the sparsity pattern of $\boldsymbol{S}$ is uniformly random, Candès et.al. show that with high probability the $n \times n$ decomposition $\boldsymbol{L} + \boldsymbol{S}$ can be exactly recovered when the rank r of $\boldsymbol{L}$ is proportional to $n/\log(n)$ and the sparsity s of $\boldsymbol{S}$ is a constant fraction of the entries, $s \leq cn^2$. This astonishing result demonstrates that the low-rank component of a data matrix can be identified even when a fixed fraction of the entries in the matrix are corrupted – and that these errors can have arbitrarily large magnitudes!

It is clear that some assumptions must be made on the individual components in the decomposition for the problem to even be well-posed. However, in many applications it may not be practical to impose such randomness in the sparsity pattern of the sparse component. We discuss this further below.

3. Compressed sensing with tight frames

In the usual CSP framework, the signal $\boldsymbol{f}$ is assumed to be sparse as in (2.3) or compressible with respect to some *orthonormal basis*. As mentioned, there are numerous applications in which the signal of interest falls into this class of signals. However, more often than not, sparsity is expressed not in terms of an orthonormal basis but in terms of an *overcomplete* dictionary. In this setting, the signal $\boldsymbol{f} = \boldsymbol{D}\boldsymbol{x}$ where $\boldsymbol{x}$ is sparse or compressible and $\boldsymbol{D}$ is an arbitrary set of column vectors which we refer to as a *dictionary* or *frame*. The dictionary need not be orthonormal or even incoherent and often it will be overcomplete, meaning it has far more columns than rows. There are numerous applications that use signals sparse in this sense, many of which are of importance to ONR. Some examples of dictionaries we encounter in practice in this setting are the following.

Oversampled DFT: Signals which are sparse with respect to the discrete Fourier matrix (DFT) are precisely those which are superpositions of sinusoids with frequencies in the lattice of those in the DFT. In practice, it is of course rare to encounter such signals. Therefore one often considers the oversampled DFT in which the sampled frequencies are taken over even smaller fixed intervals, small intervals of varying lengths, or even randomly selected intervals. This creates an overcomplete frame that may have high coherence.

Gabor frames: Radar, sonar and other imaging systems aim to recover pulse trains whose atoms have a time-frequency structure [**FS98**]. Because of this structure, Gabor frames are widely used [**Mal99**]. Gabor frames are not incoherent and often very overcomplete.

Curvelet frames: Curvelets provide a multiscale decomposition of images, and have geometric features that distinguish them from other bases like wavelets. The curvelet transform can be viewed as a multiscale pyramid with many directions at each length scale, and needle-shaped elements at fine scales [**CD04, CDDY00**]. Although the transform has many properties of an orthonormal basis, it is overcomplete, and neighboring columns have high coherence.

Wavelet Frames: The undecimated wavelet transform (UWT) is a wavelet transform with a translation invariance property that the discrete wavelet transform (DWT) does not possess [**Dut89**]. The UWT is missing the downsamplers and upsamplers in the DWT but upsamples the filter coefficients by a factor of 2^k at the $(k-1)$st level. This of course makes it very overcomplete. The Unitary Extension Principle of Ron and Shen [**RS97**] enables constructions of tight wavelet frames for $L^2(\mathbb{R}^d)$ which may also be very overcomplete. The overcompleteness has been found to be helpful in image processing [**SED04**].

Concatenations: In many applications a signal may not be sparse with respect to a single orthonormal basis, but may be a composition of sparse signals from multiple orthonormal bases. For example, a linear combination of spikes and sines is sparse with respect to the concatenation of the identity and the Fourier basis. In imaging applications one may wish to take advantage of the geometry of multiple sparsifying bases such as a combination of curvelets, wavelets, and brushlets. The concatenation of these bases is overcomplete and may be highly coherent.

Such redundant dictionaries are now used widespread in signal processing and data analysis. Often, there may simply be no good sparsifying orthonormal basis such as in the applications utilizing Gabor and Curvelet frames. In addition, researchers acknowledge and take advantage of the flexibility provided by overcomplete frames. In general linear inverse problems such as deconvolution, tomography, and signal denoising, it has been observed that using overcomplete dictionaries significantly reduces artifacts and mean squared error [**SED04, SFM07**]. Since CSP problems are special types of inverse problems it is not surprising that redundant frames are equally helpful in this setting.

3.1. The ℓ_1-analysis approach. Since in this generalized setting the sparsity is in the coefficient vector $\boldsymbol{x}$ rather than the signal $\boldsymbol{f}$, it no longer makes sense to minimize the ℓ_1-norm of the signal itself. The intuition behind the ℓ_1-analysis method is that for many dictionaries $\boldsymbol{D}$, $\boldsymbol{D}^*\boldsymbol{f}$ will have rapidly decaying coefficients and thus it becomes natural to minimize the ℓ_1-norm of this vector. Therefore for a signal $\boldsymbol{f} = \boldsymbol{D}\boldsymbol{x}$ and noisy samples $\boldsymbol{y} = \boldsymbol{\Phi}\boldsymbol{f} + \boldsymbol{e}$, the ℓ_1-analysis problem constructs an estimate $\hat{\boldsymbol{f}}$ to $\boldsymbol{f}$ as the solution to the following minimization problem:

$$\hat{\boldsymbol{f}} = \operatorname*{argmin}_{\boldsymbol{g}\in\mathbb{C}^d} \|\boldsymbol{D}^*\boldsymbol{g}\|_1 \quad \text{subject to } \|\boldsymbol{\Phi}\boldsymbol{g} - \boldsymbol{y}\|_2 \le \varepsilon,$$

where as before $\varepsilon \ge \|\boldsymbol{e}\|_2$ is a bound on the noise level.

Recently, Candès et.al. provide error bounds for ℓ_1-analysis [**CENR10**]. This result holds when the dictionary $\boldsymbol{D}$ is a *tight frame*, meaning $\boldsymbol{D}\boldsymbol{D}^*$ equals the identity. All the dictionaries mentioned above are examples of tight frames. In developing theory in this setting, an important issue that had to be addressed was the assumption on the sampling operator $\boldsymbol{\Phi}$. Since sparsity in this setting is captured in the coefficient vector rather than the signal, the following natural extension of the RIP was developed. For a given dictionary $\boldsymbol{D}$, the sampling operator $\boldsymbol{\Phi}$ satisfies the D-RIP of order s when

$$(1-\delta_s)\|\boldsymbol{D}\boldsymbol{x}\|_2^2 \le \|\boldsymbol{\Phi}\boldsymbol{D}\boldsymbol{x}\|_2^2 \le (1+\delta_s)\|\boldsymbol{D}\boldsymbol{x}\|_2^2 \quad \text{for all } s\text{-sparse vectors } \boldsymbol{x}$$

for some small δ_s, say $\delta_s \le 0.08$. Here sparsity in $\boldsymbol{x}$ is with respect to the coordinate basis. D-RIP, therefore, asks that the sampling operator $\boldsymbol{\Phi}$ be approximately orthonormal on all signals $\boldsymbol{f}$ which are sparse with respect to $\boldsymbol{D}$. Using a standard covering argument it is straightforward to show that for a dictionary $\boldsymbol{D}$ with d columns that subgaussian sampling operators satisfy the D-RIP with high probability when the number m of samples is again on the order of $s\log d$ [**CENR10**]. Moreover, if $\boldsymbol{\Phi}$ satisfies the standard RIP, then multiplying the columns by random signs yields a matrix which satisfies the D-RIP [**CENR10, KW11**]. Often, however, it may not be possible to apply random column signs to the sampling matrix. In MRI for example, one is forced to take Fourier measurements and cannot preprocess the data. Recent work by Krahmer et.al. [**KNW15**] shows that one can instead use *variable density sampling* to remove the need for these column signs. In this case, one constructs for example a randomly sub-sampled Fourier matrix by selecting the rows from the standard DFT according to some specified distribution. This shows that the same class of operators used in standard CSP can also be used in CSP with overcomplete dictionaries.

Under this assumption, the error in the estimation provided by ℓ_1-analysis is bounded by the noise level and the energy in the tail of $\boldsymbol{D}^*\boldsymbol{f}$:

THEOREM 3.1 (ℓ_1-analysis Recovery [**CENR10**]). *Let $\boldsymbol{D}$ be an arbitrary tight frame and suppose the sampling operator $\boldsymbol{\Phi}$ satisfies the D-RIP of order s. Then the solution $\hat{f}$ to the ℓ_1-analysis problem satisfies*

$$\|\hat{\boldsymbol{f}} - \boldsymbol{f}\|_2 \le C\left[\varepsilon + \frac{\|\boldsymbol{D}^*\boldsymbol{f} - (\boldsymbol{D}^*\boldsymbol{f})_s\|_1}{\sqrt{s}}\right],$$

where $(\boldsymbol{D}^*\boldsymbol{f})_s$ denotes the largest s entries in magnitude of $\boldsymbol{D}^*\boldsymbol{f}$.

This result states that ℓ_1-analysis provides robust recovery for signals $\boldsymbol{f}$ whose coefficients $\boldsymbol{D}^*\boldsymbol{f}$ decay rapidly. Observe that when the dictionary $\boldsymbol{D}$ is the identity, this recovers precisely the error bound for standard ℓ_1-minimization. Without further assumptions or modifications, this result is optimal. The bound is the natural bound one expects since the program minimizes a sparsity promoting norm over the image of $\boldsymbol{D}^*$; if $\boldsymbol{D}^*\boldsymbol{f}$ does not have decaying coefficients, there is no reason $\boldsymbol{f}$ should be close to the minimizer.

Another recent result by Gribonval et.al. analyzes this problem using a model of *cosparsity*, which captures the sparsity in $\boldsymbol{D}^*\boldsymbol{f}$ [**NDEG11**]. Their results currently only hold in the noiseless setting, and it is not known what classes of matrices satisfy the requirements they impose on the sampling operator. This alternative model deserves further analysis and future work in this direction may provide further insights.

In most of the applications discussed, namely those using curvelets, Gabor frames, and the UWT, the coefficients of $\boldsymbol{D}^*\boldsymbol{f}$ decay rapidly. Thus for these applications, ℓ_1-analysis provides strong recovery guarantees. When the dictionary is a concatenation of bases, $\boldsymbol{D}^*\boldsymbol{f}$ will not necessarily have decaying coefficients. For example, when $\boldsymbol{D}$ consists of the identity and the Fourier bases, $\boldsymbol{D}^*\boldsymbol{f}$ can be a very flat signal even when $\boldsymbol{f}$ has a sparse representation in $\boldsymbol{D}$.

Although the D-RIP is a natural extension of the standard RIP, recent work suggests that a more generalized theory may be advantageous [**CP10**]. This framework considers sampling operators whose columns are independent random vectors from an arbitrary probability distribution. Candès and Plan show that when the distribution satisfies a simple incoherence and isotropic property that ℓ_1-minimization robustly recovers signals sparse in the standard sense. A particularly useful consequence of this approach is a logarithmic reduction in the number of random Fourier samples required for reconstruction. We propose an extension of this analysis to the setting of overcomplete dictionaries which will reduce the number of samples needed and provide a framework for new sampling strategies as well.

3.2. Greedy methods. Current analysis of CSP with overcomplete dictionaries is quite limited, and what little analysis there is has focused mainly on optimization based algorithms for recovery. Recently however, Davenport et.al. analyzed a variant of the CoSaMP method [**DW11, DNW13**] summarized by the following algorithm. We use the notation $\mathcal{S}_{\boldsymbol{D}}(\boldsymbol{u}, s)$ to denote the support of the best s-sparse representation of $\boldsymbol{u}$ with respect to the dictionary $\boldsymbol{D}$, $\mathcal{R}(\boldsymbol{D}_T)$ to denote the range of the subdictionary $\boldsymbol{D}_T$, and $\mathcal{P}_{\boldsymbol{D}}(\boldsymbol{b}, s)$ to denote the signal closest to $\boldsymbol{b}$ which has an s-sparse representation in $\boldsymbol{D}$.

COSAMP WITH ARBITRARY DICTIONARIES

INPUT: Sampling operator $\boldsymbol{\Phi}$, dictionary $\boldsymbol{D}$, sample vector $\boldsymbol{y} = \boldsymbol{\Phi}\boldsymbol{f}$, sparsity level s

PROCEDURE:

Initialize: Set $\hat{\boldsymbol{f}} = \boldsymbol{0}$, $\boldsymbol{v} = \boldsymbol{y}$. Repeat the following:

Signal Proxy: Set $\boldsymbol{u} = \boldsymbol{\Phi}^*\boldsymbol{v}$, $\Omega = \mathcal{S}_{\boldsymbol{D}}(\boldsymbol{u}, 2s)$ and merge supports: $T = \Omega \cup \mathcal{S}_{\boldsymbol{D}}(\hat{\boldsymbol{f}}, 2s)$

Signal Estimation: Set $\boldsymbol{b} = \operatorname{argmin}_{\boldsymbol{z}} \|\boldsymbol{y} - \boldsymbol{\Phi}\boldsymbol{z}\|_2$ s.t. $\boldsymbol{z} \in \mathcal{R}(\boldsymbol{D}_T)$

Prune: To obtain the next approximation, set $\hat{\boldsymbol{f}} = \mathcal{P}_{\boldsymbol{D}}(\boldsymbol{b}, s)$.

Sample Update: Update the current samples: $\boldsymbol{v} = \boldsymbol{y} - \Phi\hat{\boldsymbol{f}}$.

OUTPUT: s-sparse reconstructed vector $\hat{\boldsymbol{f}}$

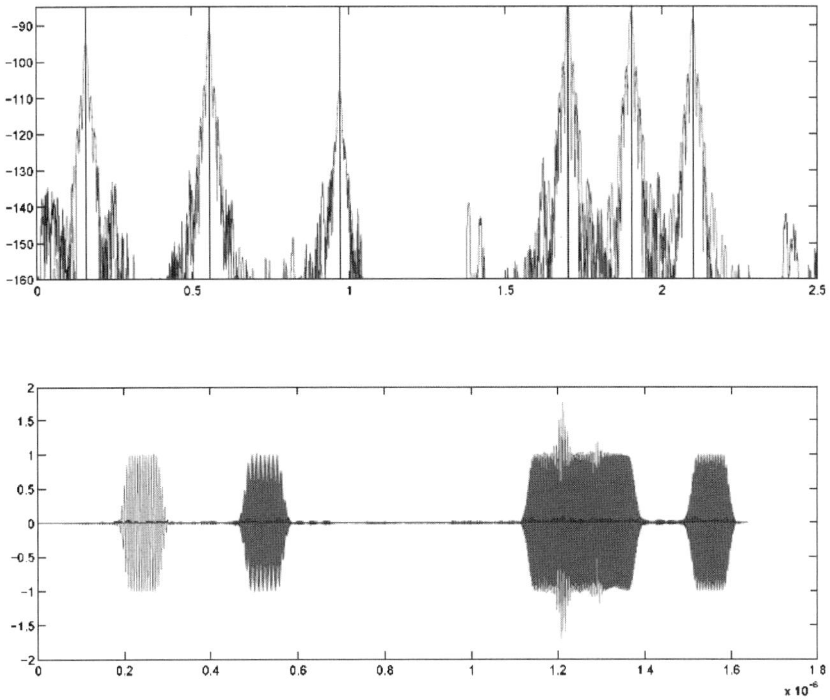

Figure 2 From [**CENR10**]. Recovery in both the time (below) and frequency (above) domains by ℓ_1-analysis after one reweighted iteration. Blue denotes the recovered signal, green the actual signal, and red the difference between the two. The RMSE is less than a third of that in Figure 2

It was recently proved that this version of CoSaMP provides robust recovery of sparse signals with respect to $\boldsymbol{D}$ when the sampling operator satisfies the $\boldsymbol{D}$-RIP [**DNW13**]. Similar results have been obtained using the co-sparse model [**GNE$^+$12**] and Iterative Hard Thresholding (IHT) [**Blu11**].

The major drawback to these results is that the projection operators $\mathcal{P}_{\boldsymbol{D}}$ and $\mathcal{S}_{\boldsymbol{D}}$ cannot in general be implemented efficiently. Indeed, Giryes and Needell [**GN13**] relax the assumptions of these operators from Davenport et al. but still require the following.

Definition 3.2. *A pair of procedures $\mathcal{S}_{\zeta k}$ and $\tilde{\mathcal{S}}_{\tilde{\zeta} k}$ implies a pair of near-optimal projections $\mathbb{P}_{\mathcal{S}_{\zeta k}(\cdot)}$ and $\mathbb{P}_{\tilde{\mathcal{S}}_{\tilde{\zeta} k}(\cdot)}$ with constants C_k and $\tilde{C}_k$ if for any $\mathbf{z} \in \mathbb{R}^d$, $|\mathcal{S}_{\zeta k}(\mathbf{z})| \leq \zeta k$, with $\zeta \geq 1$, $\left|\tilde{\mathcal{S}}_{\tilde{\zeta} k}(\mathbf{z})\right| \leq \tilde{\zeta} k$, with $\tilde{\zeta} \geq 1$, and*

$$\|\mathbb{P}_{\mathcal{S}_{\zeta k}(\mathbf{z})}\mathbf{z}\|_2^2 \leq C_k\|\mathbf{z} - \mathbb{P}_{\mathcal{S}_k^*(\mathbf{z})}\mathbf{z}\|_2^2 \quad \text{as well as} \quad \|\mathbb{P}_{\tilde{\mathcal{S}}_{\tilde{\zeta} k}(\mathbf{z})}\mathbf{z}\|_2^2 \geq \tilde{C}_k\|\mathbb{P}_{\mathcal{S}_k^*(\mathbf{z})}\mathbf{z}\|_2^2, \tag{3.1}$$

where $\mathbb{P}_{\mathcal{S}_k^*}$ denotes the optimal projection:

$$\mathcal{S}_k^*(\mathbf{z}) = \underset{|T|\leq k}{\operatorname{argmin}}\|\mathbf{z} - \mathbb{P}_T\mathbf{z}\|_2^2.$$

Under the assumption that one has access to such near-optimal projections, signal recovery can be obtained by the following result.

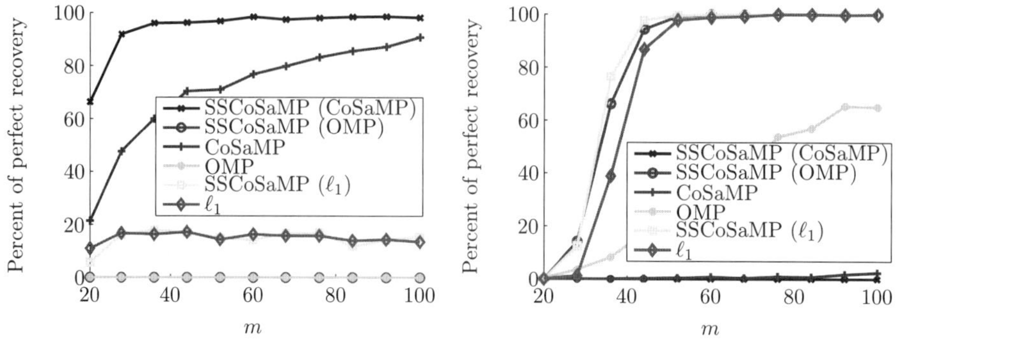

Figure 3 From [**DNW13**]. Percentage of signal recovery for CoSaMP variants and standard methods. Left: Sparse coefficients are clustered together. Right: Sparse coefficients are well separated.

THEOREM 3.3 (Signal Space CoSaMP [**GN13**]). *Let* $\mathbf{M}$ *satisfy the* $\mathbf{D}$*-RIP* (3.1) *with a constant* $\delta_{(3\zeta+1)k}$ *(*$\zeta \geq 1$*). Suppose that* $\mathcal{S}_{\zeta k}$ *and* $\tilde{\mathcal{S}}_{2\zeta k}$ *are a pair of near optimal projections (as in Definition* 3.2*) with constants* C_k *and* $\tilde{C}_{2k}$*. Apply SS-CoSaMP (with* $a = 2$*) and let* $\mathbf{x}^t$ *denote the approximation after* t *iterations. If* $\delta_{(3\zeta+1)k} < \epsilon^2_{C_k,\tilde{C}_{2k},\gamma}$ *and*

$$(1 + C_k)\left(1 - \frac{\tilde{C}_{2k}}{(1+\gamma)^2}\right) < 1, \tag{3.2}$$

then after a constant number of iterations t^* *it holds that*

$$\|\mathbf{x}^{t^*} - \mathbf{x}\|_2 \leq \eta_0 \|\mathbf{e}\|_2, \tag{3.3}$$

where γ *is an arbitrary constant, and* η_0 *is a constant depending on* $\delta_{(3\zeta+1)k}$*,* C_k*,* $\tilde{C}_{2k}$ *and* γ*. The constant* $\epsilon_{C_k,\tilde{C}_{2k},\gamma}$ *is greater than zero if and only if* (3.2) *holds.*

Unfortunately, it is unknown whether there exist efficient approximate-projections for general redundant frames that satisfy Definition 3.2. Empirically, however, traditional CSP methods like OMP, CoSaMP, or ℓ_1-minimization often provide accurate recovery [**DW11, DNW13, GNE$^+$12**]. We also find that the method used to solve the projection may have a significant impact. For example, Figure 3 (left) shows the percentage of correctly recovered signals (as a function of the number of measurements m) with a $256 \times 4(256)$ oversampled Fourier dictionary in which CoSaMP with projection approximated by CoSaMP clearly outperforms the other CoSaMP methods as well as the standard methods.

On the other hand, if one employs the co-sparse or "analysis-sparse" model, a need for such projections can be eliminated. Rather than assuming the signal is sparse in the overcomplete frame $\mathbf{D}$ (that is, $\mathbf{f} = \mathbf{D}\mathbf{x}$ for sparse $\mathbf{x}$, the analysis-sparse model assumes that the analysis coefficients $\mathbf{D}^*\mathbf{f}$ are sparse (or approximately sparse). Foucart, for example, shows that under the analysis-sparse model hard thresholding algorithms provide the same guarantees as ℓ_1-minimization without the need for approximate projections [**F15**]. Of course, these two models of sparsity can be quite different for frames of interst, and the practicality of either may vary from application to application.

4. Dictionary Learning: An introduction

In the CS framework, the dictionary $\mathbf{D}$ is often chosen as a random matrix satisfying the Restricted Isometry Property (e.g., Subgaussian or partial bounded orthogonal matrices), or designed based on intuitive expectations of the signal of interest (such as the oversampled DFT, Gabor frames, wavelets, and curvelets). The resulting dictionary is thus not directly related to the observed signals. However, in reality, the observed data often do not obey those model assumptions, so that pre-designed dictionaries would not work well. Accordingly, it is important to consider dictionaries which adapt to the observed data, often called *data-dependent dictionaries*. Starting this section, we talk about how to learn such dictionaries directly from data and apply them to image processing tasks.

Notation. We often use boldface lowercase Roman letters to represent vectors (for example $\mathbf{x}, \mathbf{y}, \mathbf{e}$) and boldface uppercase Roman letters for matrices (e.g., $\mathbf{A}, \mathbf{D}$). For any such vector (e.g. $\mathbf{x}$), we use the plain version of the letter plus a subscript (i.e., x_i) to refer to the specific entry of the vector. Meanwhile, for some vectors and matrices that are interpreted as coefficients, we will use Greek letters to denote them. In these cases, we use lowercase letters with subscripts to represent vectors (e.g., γ_i) and uppercase letters for matrices (e.g. Γ). For a matrix $\mathbf{A}$, we write A_{ij} to denote its (i, j) entry; we use $\mathbf{A}(:, j)$ to denote the jth column of $\mathbf{A}$ and $\mathbf{A}(i, :)$ its ith row. For any $0 < p < \infty$, the p-norm of a vector $\mathbf{x} \in \mathbb{R}^L$ is defined as

$$\|\mathbf{x}\|_p = \left(\sum_{i=1}^{L} |x_i|^p \right)^{1/p}.$$

If $p = 0$, then $\|\mathbf{x}\|_0$ counts the number of its nonzero entries:

$$\|\mathbf{x}\|_0 = \#\{i \mid x_i \neq 0\}.$$

The Frobenius norm of a matrix $\mathbf{A}$ is

$$\|\mathbf{A}\|_F = \sqrt{\sum_{i,j} A_{ij}^2},$$

and its $\ell_{1,1}$ norm is

$$\|\mathbf{A}\|_{1,1} = \sum_j \|\mathbf{A}(:, j)\|_1$$

If $\mathbf{A}$ is a square matrix, then its trace is defined as

$$\mathrm{trace}(\mathbf{A}) = \sum_i A_{ii}.$$

4.1. The Dictionary Learning Problem. Suppose we are given a finite set of training signals in $\mathbb{R}^L$, for example, $\sqrt{L} \times \sqrt{L}$ pixel images or $\sqrt{L} \times \sqrt{L}$ patches taken from a large digital image. We want to to learn a collection of atomic signals called *atoms*, directly from the given signals so that they can be represented as, or closely approximated by, linear combinations of few atoms. A good analog of this problem is the construction of the English dictionary from many sentences or the recovery of the periodic table of chemical elements from a large variety of materials.

Specifically, given the training data $\mathbf{x}_1, \ldots, \mathbf{x}_n \in \mathbb{R}^L$, and positive integers m, s, we wish to find an $L \times m$ matrix $\mathbf{D}$ and s-sparse vectors $\gamma_1, \ldots, \gamma_n \in \mathbb{R}^m$ such that

$\mathbf{D}\gamma_i$ is "close" to $\mathbf{x}_i$ for all i. Using the ℓ_2 norm to quantify the error, we formulate the following dictionary learning problem:

$$(4.1)\qquad \min_{\mathbf{D},\gamma_1,\ldots,\gamma_n} \sum_{i=1}^{n} \|\mathbf{x}_i - \mathbf{D}\gamma_i\|_2^2 \quad \text{such that} \quad \|\gamma_i\|_0 \le s, \text{ for all } i.$$

Here, $\mathbf{D} = [\mathbf{d}_1, \ldots, \mathbf{d}_m] \in \mathbb{R}^{L\times m}$ is called the *dictionary*, and its columns represent atoms. The vector $\gamma_i \in \mathbb{R}^m$, with at most s nonzero entries, contains the coefficients needed by the columns of $\mathbf{D}$ to linearly represent $\mathbf{x}_i$. To make the choices of $\mathbf{D}$ and γ_i unique, we constrain the columns of $\mathbf{D}$ to be on the unit sphere in $\mathbb{R}^L$, i.e., $\|\mathbf{d}_i\|_2 = 1$. The dictionary size m is allowed to exceed the ambient dimension L in order to exploit redundancy. In contrast, the sparsity parameter often satisfies $s \ll L$.

In the special case where each γ_i is enforced to be 1-sparse (i.e., $s = 1$) with the only nonzero entry being 1, the problem in (4.1) aims to use the most similar atom to represent each signal. This corresponds to the Kmeans clustering problem [**Mac67**], where the training data are divided into n disjoint subsets, each surrounding a unique atom as its center, such that points in each subset are closer to the corresponding center than to other centers. Here, we mention a recent paper by Awasthi et al. [**ABC$^+$15**] which provides global recovery guarantees for an SDP relaxation of the Kmeans optimization problem.

Let us look at an example. Suppose we extract all 8×8 patches from a 512×512 digital image and consider them as our signal data. Here, the signal dimension $L = 64$, but the number of signals is very large ($n \approx 512^2$). A typical choice of the dictionary size is $m = 256$, which is four times as large as the signal dimension L so that $\mathbf{D}$ is overcomplete. Lastly, s is often set to some positive integer not more than 10. Performing dictionary learning in this setting is thus equivalent to finding 256 elementary image patches so that each original patch can be most closely approximated by a linear combination of at most 10 elementary patches.

Note that in (4.1) we used the square loss function to quantify the representation error

$$\ell(\mathbf{x}_i, \mathbf{D}) = \|\mathbf{x}_i - \mathbf{D}\gamma_i\|_2^2,$$

but this can be replaced by any other loss function, for example ℓ_1. The dictionary is considered "good" at representing the signals if the total loss is "small". Furthermore, the fewer columns $\mathbf{D}$ has, the more efficient it is.

If we let $\mathbf{X} = [\mathbf{x}_1, \ldots, \mathbf{x}_n]$ and $\Gamma = [\gamma_1, \ldots, \gamma_n]$ be two matrices representing respectively the signals and the coefficients in columns, then the dictionary learning problem in (4.1) can be readily rewritten as a matrix factorization problem

$$(4.2)\qquad \min_{\mathbf{D},\Gamma} \|\mathbf{X} - \mathbf{D}\Gamma\|_F^2 \quad \text{such that} \quad \|\gamma_i\|_0 \le k, \text{ for all } i.$$

Here, the matrix $\mathbf{D}$ is required to have unit-norm columns while Γ must be column-sparse.

In some cases we are not given the signal sparsity s but a precision requirement ϵ on the approximation error for individual signals. We then reformulate the above problem as follows:

$$(4.3)\qquad \min_{\mathbf{D},\gamma_1,\ldots,\gamma_n} \sum_{i=1}^{n} \|\gamma_i\|_0 \quad \text{such that} \quad \|\mathbf{x}_i - \mathbf{D}\gamma_i\|_2 \le \varepsilon, \text{ for all } i$$

Here, the objective function can be thought of the total cost for representing the signals with respect to a dictionary.

The two formulations of the dictionary learning problem in (4.1) and (4.3) can be unified into a single problem without mentioning s or ϵ:

$$\min_{\mathbf{D},\Gamma} \sum_{i=1}^{n} \|\mathbf{x}_i - \mathbf{D}\gamma_i\|_2^2 + \lambda\|\gamma_i\|_0. \tag{4.4}$$

Here, λ is a regularization parameter whose role is to balance between representation error and cost (i.e., sparsity). That is, large values of λ force the ℓ_0 penalty term to be small, leading to very sparse representations. On the other hand, smaller values of λ place a smaller weight on sparsity and correspondingly enforce the program to significantly reduce the total error.

Unfortunately, the combinatorial nature of the ℓ_0 penalty requires an exhaustive search for the support set of each coefficient vector γ_i, making none of the problems (4.1)-(4.4) practically tractable (in fact, they are all NP-hard). One often replaces it by the ℓ_1 penalty (which is the closest convex norm) and considers instead

$$\min_{\mathbf{D},\gamma_1,\ldots,\gamma_n} \sum_{i=1}^{n} \|\mathbf{x}_i - \mathbf{D}\gamma_i\|_2^2 + \lambda\|\gamma_i\|_1, \tag{4.5}$$

or its matrix version

$$\min_{\mathbf{D},\Gamma} \|\mathbf{X} - \mathbf{D}\Gamma\|_F^2 + \lambda\|\Gamma\|_{1,1}, \tag{4.6}$$

hoping that the new problem still preserves, at least approximately, the solution of the original problem. The problem in (4.6) is now convex in each of the variables $\mathbf{D}, \Gamma$, but not jointly convex. It is thus often solved by fixing one of $\mathbf{D}$ and Γ and updating the other in an alternating fashion. From now on, we will focus on (4.5) and its matrix version (4.6) due to its tractability and unifying nature.

4.2. Connections to several other fields. Dictionary learning (DL) is closely related to the following fields.

4.2.1. *Compressive sensing (CS).* In DL both the dictionary and sparse coefficients are simultaneously learned from the training data. When the dictionary $\mathbf{D}$ is fixed, the optimization problem in (4.5) is over the coefficients $\gamma_1, \ldots, \gamma_n$, in which case the n terms in the sum of (4.5) can be decoupled, leading to n similar problems:

$$\min_{\gamma} \|\mathbf{x}_i - \mathbf{D}\gamma\|_2^2 + \lambda\|\gamma\|_1. \tag{4.7}$$

This is exactly the *sparse coding* problem, studied extensively in the CS framework [**BDE09**]. Indeed, the CS research has shown that the relaxation to the $\|\gamma\|_1$ penalty (from $\|\gamma\|_0$) preserves the sparse solution, at least when $\mathbf{D}$ satisfies the RIP condition [**CRT06b**]. Additionally, there are efficient pursuit algorithms for solving this problem, such as the OMP [**TG07**], Basis Pursuit [**CRT06b**], and CoSamp [**NT08b, NT08a**]. Thus, one may solve this problem by using any of these pursuit algorithms, which are described in the first part of the lecture notes.

Although both CS and DL contain the same coding problem (4.7), the interpretations of the variables in (4.7) are markedly different. In the CS setting the matrix $\mathbf{D}$ serves as the sensing matrix whose rows are carefully picked to linearly interact with the unknown signal γ, and $\mathbf{x}$ represents the vector of compressed measurements. The main goal of solving (4.7) is to recover both the support set and entries of the sparse signal γ. In contrast, in the DL framework the emphasis is placed on the columns of $\mathbf{D}$ which are regarded as prototype signals and used to linearly represent the training signals of the same dimension. Accordingly, $\mathbf{x}$ should not be regarded as the measurement vector, but just a training signal. Moreover, the vector γ no longer represents the sparse signal to be recovered, but indicates the sparse linear representation of the training signal $\mathbf{x}$ with respect to the dictionary $\mathbf{D}$.

4.2.2. *Frame theory.* Frame design has been a very active research field for decades, and it lies at the intersection of many subjects, theoretical or applied, such as pure mathematics, harmonic analysis, compressive sensing, dictionary learning, and signal processing. Specifically, a frame for a finite dimensional Hilbert space (i.e., $\mathbb{R}^L$) is a spanning set $\{\mathbf{e}_k\}$ for the space, without requiring linear independence among them, that satisfies the following frame condition [**CK**]: There exist two fixed constants $B \geq A > 0$ such that for every $\mathbf{x} \in \mathbb{R}^L$,

$$A\|\mathbf{x}\|_2^2 \leq \sum_k |\langle \mathbf{x}, \mathbf{e}_k \rangle|^2 \leq B\|\mathbf{x}\|_2^2.$$

The central problem in frame theory is signal representation and reconstruction by using the frame $\{\mathbf{e}_k\}$ and its dual $\{\tilde{\mathbf{e}}_k\}$:

$$\mathbf{x} = \sum_k \langle \mathbf{x}, \tilde{\mathbf{e}}_k \rangle \mathbf{e}_k = \sum_k \langle \mathbf{x}, \mathbf{e}_k \rangle \tilde{\mathbf{e}}_k.$$

The concept of frames occurred much earlier than that of dictionaries, representing an important intermediate step from orthogonal bases modeling to sparse and redundant modeling. Like dictionaries, frames are overcomplete systems of signals that can represent other signals. Because of the redundancy every vector $\mathbf{x} \in \mathbb{R}^L$ has infinitely many representations, and this great flexibility is what makes both of them (frames and dictionaries) so useful in many applications: By representing a signal in many different ways, we are better able to sustain losses and noise while still having accurate and robust reconstructions.

Though both frames and dictionaries depend on the same notion of redundancy, they use it in different ways. The vectors in a frame must satisfy a frame condition which enables rigorous analysis of the system and guarantees many attractive theoretical properties. For example, though no longer an orthogonal basis, the linear coefficients can still be obtained through dot products between the (dual) frame and the signal. In contrast, dictionaries, especially data-dependent ones, do not require such a condition for its elements, but introduce a new notion of sparsity for the representation coefficients. That is, it places a small upper bound on the number of its elements that can be used for representing any given signal, so as to promote simple and interpretable representations. While the sparsity concept is quite easy to understand, its discrete nature makes it extremely difficult to analyze and often one can only consider a convex relaxation of it. Accordingly, comparing with frames, there is much less theory for dictionaries. Despite the theoretical challenge, dictionaries have proven to improve over frames in many applications, because of its greater flexibility and better ability to adapt to real data. Furthermore, the elements of a dictionary represent prototype signals and thus have a more clear physical interpretation.

4.2.3. *Subspace clustering.* Subspace clustering extends the classic Principle Component Analysis (PCA) to deal with hybrid linear data. The PCA is a linear transform which adapts to signals sampled from a Gaussian distribution, by fitting a low-dimensional subspace to the data with the lowest L_2 approximation error [**Jol02**]. Subspace clustering is a natural extension of PCA by using more than one subspace. Specifically, given a set of signals $\mathbf{x}_1, \ldots, \mathbf{x}_n$ in $\mathbb{R}^L$ which are sampled from a mixture of unknown number of subspaces with unknown dimensions, the goal of subspace clustering is to estimate the parameters of the model planes and their bases, and then cluster data according to the identified planes. It has been a very hot topic since the beginning of this century. We refer the reader to [**Vid11**] for a tutorial on this field and for an introduction to state-of-the art algorithms such as SCC [**CL09a**, **CL09b**].

From a DL perspective, the overall collection of the subspace bases forms a dictionary, and every given signal is expressed as a linear combination of several basis vectors, depending on the subspace it belongs to. In other words, the dictionary here consists of a few subdictionaries, each one reserved for a particular group of signals, and the different subdictionaries are not allowed to be mixed to form other kinds of linear representations. Clearly, such a dictionary is a lot more restrictive. Though sparsity is still enforced here, redundancy is not exploited because the dictionary is small for low dimensional subspaces. Finally, the subspace bases, which must be linearly independent, may not have the interpretation of atomic signals.

In (4.2)-(4.4), the signals all have at most s non-zeros in its representation. In other words, the size of the support set of each coefficient is no bigger than s. More importantly, there is no restriction on which combination of atoms can be used for representing a signal. Thus, there are a total of $\binom{m}{s} + \binom{m}{s-1} + \cdots + \binom{m}{1} = \binom{m+1}{s}$ possibilities for a support, where m is the dictionary size. Each such support defines a unique subspace of dimension (at most) s in $\mathbb{R}^L$, and the overall signal model is therefore a union of a large number of subspaces, to one of which each signal is believed to belong. Consequently, this model is a relaxation of the mixture of subspaces model mentioned above, and the representations here are thus more flexible and efficient.

4.3. Applications to image processing tasks. Sparse and redundant dictionaries offer a new way for modeling complex image content, by representing images as linear combinations of few atomic images chosen from a large redundant collection (i.e., dictionary). Because of the many advantages associated with dictionaries, dictionary-based methods tend to improve over traditional image processing algorithms, leading to state-of-the-art results in practice. Below we briefly survey some of the most common imaging applications and their solution by dictionary learning. To gain a deeper and more thorough understanding of the field, we refer the reader to [**EFM10**], which is also our main reference for writing this part.

4.3.1. *Introduction.* Consider a clean image or a patch taken from it, I, of size $\sqrt{L} \times \sqrt{L}$, where $\sqrt{L}$ is a positive integer. Typically, $\sqrt{L} = 512$ (for full digital images), or $\sqrt{L} = 8$ (when operating on patches taken from a full image). We vectorize the image I to obtain $\mathbf{t} \in \mathbb{R}^L$, by following some fixed order (e.g., lexicographical order). Normally we do not observe the clean image $\mathbf{t}$, but rather a noisy measurement of it (see Fig. 4):

$$\mathbf{x} = \mathbf{t} + \mathbf{e}.$$

Here, $\mathbf{e}$ represents an (unknown) additive noise contaminating the image. Naturally, given $\mathbf{x}$, we would like to recover the true image $\mathbf{t}$, at least as closely as possible. This is the *image denoising* problem [**BCM05**, **EA06**, **MSE**].

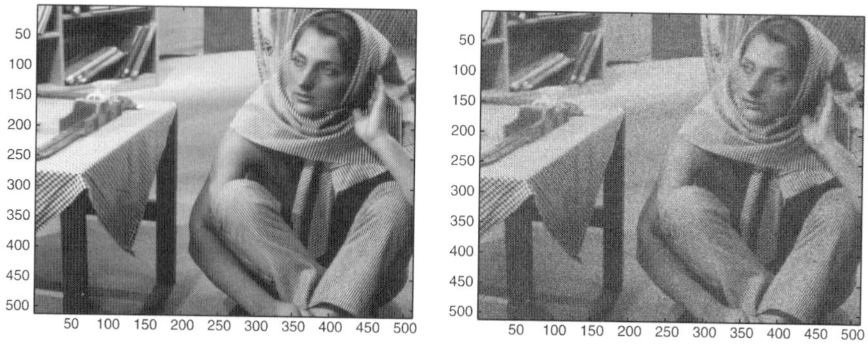

Figure 4 A clean image and its noisy version (by adding zero-mean Gaussian noise with standard deviation $\sigma = 25$). We assume that only the noisy image is given to us, and we wish to recover the clean image.

Assuming that the noise $\mathbf{e}$ has bounded norm ($\|\mathbf{e}\|_2 \le \delta$), the true image $\mathbf{t}$ and its noisy realization $\mathbf{x}$ are within a distance of δ from each other. In theory, if we know the value of δ, then we may search the δ-ball centered at $\mathbf{x}$ for the clean image $\mathbf{t}$. However, we cannot use the concept of cleanness directly. In addition, this space is prohibitively large for performing any practical task. So we need to choose a model for the clean image $\mathbf{t}$ and correspondingly focus on a smaller class of images. The "best" image in that class is then used as an estimator for the clean image $\mathbf{t}$:

$$\hat{\mathbf{t}} = \operatorname*{argmin}_{\mathbf{y}} C(\mathbf{y}) \qquad \text{subject to} \qquad \|\mathbf{x} - \mathbf{y}\|_2 \le \delta.$$

In the above, $C(\cdot)$ represents a cost function, often naturally associated with the selected model, such that smaller cost means better estimation.

For example, if we let $C(\mathbf{y}) = \|\mathbf{Ly}\|_2^2$, where $\mathbf{L}$ is a Laplacian matrix representing the operation of applying the Laplacian filter to the image $\mathbf{y}$, then the cost is the deviation of $\mathbf{t}$ from spatial smoothness. In other words, the class of spatially smooth images that lie in the δ-ball around $\mathbf{t}$ is considered and the most spatially smooth image is selected to estimate $\mathbf{t}$. A second example is $C(\mathbf{y}) = \|\mathbf{Wy}\|_1$, where $\mathbf{W}$ is a matrix representing the orthogonal wavelet transform, and the ℓ_1 norm measures the sparsity of the wavelet coefficients. This corresponds to wavelet denoising, which combines spatial smoothness (of a lower order) and a robust measure in the cost function. There are also many other choices of $C(\mathbf{y})$, e.g., the total variation measure [**ROF92**].

Recently, inspired by sparse and redundant modeling, the sparsity of the coefficient of $\mathbf{y}$ with respect to an overcomplete dictionary $\mathbf{D}$ is adopted as the cost function:

$$\min \|\gamma\|_1 \qquad \text{subject to} \qquad \|\mathbf{x} - \mathbf{D}\gamma\|_2 \le \delta.$$

Here, $\mathbf{D}$ represents a global dictionary, learned in advance from many image examples of the same size. In this case, the dictionary learning and sparse coding parts are actually decoupled from each other, and one thus solves them separately. The minimizer $\hat{\gamma}$ of the sparse coding problem corresponds to the "simplest" image with respect to the global dictionary $\mathbf{D}$, and the clean image estimate is $\hat{\mathbf{t}} = \mathbf{D}\hat{\gamma}$.

In practice, instead of working on the full image and using many similar examples, one often extracts the patches $\mathbf{p}_i$, at a fixed size, of the noisy image $\mathbf{x}$ as training signals. We then learn a (patch) dictionary $\mathbf{D}$, along with coefficients $\hat{\gamma}_i$, directly from those patches:

$$\min_{\mathbf{D},\{\gamma_i\}} \sum_i \|\gamma_i\|_1 \qquad \text{subject to} \quad \|\mathbf{p}_i - \mathbf{D}\gamma_i\|_2 \leq \delta \ \text{ for all } i.$$

The denoised patches are given by $\mathbf{D}\hat{\gamma}_i$, from which we reconstruct the clean image. Such a patch-based approach has two immediate advantages: First, the signal dimension becomes much smaller, which greatly mitigates the computational burden. Secondly, since the dictionary is self-learned, there is no need to use other exemplary images. Fig. 5 displays both a dictionary trained on patches of size 8×8 taken from the noisy image in Fig. 4 and the corresponding denoised image.

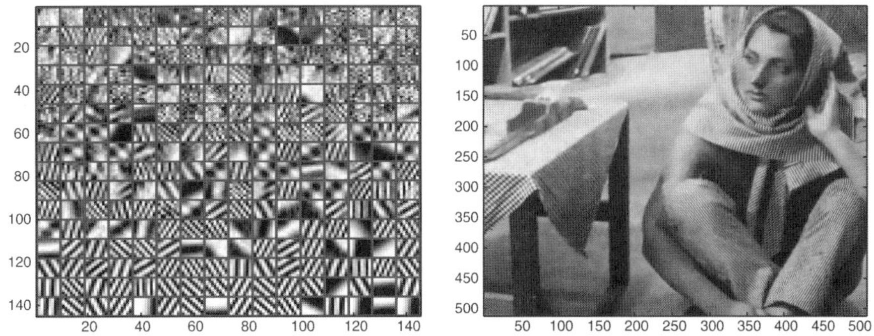

Figure 5 Trained dictionary (left) and corresponding denoised result (right), using the K-SVD algorithm [**AEB05**, **AEB06**].

More generally, we assume that we observe a noisy degraded version of $\mathbf{x}$:

$$\mathbf{x} = \mathbf{H}\mathbf{t} + \mathbf{e}$$

where $\mathbf{H}$ is a linear operator representing some kind of degradation of the signal, such as

- a blur,
- the masking of some pixels,
- the downsampling, and
- a random set of projections.

Our goal is still to recover the true image $\mathbf{t}$ from its noisy observation $\mathbf{x}$. The corresponding problems are respectively referred to as

- image deblurring [**HX13**],
- image inpainting [**MSE**],
- image super-resolution [**PETM09**], and
- compressive sensing.

When $\mathbf{H}$ is taken to be the identity operator, then the problem reduces to image denoising. These problems are all special types of inverse problems in image processing.

Similarly, if we adopt the ℓ_1 cost function and learn a dictionary $\mathbf{D}$ elsewhere from many image examples, we may then consider the following problem:

$$\min \|\gamma\|_1 \qquad \text{subject to} \qquad \|\mathbf{x} - \mathbf{H}\mathbf{D}\gamma\|_2 \leq \delta.$$

Here, we assume $\mathbf{H}$ is known to us. We solve it by regarding $\mathbf{HD}$ as a whole. The minimizer $\hat{\gamma}$ of the above problem then gives the clean image estimate $\hat{\mathbf{t}} = \mathbf{D}\hat{\gamma}$. We refer the reader to the above references (corresponding to the specific applications) for more details as well as experimental results.

5. Dictionary Learning: Algorithms

Since the beginning of this century, many algorithms have been proposed for solving the dictionary learning problem, most of which use the formulation (4.6) and have an iterative fashion. That is, by fixing one of the matrices $\mathbf{D}$ and Γ, they consider the optimization over the other variable and strive to find the best update for it; such an alternating procedure is repeated until convergence. In the following, we review three state-of-the-art dictionary learning algorithms, K-SVD [**AEB05**, **AEB06**], Geometric Multi-Resolution Analysis (GRMA) [**CM10**, **CM11b**, **ACM12**], and Online Dictionary Learning (ODL) [**MBPS09**, **MBPS10**], which have very different flavors and adequately represent their own categories. The review will also enable the reader to learn the different rationals and ideas used in the data-dependent dictionary learning research. For a more complete survey on dictionary learning approaches, we refer the reader to [**RBE10**].

5.1. K-SVD. The K-SVD algorithm is an iterative algorithm, developed by Elad et al. [**AEB05**, **AEB06**], that minimizes the expression in (4.1), or its matrix form (4.2). It consists of two stages, similarly to the Kmeans algorithm [**Mac67**].

First, at any iteration, the dictionary $\mathbf{D}$ is held fixed and the best coefficient matrix Γ is seeked. In this case, the n terms in the sum of (4.1) can be decoupled, leading to n similar problems:

$$\min_{\gamma_i} \|\mathbf{x}_i - \mathbf{D}\gamma_i\|_2^2 \quad \text{subject to} \quad \|\gamma_i\|_0 \le s, \tag{5.1}$$

where i is taken to be from 1 to n. This is exactly the sparse coding problem, being solved n times. Therefore, any of the pursuit algorithms (such as OMP and CoSamp) that are mentioned in the first half of this paper may be used at this stage.

At the second stage of the same iteration, K-SVD then fixes the new coefficient matrix Γ and searches for a better dictionary $\mathbf{D}$ relative to the coefficients. However, unlike some of the approaches described in [**RBE10**] which update the whole matrix $\mathbf{D}$ by treating it as a single variable, the K-SVD algorithm updates one column at a time, fixing the other columns of $\mathbf{D}$. Meanwhile, as a byproduct, new coefficient corresponding to the updated column is also obtained. Such adoptions have at least two important advantages. First, as we shall see, the process of updating only one column of $\mathbf{D}$ at a time is a simple problem with a straightforward solution based on the singular value decomposition (SVD). Second, allowing a change in the coefficient values while updating the dictionary columns accelerates convergence, since the subsequent column updates will be based on the more relevant coefficients.

5.1.1. *Detailed description of the KSVD algorithm.* Let us present such ideas more carefully. Assume that at some iteration both Γ and all columns of $\mathbf{D}$ except one $\mathbf{d}_k$ are fixed. The goal is to update $\mathbf{d}_k$ and $\gamma^{(k)}$ simultaneously so as to reduce the overall representation error. Denote by $\gamma^{(i)}$ the ith row of Γ for all i (note the the ith column of Γ is denoted by γ_i). Then, by writing out the individual rank-1 matrices in the product $\mathbf{D}\Gamma$ and regrouping terms, we obtain from the objective function in (4.2) the following

$$\|\mathbf{X} - \mathbf{D}\Gamma\|_{\mathrm{F}}^2 = \left\|\mathbf{X} - \sum_j \mathbf{d}_j \gamma^{(j)}\right\|_{\mathrm{F}}^2 = \left\|\left(\mathbf{X} - \sum_{j\neq k} \mathbf{d}_j \gamma^{(j)}\right) - \mathbf{d}_k \gamma^{(k)}\right\|_{\mathrm{F}}^2. \tag{5.2}$$

Denoting

$$\mathbf{E}_k = \mathbf{X} - \sum_{j\neq k} \mathbf{d}_j \gamma^{(j)}, \tag{5.3}$$

which stores the errors for all the training data when the kth atom is omitted, the optimization problem in (4.2) becomes

$$\min_{\mathbf{d}_k, \gamma^{(k)}} \left\|\mathbf{E}_k - \mathbf{d}_k \gamma^{(k)}\right\|_{\mathrm{F}}^2. \tag{5.4}$$

Note that in the above equation the matrix $\mathbf{E}_k$ is considered fixed. The problem thus tries to find the closest rank-1 approximation to $\mathbf{E}_k$, expressing each of its columns as a constant multiple of $\mathbf{d}_k$. A seemingly natural solution would be to perform a rank-1 SVD of $\mathbf{E}_k$ to update both $\mathbf{d}_k$ and $\gamma^{(k)}$. However, this disregards any sparsity structure that $\gamma^{(k)}$ presents[1] and the SVD will very likely fill all its entries to minimize the objective function. Collectively, when all atoms along with their coefficients are sequentially updated, such a method would destroy the overall sparsity pattern of the coefficient matrix Γ. As a result, the convergence of the algorithm will be significantly impaired.

It is thus important to preserve the support of $\gamma^{(k)}$, when solving the above problem, to ensure convergence. The K-SVD algorithm introduces the following simple solution to address the issue. Let the support set of $\gamma^{(k)}$ be denoted by

$$\Omega_k = \{i \mid \gamma^{(k)}(i) \neq 0\},$$

and its reduced version by

$$\gamma_\Omega^{(k)} = \gamma^{(k)}(\Omega_k) = \left(\gamma^{(k)}(i)\right)_{i\in\Omega_k}.$$

We also restrict our attention to the same subset of columns of $\mathbf{E}_k$:

$$\mathbf{E}_k^\Omega = \mathbf{E}_k(:, \Omega_k).$$

Using such notation, we may rewrite the above problem as

$$\min_{\mathbf{d}_k, \gamma_\Omega^{(k)}} \left\|\mathbf{E}_k^\Omega - \mathbf{d}_k \gamma_\Omega^{(k)}\right\|_{\mathrm{F}}^2, \tag{5.5}$$

in which $\gamma_\Omega^{(k)}$ now has a full support. Since the sparsity constraint has been removed, this problem bears a simple and straightforward solution, computable from rank-1 SVD. Specifically, if the SVD of $\mathbf{E}_k^\Omega$ is given by

$$\mathbf{E}_k^\Omega = U\Sigma V^T$$

where U, V are orthonormal and Σ is diagonal, then the solution of the above problem is

$$\widetilde{\mathbf{d}}_k = U(:, 1), \quad \gamma_\Omega^{(k)} = \Sigma_{11} V(:, 1).$$

[1] Recall that $\gamma^{(k)}$ is the kth row of the coefficient matrix Γ which has sparse columns. So it is very likely that each row of Γ also contains many zeros and thus has a (nearly) sparse pattern.

One immediate benefit of such a solution is that the new atom $\widetilde{\mathbf{d}}_k$ remains normalized.

We now summarize the steps of K-SVD in Algorithm 1.

Algorithm 1 Pseudocode for the K-SVD Algorithm

Input: Training data $\mathbf{X} = \{\mathbf{x}_1, \ldots, \mathbf{x}_n\}$, sparsity parameter s, initial dictionary $\mathbf{D}^{(0)}$

Output: Dictionary $\mathbf{D}$, sparse coefficients Γ

Steps:

Initialization: $J \leftarrow 1$ (iteration index)

WHILE *stopping criterion not met*

- **Sparse coding stage**: For each data point $\mathbf{x}_i, i = 1, \ldots, n$, solve
$$\min_{\gamma_i} \|\mathbf{x}_i - \mathbf{D}^{(J-1)}\gamma_i\|_2^2 \quad \text{subject to} \quad \|\gamma_i\|_0 \leq s,$$
using any pursuit algorithm (e.g. OMP). Denote the resultant coefficient matrix by Γ.
- **Dictionary update stage**: For each dictionary atom $\mathbf{d}_i$ of $D^{(J-1)}, i = 1, \ldots, n$,
  - Identify the support set Ω_i of $\gamma^{(i)}$, the ith row of the *current* matrix Γ.
  - Compute
  $$\mathbf{E}_i = \mathbf{X} - \sum_{j \neq i} \mathbf{d}_j \gamma^{(j)},$$
  and restrict it to the subset Ω_i of columns of $\mathbf{E}_i$ to form $\mathbf{E}_i^{\Omega}$.
  - Apply rank-1 SVD to $\mathbf{E}_i^{\Omega}$ to update $\mathbf{d}_i$ and $\gamma_{\Omega}^{(i)}$ simultaneously
- $J \leftarrow J + 1$

ENDWHILE

Return $\mathbf{D}, \Gamma$

5.1.2. *A few remarks about K-SVD.* We make the following comments on K-SVD.

- The K-SVD algorithm has many advantages. For example, it is simple to implement, fast to run, and converges (assuming the pursuit algorithm used for sparse coding is accurate). It has been successfully applied to many imaging applications (e.g., [**EA06**]).
- However, the success of K-SVD depends on the choice of the initial dictionary. In other words, though it converges, it might be trapped in a suboptimal solution. Its performance also depends on the pursuit algorithm used. For example, convergence of K-SVD is guaranteed only if the pursuit algorithm solves the sparse coding problems accurately.
- The K-SVD algorithm closely resembles the Kmeans algorithm, and can be viewed as a natural extension of it. This explains why K-SVD shares the same advantages and drawbacks with Kmeans, like those mentioned above.

- The dictionary built by K-SVD is completely unstructured, making sparse representation of any new signal (relative to the trained dictionary) a nontrivial task, which requires to use one of the pursuit algorithms.

5.2. Geometric Multi-Resolution Analysis (GMRA). The GMRA [**CM10, CM11b, ACM12**] is a wavelet-like algorithm based on a geometric multiresolution analysis of the data. It builds data-dependent dictionaries that are structured and multiscale. When the data is sampled from a manifold, there are theoretical guarantees for both the size of the dictionary and the sparsity of the coefficients.

Specifically, let $(\mathcal{M}, \rho, \mu)$ be a metric measure space with μ a Borel probability measure, ρ a metric function, and $\mathcal{M} \subseteq \mathbb{R}^D$ a set. For example, $(\mathcal{M}, \rho, \mu)$ can be a smooth compact Riemannian manifold of dimension d isometrically embedded in $\mathbb{R}^D$, endowed with the natural volume measure. The GMRA construction consists of three steps. First, it performs a nested geometric decomposition of the set $\mathcal{M}$ into dyadic cubes at a total of J scales, arranged in a tree. Second, it obtains an affine approximation in each cube, generating a sequence of piecewise linear sets $\{\mathcal{M}_j\}_{1\le j\le J}$ approximating $\mathcal{M}$. Lastly, it constructs low-dimensional affine difference operators that efficiently encode the differences between $\mathcal{M}_j$ and $\mathcal{M}_{j+1}$, producing a hierarchically organized dictionary that is adapted to the data. Associated to this dictionary, there exist efficient geometric wavelet transforms, an advantage not commonly seen in the current dictionary learning algorithms.

5.2.1. *Multiscale Geometric Decomposition.* For any $\mathbf{x} \in \mathcal{M}$ and $r > 0$, we use $B_r(\mathbf{x})$ to denote the ball in the set $\mathcal{M}$ of radius r centered at $\mathbf{x}$. We start by a spatial multiscale decomposition of $\mathcal{M}$ into *dyadic cubes*, $\{\mathbb{C}_{j,k}\}_{k\in\Gamma_j, j\in\mathbb{Z}}$, which are open sets in $\mathcal{M}$ such that

(i) for every $j \in \mathbb{Z}$, $\mu(\mathcal{M} \setminus \cup_{k\in\Gamma_j}\mathbb{C}_{j,k}) = 0$;

(ii) for $j' \ge j$ either $\mathbb{C}_{j',k'} \subseteq \mathbb{C}_{j,k}$ or $\mu(\mathbb{C}_{j,k} \cap \mathbb{C}_{j',k'}) = 0$;

(iii) for any $j < j'$ and $k' \in \Gamma_{j'}$, there exists a unique $k \in \Gamma_j$ such that $\mathbb{C}_{j',k'} \subseteq \mathbb{C}_{j,k}$;

(iv) each $\mathbb{C}_{j,k}$ contains a point $\mathbf{c}_{j,k}$, called center of $\mathbb{C}_{j,k}$, such that

$$B_{c_1\cdot 2^{-j}}(\mathbf{c}_{j,k}) \subseteq \mathbb{C}_{j,k} \subseteq B_{\min\{c_2\cdot 2^{-j}, \operatorname{diam}(\mathcal{M})\}}(\mathbf{c}_{j,k}),$$

for fixed constants c_1, c_2 depending on intrinsic geometric properties of $\mathcal{M}$. In particular, we have $\mu(\mathbb{C}_{j,k}) \sim 2^{-dj}$;

(v) the boundary of each $\mathbb{C}_{j,k}$ is piecewise smooth.

The properties above imply that there is a natural tree structure $\mathcal{T}$ associated to the family of dyadic cubes: for any (j, k), we let

$$\text{children}(j,k) = \{k' \in \Gamma_{j+1} : \mathbb{C}_{j+1,k'} \subseteq \mathbb{C}_{j,k}\}.$$

Note that $\mathbb{C}_{j,k}$ is the disjoint union of its children. For every $\mathbf{x} \in \mathcal{M}$, with abuse of notation we let $(j, \mathbf{x})$ be the unique $k \in \Gamma_j$ such that $\mathbf{x} \in \mathbb{C}_{j,k}$.

5.2.2. *Multiscale Singular Value Decomposition (MSVD).* We start with some geometric objects that are associated to the dyadic cubes. For each $\mathbb{C}_{j,k}$ we define the mean

$$\mathbf{c}_{j,k} := \mathbb{E}_\mu[\mathbf{x}|\mathbf{x} \in \mathbb{C}_{j,\mathbf{x}}] = \frac{1}{\mu(\mathbb{C}_{j,k})}\int_{\mathbb{C}_{j,k}} \mathbf{x}\, d\mu(\mathbf{x}), \tag{5.6}$$

and the covariance operator restricted to $C_{j,k}$,

$$\operatorname{cov}_{j,k} = \mathbb{E}_\mu[(\mathbf{x} - \mathbf{c}_{j,k})(\mathbf{x} - \mathbf{c}_{j,k})^*|\mathbf{x} \in \mathbb{C}_{j,k}]. \tag{5.7}$$

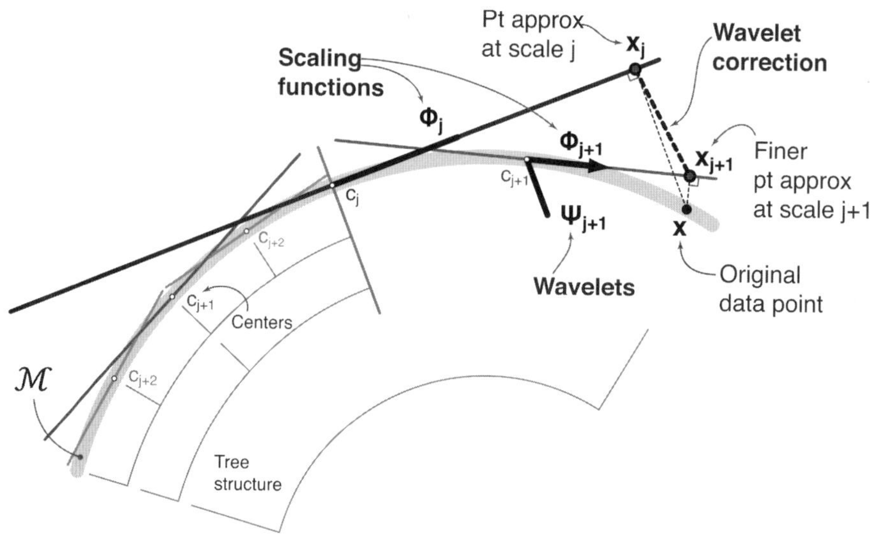

Figure 6 Construction of GMRA.

Let τ_0 be some method that chooses local dimensions $d_{j,k}$ at the dyadic cubes $\mathbb{C}_{j,k}$. For example, when the data is sampled from a manifold of dimension d, τ_0 assigns $d_{j,k} = d$ for all (j, k). In the setting of nonmanifold data, τ_0 can instead picks $d_{j,k}$ so that certain (absolute/relative) error criterion is met. We then compute the rank-$d_{j,k}$ Singular Value Decomposition (SVD) of the above covariance matrix

$$\text{cov}_{j,k} \approx \Phi_{j,k}\Sigma_{j,k}\Phi_{j,k}^*, \tag{5.8}$$

and define the approximate local tangent space

$$\mathbb{V}_{j,k} := \mathcal{V}_{j,k} + \mathbf{c}_{j,k}, \quad \mathcal{V}_{j,k} = \langle \Phi_{j,k} \rangle, \tag{5.9}$$

where $\langle \Phi_{j,k} \rangle$ denotes the span of the columns of $\Phi_{j,k}$. Let $\mathbb{P}_{j,k}$ be the associated affine projection onto $\mathbb{V}_{j,k}$: for any $\mathbf{x} \in \mathbb{C}_{j,k}$,

$$\mathbb{P}_{j,k}(\mathbf{x}) := \mathbf{P}_{j,k} \cdot (\mathbf{x} - \mathbf{c}_{j,k}) + \mathbf{c}_{j,k}, \quad \mathbf{P}_{j,k} = \Phi_{j,k}\Phi_{j,k}^*, \tag{5.10}$$

and define a coarse approximation of $\mathcal{M}$ at scale j,

$$\mathcal{M}_j := \cup_{k\in\Gamma_j} \mathbb{P}_{j,k}(\mathbb{C}_{j,k}). \tag{5.11}$$

When $\mathcal{M}$ is a manifold and $d_{j,k} = d$, $\mathcal{M}_j \to \mathcal{M}$ in the Hausdorff distance, as $J \to +\infty$.

5.2.3. *Construction of Geometric Wavelets.* We introduce our wavelet encoding of the difference between $\mathcal{M}_j$ and $\mathcal{M}_{j+1}$, for $j \geq 0$. Fix a point $\mathbf{x} \in \mathbb{C}_{j+1,k'} \subset \mathbb{C}_{j,k}$. There are two ways to define its approximation at scale j, denoted by $\mathbf{x}_j$ or equivalently by $P_{\mathcal{M}_j}(\mathbf{x})$:

$$\mathbf{x}_j := \mathbb{P}_{j,k}(\mathbf{x}); \tag{5.12}$$

and

$$\mathbf{x}_j := \mathbb{P}_{j,k}(\mathbf{x}_{j+1}), \quad \text{for } j < J; \quad \text{and } \mathbf{x}_J = \mathbb{P}_{J,\mathbf{x}}(\mathbf{x}). \tag{5.13}$$

Clearly, the first definition is the direct projection of the point $\mathbf{x}$ onto the approximate local tangent subspace $\mathbb{V}_{j,k}$ (thus it is the closest approximation to x from

$\mathbb{V}_{j,k}$ in the least-squares sense). In contrast, the second definition is the successive projection of $\mathbf{x}$ onto the sequence of tangent spaces $\mathbb{V}_{J,\mathbf{x}}, \ldots, \mathbb{V}_{j,\mathbf{x}}$.

Regardless of which definition, we will see that the difference $\mathbf{x}_{j+1} - \mathbf{x}_j$ is a high-dimensional vector in $\mathcal{R}^D$, however it may be decomposed into a sum of vectors in certain well-chosen low-dimensional spaces, shared across multiple points, in a multiscale fashion.

For reasons that will become obvious later, we define the geometric wavelet subspaces and translations as

$$W_{j+1,k'} := (\mathbf{I} - \mathbf{P}_{j,k})\,\mathcal{V}_{j+1,k'}; \tag{5.14}$$

$$\mathbf{w}_{j+1,k'} := (\mathbf{I} - \mathbf{P}_{j,k})(\mathbf{c}_{j+1,k'} - \mathbf{c}_{j,k}), \tag{5.15}$$

and let $\Psi_{j+1,k'}$ be an orthonormal basis for $W_{j+1,k'}$ which we call a *geometric wavelet basis* (see Fig. 6).

We proceed using the two definitions of $\mathbf{x}_j$ separately. First, with (5.12), we have for $j \leq J-1$

$$\begin{aligned} Q_{\mathcal{M}_{j+1}}(\mathbf{x}) &:= \mathbf{x}_{j+1} - \mathbf{x}_j \\ &= \mathbf{x}_{j+1} - \mathbb{P}_{j,k}(\mathbf{x}_{j+1}) + \mathbb{P}_{j,k}(\mathbf{x}_{j+1}) - \mathbb{P}_{j,k}(\mathbf{x}) \\ &= (\mathbf{I} - \mathbf{P}_{j,k})(\mathbf{x}_{j+1} - \mathbf{c}_{j,k}) + \mathbf{P}_{j,k}(\mathbf{x}_{j+1} - \mathbf{x}). \end{aligned} \tag{5.16}$$

Since $\mathbf{x}_{j+1} - \mathbf{c}_{j,k} = \mathbf{x}_{j+1} - \mathbf{c}_{j+1,k'} + \mathbf{c}_{j+1,k'} - \mathbf{c}_{j,k}$ and $\mathbf{x}_{j+1} - \mathbf{c}_{j+1,k'} \in \mathcal{V}_{j+1,k'}$, we obtain from (5.16), (5.14), (5.15)

$$\begin{aligned} Q_{\mathcal{M}_{j+1}}(\mathbf{x}) = \Psi_{j+1,k'}\Psi^*_{j+1,k'}(\mathbf{x}_{j+1} - \mathbf{c}_{j+1,k'}) + \mathbf{w}_{j+1,k'} \\ - \Phi_{j,k}\Phi^*_{j,k}(\mathbf{x} - \mathbf{x}_{j+1}). \end{aligned} \tag{5.17}$$

Note that the last term $\mathbf{x} - \mathbf{x}_{j+1}$ can be closely approximated by $\mathbf{x}_J - \mathbf{x}_{j+1} = \sum_{l=j+1}^{J-1} Q_{\mathcal{M}_{l+1}}(\mathbf{x})$ as the finest scale $J \to +\infty$, under general conditions. This equation splits the difference $\mathbf{x}_{j+1} - \mathbf{x}_j$ into a component in $W_{j+1,k'}$, a translation term that only depends on the cube (j,k) (and not on individual points), and a projection onto $V_{j,k}$ of a sum of differences $\mathbf{x}_{l+1} - \mathbf{x}_l$ at finer scales.

Second, with (5.13), we may obtain a simpler representation of the difference

$$\begin{aligned} Q_{\mathcal{M}_{j+1}}(\mathbf{x}) &= \mathbf{x}_{j+1} - (\mathbf{P}_{j,k}(\mathbf{x}_{j+1} - \mathbf{c}_{j,k}) + \mathbf{c}_{j,k}) \\ &= (\mathbf{I} - \mathbf{P}_{j,k})(\mathbf{x}_{j+1} - \mathbf{c}_{j+1,k'} + \mathbf{c}_{j+1,k'} - \mathbf{c}_{j,k}) \\ &= \Psi_{j+1,k'}\Psi^*_{j+1,k'}(\mathbf{x}_{j+1} - \mathbf{c}_{j+1,k'}) + \mathbf{w}_{j+1,k'}. \end{aligned} \tag{5.18}$$

The term $\mathbf{x} - \mathbf{x}_{j+1}$ no longer appears in this equation and the difference depends only on a component in $W_{j+1,k'}$ and a translation term.

Comparing (5.17) and (5.18) we see that the main advantage of the construction in (5.17) is that the approximations $\mathbf{x}_j$ have clear-cut interpretations as the best least-squares approximations. However, it is at the expense of the size of the dictionary which must contain the scaling functions $\Phi_{j,k}$. The construction in (5.18), leading to a smaller dictionary, is particularly useful when one does not care about the intermediate approximations, for example, in data compression tasks.

It is also worth mentioning that the definition of wavelet subspaces and translations (see (5.14), (5.15)) is independent of that of the x_j. We present their construction in Alg. 2. Moreover, regardless of the definition of the approximations, we have the following two-scale relationship (by definition of $Q_{\mathcal{M}_{j+1}}$)

$$P_{\mathcal{M}_{j+1}}(\mathbf{x}) = P_{\mathcal{M}_j}(\mathbf{x}) + Q_{\mathcal{M}_{j+1}}(\mathbf{x}), \tag{5.19}$$

Algorithm 2 Pseudocode for the construction of geometric wavelets

Input: $\mathbb{X}$: a set of n samples from $\mathcal{M} \subset \mathbb{R}^D$;
τ_0: some method for choosing local dimensions;
ϵ: a precision parameter
Output: A tree $\mathcal{T}$ of dyadic cubes $\{\mathbb{C}_{j,k}\}$, with local means $\{\mathbf{c}_{j,k}\}$ and SVD bases $\{\Phi_{j,k}\}$, as well as a family of geometric wavelets $\{\Psi_{j,k}\}$, $\{\mathbf{w}_{j,k}\}$
Steps:
1: Construct a tree $\mathcal{T}$ of dyadic cubes $\{\mathbb{C}_{j,k}\}$ with centers $\{\mathbf{c}_{j,k}\}$.
2: $J \leftarrow$ finest scale with the ϵ-approximation property.
3: Let $\text{cov}_{J,k} = |C_{J,k}|^{-1} \sum_{\mathbf{x} \in C_{J,k}} (\mathbf{x} - \mathbf{c}_{J,k})(\mathbf{x} - \mathbf{c}_{J,k})^*$, for all $k \in \Gamma_J$, and compute $\text{SVD}(\text{cov}_{J,k}) \approx \Phi_{J,k} \Sigma_{J,k} \Phi_{J,k}^*$ (where the rank of $\Phi_{J,k}$ is determined by τ_0).
4: **FOR** $j = J - 1$ **down to** 0
  FOR $k \in \Gamma_j$
  - Compute $\text{cov}_{j,k}$ and $\Phi_{j,k}$ as above
  - For each $k' \in \text{children}(j,k)$, construct the wavelet basis $\Psi_{j+1,k'}$ and translation $\mathbf{w}_{j+1,k'}$ using (5.14) and (5.15)

  ENDFOR
ENDFOR
5: **Return** $\Psi_{0,k} := \Phi_{0,k}$ and $\mathbf{w}_{0,k} := \mathbf{c}_{0,k}$ for $k \in \Gamma_0$.

and it may be iterated across scales:

$$\mathbf{x} = P_{\mathcal{M}_j}(\mathbf{x}) + \sum_{l=j}^{J-1} Q_{\mathcal{M}_{l+1}}(\mathbf{x}) + (\mathbf{x} - P_{\mathcal{M}_J}(\mathbf{x})). \tag{5.20}$$

The above equations allow to efficiently decompose each step along low dimensional subspaces, leading to efficient encoding of the data. We have therefore constructed a multiscale family of projection operators $P_{\mathcal{M}_j}$ (one for each node $\mathbb{C}_{j,k}$) onto approximate local tangent planes and detail projection operators $Q_{\mathcal{M}_{j+1}}$ (one for each edge) encoding the differences, collectively referred to as a GMRA structure. The cost of encoding the GMRA structure is at most $O(dD\epsilon^{-\frac{d}{2}})$ (when also encoding the scaling functions $\{\Phi_{j,k}\}$), and the time complexity of the algorithm is $O(Dn\log(n))$ [**ACM12**].

Finally, we mention that various other variations, optimizations, and generalizations of the construction, such as orthogonalization, splitting, pruning, out-of-sample extension, etc., can be found in [**ACM12**]. Due to space considerations, we omit their details here.

5.2.4. *Associated Geometric Wavelet Transforms (GWT).* Given a GMRA structure, we may compute a Discrete Forward GWT for a point $\mathbf{x} \in \mathcal{M}$ that maps it to a sequence of wavelet coefficient vectors:

$$\mathbf{q}_\mathbf{x} = (\mathbf{q}_{J,\mathbf{x}}, \mathbf{q}_{J-1,\mathbf{x}}, \ldots, \mathbf{q}_{1,\mathbf{x}}, \mathbf{q}_{0,\mathbf{x}}) \in \mathcal{R}^{d+\sum_{j=1}^{J} d_{j,\mathbf{x}}^w} \tag{5.21}$$

where $\mathbf{q}_{j,\mathbf{x}} := \Psi_{j,\mathbf{x}}^*(\mathbf{x}_j - \mathbf{c}_{j,\mathbf{x}})$, and $d_{j,\mathbf{x}}^w := \text{rank}(\Psi_{j,\mathbf{x}})$. Note that, in the case of a d-dimensional manifold and for a fixed precision $\epsilon > 0$, $\mathbf{q}_\mathbf{x}$ has a maximum possible length $(1 + \frac{1}{2}\log_2 \frac{1}{\epsilon})d$, which is independent of D and nearly optimal in d [**CM10**]. On the other hand, we may easily reconstruct the point $\mathbf{x}$ using the

```
{q_{j,x}} =FGWT(GMRA, x)

// Input: GMRA structure, x ∈ M
// Output: Wavelet coefficients {q_{j,x}}
for j = J down to 0
        x_j = Φ_{j,x} Φ*_{j,x} (x − c_{j,x}) + c_{j,x}
        q_{j,x} = Ψ*_{j,x} (x_j − c_{j,x})
end
```

```
{q_{j,x}} =FGWT(GMRA, x)

// Input: GMRA structure, x ∈ M
// Output: Wavelet coefficients {q_{j,x}}
for j = J down to 0
        q_{j,x} = Ψ*_{j,x} (x − c_{j,x})
        x = x − (Ψ_{j,x} q_{j,x} + w_{j,x})
end
```

```
x =IGWT(GMRA,{q_{j,x}})

// Input: GMRA structure, wavelet coefficients {q_{j,x}}
// Output: Reconstruction x

Q_{M_J}(x) = Ψ_{J,x} q_{J,x} + w_{J,x}
for j = J − 1 down to 1
        Q_{M_j}(x) = Ψ_{j,x} q_{j,x} + w_{j,x} − P_{j−1,x} Σ_{ℓ>j} Q_{M_ℓ}(x)
end
x = Ψ_{0,x} q_{0,x} + w_{0,x} + Σ_{j>0} Q_{M_j}(x)
```

```
x =IGWT(GMRA,{q_{j,x}})

// Input: GMRA structure, wavelet coefficients {q_{j,x}}
// Output: Reconstruction x

for j = J down to 0
        Q_{M_j}(x) = Ψ_{j,x} q_{j,x} + w_{j,x}
end
x = Σ_{0≤j≤J} Q_{M_j}(x)
```

Figure 7 Pseudocodes for the Forward (top row) and Inverse (bottom row) GWTs corresponding to different wavelet constructions (5.17) (left column) and (5.18) (right column)

GMRA structure and the wavelet coefficients, by a Discrete Inverse GWT. See Fig. 7 for the pseudocodes of both transforms.

5.2.5. *A toy example.* We consider a 2-dimensional *SwissRoll* manifold in $\mathbb{R}^{50}$ and sample 2000 points from it without adding any noise. We apply the GMRA to this synthetic data set to illustrate how the GMRA works in general. The corresponding results are shown in Fig. 8.

5.2.6. *A few remarks about GMRA.* We make the following comments.

- The GMRA algorithm presents an appealing framework for constructing data-dependent dictionaries using a geometric multiresolution analysis. Unlike the K-SVD dictionaries, the GMRA outputs dictionaries that are structured and hierarchically organized. Moreover, the different subgroups of such a dictionary correspond to different scales and have clear interpretations as detail operators.
- It has many other advantages, for example the construction is based on many local SVD and thus is fast to execute. In addition, there are theoretical guarantees on the size of the dictionary and the sparsity of the representation, at least when the data follows a manifold model. It is also associated with fast transforms, making the sparse coding component extremely simple and fast, which is typically unavailable for other algorithms.
- The GMRA algorithm naturally extends the wavelet transform for 1-dimensional signals to efficient multiscale transforms for higher dimensional data. The nonlinear space $\mathcal{M}$ replaces the classical function spaces,

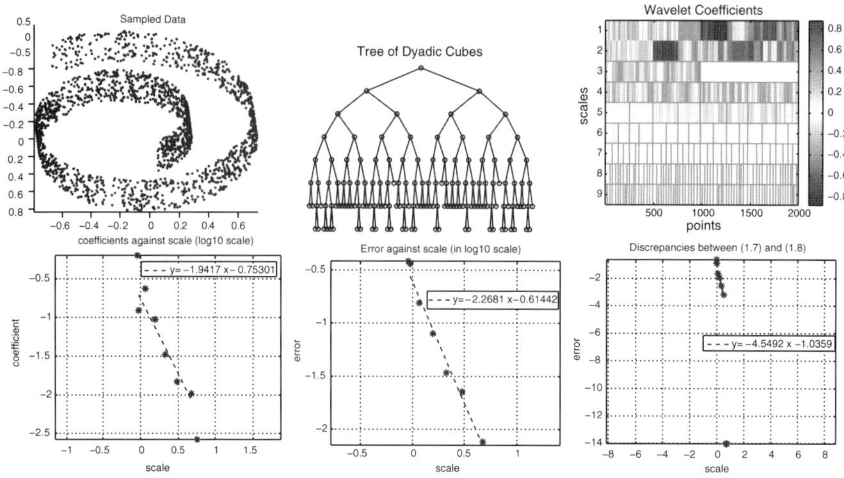

Figure 8 Illustration of the GMRA (with the projection defined in (5.13)) on a data set of 2000 points sampled from a 2-dimensional SwissRoll manifold in $\mathbb{R}^{50}$. Top left: sampled data; top middle: the tree of dyadic cubes obtained by the METIS algorithm [**KK99**]; top right: matrix of wavelet coefficients. The x-axis indexes the points (arranged according to the tree), and the y axis indexes the scales from coarse (top) to fine (bottom). Note that each block corresponds to a different dyadic cube. Bottom left: (average) magnitude of wavelet coefficients versus scale; bottom middle: approximation error of the projection (5.13) to the data as a function of scale; bottom right: deviation of the projection (5.13) from the best possible one (5.12) at each scale (also in $\log_{10}$ scale). The last plot shows that the projection (5.13) deviates from (5.12) at a rate of 4 and in particular, the two almost coincide with each other at fine scales.

the piecewise affine approximation at each scale substitutes the linear projection on scaling function spaces, and the difference operators play the role of the linear wavelet projections. But it is also quite different in many crucial aspects. It is nonlinear, as its adapts to the nonlinear manifolds modeling the data space, but every scale-to-scale step is linear. Translations or dilations do not play any role here, while they are often considered crucial in classical wavelet constructions.

5.3. Online dictionary learning (ODL). The ODL algorithm, developed by Mairal et al. [**MBPS09, MBPS10**], is an online algorithm that is designed to handle extremely large data sets. It starts by assuming a generative model $\mathbf{x}_t \sim p(\mathbf{x})$, where $p(\mathbf{x})$ represents a probability density function governing the data distribution.[2] At each time $t = 1, 2, \ldots$, it draws a new sample $\mathbf{x}_t$ from the distribution and uses it to refine the dictionary $\mathbf{D}_{t-1}$ obtained at time $t-1$. This procedure is repeated until a stopping criterion has been met (for example, t has reached an upper bound T, or the dictionary $\mathbf{D}_t$ no longer changes noticeably).

[2]For real data sets one does not know the underlying probability distribution; in this case, the uniform discrete measure can be used. This is equivalent to first randomly permuting the data points and then sequentially processing them, one at a time.

5.3.1. *Detailed description of the ODL algorithm.* To present the specific ideas, we consider iteration t when a new sample $\mathbf{x}_t$ arrives. By this time, the first $t-1$ samples $\mathbf{x}_1, \ldots, \mathbf{x}_{t-1}$ have already been drawn from the distribution and used to train a dictionary $\mathbf{D}_{t-1}$. Of course, if $t = 1$, then $\mathbf{D}_0$ represents an initial guess of the dictionary provided by the user to start with. We now would like to use $\mathbf{x}_t$ to update the dictionary $\mathbf{D}_{t-1}$ to $\mathbf{D}_t$. This is achieved in two steps: First, we find the sparse coefficient of $\mathbf{x}_t$ relative to $\mathbf{D}_{t-1}$ by solving the sparse coding problem

$$\gamma_t = \operatorname*{argmin}_{\gamma} \frac{1}{2}\|\mathbf{x}_t - \mathbf{D}_{t-1}\gamma\|^2 + \lambda\|\gamma\|_1$$

where λ is the user-specified tuning parameter, fixed throughout all iterations. We have repeatedly encountered this problem and recall that it can be easily solved by any of those pursuit algorithms mentioned in the CS framework. Next, we fix the coefficient γ_t obtained above, together with the previous ones $\gamma_1, \ldots, \gamma_{t-1}$, and minimize the same objective function, this time with respect to the dictionary $\mathbf{D}$ (constrained to have at most unit-norm columns), hoping to find a new dictionary $\mathbf{D}_t$ that is better suited to all coefficients $\gamma_i, 1 \le i \le t$:

$$\mathbf{D}_t = \operatorname*{argmin}_{\mathbf{D}} \frac{1}{2t}\sum_{i=1}^{t} \|\mathbf{x}_i - \mathbf{D}\gamma_i\|^2 + \lambda\|\gamma_i\|_1.$$

In this step, since the γ_i are fixed, we may remove the second term from the objective function and consider instead

$$\mathbf{D}_t = \operatorname*{argmin}_{\mathbf{D}} \sum_{i=1}^{t} \|\mathbf{x}_i - \mathbf{D}\gamma_i\|^2.$$

To see how the square-loss objective is minimized, we rewrite it using matrix notation:

$$\begin{aligned}
\sum_{i=1}^{t} \|\mathbf{x}_i - \mathbf{D}\gamma_i\|^2 &= \sum_{i=1}^{t} \operatorname{trace}\left((\mathbf{x}_i - \mathbf{D}\gamma_i)(\mathbf{x}_i - \mathbf{D}\gamma_i)^T\right) \\
&= \sum_{i=1}^{t} \operatorname{trace}\left(\mathbf{x}_i\mathbf{x}_i^T - \mathbf{x}_i\gamma_i^T\mathbf{D}^T - \mathbf{D}\gamma_i\mathbf{x}_i^T + \mathbf{D}\gamma_i\gamma_i^T\mathbf{D}^T\right) \\
&= \sum_{i=1}^{t} \left(\operatorname{trace}\left(\mathbf{x}_i\mathbf{x}_i^T\right) - 2\operatorname{trace}\left(\mathbf{x}_i\gamma_i^T\mathbf{D}^T\right) + \operatorname{trace}\left(\mathbf{D}^T\mathbf{D}\gamma_i\gamma_i^T\right)\right) \\
&= \operatorname{trace}\left(\sum_{i=1}^{t} \mathbf{x}_i\mathbf{x}_i^T\right) - 2\operatorname{trace}\left(\mathbf{D}^T\sum_{i=1}^{t} \mathbf{x}_i\gamma_i^T\right) \\
&\quad + \operatorname{trace}\left(\mathbf{D}^T\mathbf{D}\sum_{i=1}^{t} \gamma_i\gamma_i^T\right)
\end{aligned}$$

Letting

$$\mathbf{A}_t = \sum_{i=1}^{t} \gamma_i\gamma_i^T \in \mathbb{R}^{m\times m}, \quad \mathbf{B}_t = \sum_{i=1}^{t} \mathbf{x}_i\gamma_i^T \in \mathbb{R}^{L\times m}$$

and discarding the first term that does not depend on $\mathbf{D}$, we arrive at the final form

$$\mathbf{D}_t = \operatorname*{argmin}_{\mathbf{D}} \operatorname{trace}(\mathbf{D}^T\mathbf{D}\mathbf{A}_t) - 2\operatorname{trace}(\mathbf{D}^T\mathbf{B}_t)$$

This problem can be solved by block coordinate descent, using $\mathbf{D}_{t-1}$ as an initial dictionary. Specifically, we hold all except the kth column $\mathbf{d}_k$ of $\mathbf{D} = \mathbf{D}_{t-1}$ fixed and consider the resulting optimization problem, which is only over $\mathbf{d}_k$. Using straightforward multivariable calculus, we obtain the gradient of the objective function as follows:

$$2(\mathbf{D}\,\mathbf{A}_t(:,k) - \mathbf{B}_t(:,k)).$$

Setting it equal to zero and solving the corresponding equation yields a unique critical point

$$\frac{1}{(\mathbf{A}_t)_{kk}}(\mathbf{B}_t(:,k) - \sum_{j\neq k} \mathbf{d}_j\,(\mathbf{A}_t)_{jk}) = \mathbf{d}_k - \frac{1}{(\mathbf{A}_t)_{kk}}\left(\mathbf{D}\,\mathbf{A}_t(:,k) - \mathbf{B}_t(:,k)\right).$$

The Hessian matrix of the objective function, $(\mathbf{A}_t)_{kk}\,\mathbf{I}$, is strictly positive definite (because $\mathbf{A}_t$ also is), implying that the critical point is a global minimizer. Thus, we update $\mathbf{d}_k$ by setting

$$\mathbf{d}_k \leftarrow \mathbf{d}_k - \frac{1}{(\mathbf{A}_t)_{kk}}\left(\mathbf{D}\,\mathbf{A}_t(:,k) - \mathbf{B}_t(:,k)\right),$$

in order to reduce the objective function as much as possible (if $\|\mathbf{d}_k\|_2 > 1$, we then need to normalize it to unit norm). We sequentially update all the columns of $\mathbf{D}$ by varying k, always including the updated columns in $\mathbf{D}$ for updating the remaining columns, in order to accelerate convergence. We repeat this procedure until the objective function no longer decreases and denote the corresponding dictionary by $\mathbf{D}_t$.

We now summarize the steps of ODL in Algorithm 3. We refer the reader to [**MBPS09**, **MBPS10**] for convergence analysis and performance evaluations.

5.3.2. *A few remarks about the ODL algorithm.*

- ODL and K-SVD are similar in several ways: (1) Both involve solving the sparse coding problem (4.7) many times (once per signal, per iteration). (2) The columns of $\mathbf{D}$ are updated sequentially in both algorithms, which speeds up convergence. (3) Both require an initial guess of the dictionary, which affects convergence.
- Both ODL and GMRA assume a generative model for the data, so they both aim to build dictionaries for some probability distribution, not just for a particular data set. In contrast, K-SVD only aims to achieve the minimal empirical cost on the training data.
- The ODL is also quite different from K-SVD and GMRA, in the sense that the latter two are batch methods which must access the whole training set in order to learn the dictionary. In contrast, the ODL uses one sample at a time, without needing to store or access the entire data set. This advantage makes ODL particularly suitable for working with very large data sets, which might have millions of samples, and with dynamic training data changing over time, such as video sequences. Additionally, since it does not need to store the entire data set, it has a low memory consumption and lower computational cost than classical batch algorithms.

5.4. Future directions in dictionary learning. Despite all the impressive achievements by sparse and redundant modeling, many questions remain to be answered and there are a large number of future research directions. We list only

Algorithm 3 Pseudocode for the Online Dictionary Learning (ODL) Algorithm

Input: Density $p(\mathbf{x})$, tuning parameter λ, initial dictionary $\mathbf{D}_0$, and number of iterations T

Output: Final dictionary $\mathbf{D}_T$

Steps:

1: **Initialization**: Set $\mathbf{A}_0 \leftarrow 0$ and $\mathbf{B}_0 \leftarrow 0$

2: **FOR** $t = 1 : T$

- **Sampling:** Draw a new sample $\mathbf{x}_t$ from $p(\mathbf{x})$
- **Sparse coding:** Find the sparse coefficient of $\mathbf{x}_t$ relative to $\mathbf{D}_{t-1}$ by solving

$$\gamma_t = \underset{\gamma}{\operatorname{argmin}} \ \|\mathbf{x}_t - \mathbf{D}_{t-1}\gamma\|_2^2 + \lambda\|\gamma\|_1.$$

  using any pursuit algorithm (e.g., OMP).
- **Recording:** Update $\mathbf{A}_t$ and $\mathbf{B}_t$ to include $\mathbf{x}_t$ and γ_t:

$$\mathbf{A}_t \leftarrow \mathbf{A}_{t-1} + \gamma_t\gamma_t^T, \quad \mathbf{B}_t \leftarrow \mathbf{B}_{t-1} + \mathbf{x}_t\gamma_t^T$$

- **Dictionary update:** Update the columns $\mathbf{d}_k$ of $\mathbf{D}_{t-1}$ sequentially, using

$$\mathbf{d}_k \leftarrow \mathbf{d}_k - \frac{1}{(\mathbf{A}_t)_{kk}} \left(\mathbf{D}\mathbf{A}_t(:,k) - \mathbf{B}_t(:,k)\right).$$

  If $\|\mathbf{d}_k\|_2 > 1$, then normalize it to have unit norm. Repeat this procedure until convergence.

ENDFOR

3: Return $\mathbf{D}_t$

a few below, while referring the reader to [**Ela12**] for a detailed discussion on this topic.

- **Theoretical justification of dictionary learning algorithms**. So far most of the dictionary learning algorithms are essentially empirical methods (e.g., K-SVD [**AEB05, AEB06**]). Little is known about their stability, especially in the presence of noise. Furthermore, it is unclear which conditions would guarantee the algorithm to succeed. Finally, new measures of dictionary quality (not just worst-case conditions) need to be developed in order to study the goodness of the learned dictionary.
- **Introducing structure to dictionaries**. Currently, most learned dictionaries (such as the K-SVD dictionary) are completely unstructured. Finding the sparse representation of a signal relative to such a dictionary is a nontrivial task. This causes a great computational burden when dealing with large data sets. It is thus desirable to impose structures to the dictionary atoms to simplify the sparse coding task. It is noteworthy to mention that the GMRA dictionary represents an advancement in this direction, as it organizes its atoms into a tree which makes the coding part extremely simple and fast. It will be interesting to explore other ways to construct structured dictionaries.
- **Developing next generation models**. Sparse and redundant modeling represents the current state of the art, having evolved from transforms and

union-of-subspaces models. Inevitably such a model will be replaced by newer, more powerful models, just like its predecessors. It will be exciting to see any new research devoted in this direction.

Acknowledgment

The images in Fig. 5 are generated by the K-SVD software available at `http://www.cs.technion.ac.il/~elad/software/`.

References

[ABC+15] P. Awasthi, and A. S. Bandeira, and M. Charikar, and R. Krishnaswamy, and S. Villar, and R. Ward. Relax, No Need to Round: Integrality of Clustering Formulations. In *Proceedings of the 2015 Conference on Innovations in Theoretical Computer Science*, pages 191–20, January 2015.

[ACM12] William K. Allard, Guangliang Chen, and Mauro Maggioni, *Multi-scale geometric methods for data sets II: Geometric multi-resolution analysis*, Appl. Comput. Harmon. Anal. **32** (2012), no. 3, 435–462, DOI 10.1016/j.acha.2011.08.001. MR2892743

[AEB05] M. Aharon, M. Elad, and A. M. Bruckstein. The k-svd algorithm. In *SPARSE*, 2005.

[AEB06] M. Aharon, M. Elad, and A. M. Bruckstein. K-svd: An algorithm for designing of overcomplete dictionaries for sparse representation. *IEEE T. Signal Proces.*, 54(11):4311–4322, November 2006.

[AR77] J. B. Allen and L. R. Rabiner. A unified approach to short-time fourier analysis and synthesis. In *Proceedings of the IEEE*, volume 65, pages 1558–1564, 1977.

[Bas80] M. J. Bastiaans. Gabor's expansion of a signal into gaussian elementary signals. In *Proceedings of the IEEE*, volume 68, pages 538–539, 1980.

[BCM05] A. Buades, B. Coll, and J. M. Morel, *A review of image denoising algorithms, with a new one*, Multiscale Model. Simul. **4** (2005), no. 2, 490–530, DOI 10.1137/040616024. MR2162865

[BD09] Thomas Blumensath and Mike E. Davies, *Iterative hard thresholding for compressed sensing*, Appl. Comput. Harmon. Anal. **27** (2009), no. 3, 265–274, DOI 10.1016/j.acha.2009.04.002. MR2559726 (2010i:94048)

[BDE09] Alfred M. Bruckstein, David L. Donoho, and Michael Elad, *From sparse solutions of systems of equations to sparse modeling of signals and images*, SIAM Rev. **51** (2009), no. 1, 34–81, DOI 10.1137/060657704. MR2481111 (2010d:94012)

[BE08] O. Bryt and M. Elad. Compression of facial images using the k-svd algorithm. *Journal of Visual Communication and Image Representation*, 19:270–283, 2008.

[Blu11] Thomas Blumensath, *Sampling and reconstructing signals from a union of linear subspaces*, IEEE Trans. Inform. Theory **57** (2011), no. 7, 4660–4671, DOI 10.1109/TIT.2011.2146550. MR2840482 (2012i:94102)

[BS07] R. Baraniuk and P. Steeghs. Compressive radar imaging. In *Proc. IEEE Radar Conf*, pages 128–133. Ieee, 2007.

[Can06] Emmanuel J. Candès, *Compressive sampling*, International Congress of Mathematicians. Vol. III, Eur. Math. Soc., Zürich, 2006, pp. 1433–1452. MR2275736 (2008e:62033)

[CD04] Emmanuel J. Candès and David L. Donoho, *New tight frames of curvelets and optimal representations of objects with piecewise C^2 singularities*, Comm. Pure Appl. Math. **57** (2004), no. 2, 219–266, DOI 10.1002/cpa.10116. MR2012649 (2004k:42052)

[CDD09] Albert Cohen, Wolfgang Dahmen, and Ronald DeVore, *Compressed sensing and best k-term approximation*, J. Amer. Math. Soc. **22** (2009), no. 1, 211–231, DOI 10.1090/S0894-0347-08-00610-3. MR2449058 (2010d:94024)

[CDDY00] E. J. Candès, L. Demanet, D. L. Donoho, and L. Ying. Fast discrete curvelet transforms. *Multiscale Model. Simul.*, 5:861–899, 2000.

[CSPW11] Venkat Chandrasekaran, Sujay Sanghavi, Pablo A. Parrilo, and Alan S. Willsky, *Rank-sparsity incoherence for matrix decomposition*, SIAM J. Optim. **21** (2011), no. 2, 572–596, DOI 10.1137/090761793. MR2817479 (2012m:90128)

[CDS98] Scott Shaobing Chen, David L. Donoho, and Michael A. Saunders, *Atomic decomposition by basis pursuit*, SIAM J. Sci. Comput. **20** (1998), no. 1, 33–61, DOI 10.1137/S1064827596304010. MR1639094 (99h:94013)

[CENR10] Emmanuel J. Candès, Yonina C. Eldar, Deanna Needell, and Paige Randall, *Compressed sensing with coherent and redundant dictionaries*, Appl. Comput. Harmon. Anal. **31** (2011), no. 1, 59–73, DOI 10.1016/j.acha.2010.10.002. MR2795875 (2012d:94014)

[CK] *Finite frames*, Applied and Numerical Harmonic Analysis, Birkhäuser/Springer, New York, 2013. Theory and applications; Edited by Peter G. Casazza and Gitta Kutyniok. MR2964005

[CL09a] Guangliang Chen and Gilad Lerman, *Foundations of a multi-way spectral clustering framework for hybrid linear modeling*, Found. Comput. Math. **9** (2009), no. 5, 517–558, DOI 10.1007/s10208-009-9043-7. MR2534403 (2010k:62299)

[CL09b] G. Chen and G. Lerman. Spectral curvature clustering (SCC). *Int. J. Comput. Vision*, 81(3):317–330, 2009.

[CM10] G. Chen and M. Maggioni. Multiscale geometric wavelets for the analysis of point clouds. In *Proc. of the 44th Annual Conference on Information Sciences and Systems (CISS)*, Princeton, NJ, March 2010.

[CM11a] G. Chen and M. Maggioni. Multiscale geometric and spectral analysis of plane arrangements. In *IEEE Conference on Computer Vision and Pattern Recognition*, 2011.

[CM11b] G. Chen and M. Maggioni. Multiscale geometric dictionaries for point-cloud data. In *Proc. of the 9th International Conference on Sampling Theory and Applications (SampTA)*, Singapore, May 2011.

[CMW92] Ronald R. Coifman, Yves Meyer, and Victor Wickerhauser, *Wavelet analysis and signal processing*, Wavelets and their applications, Jones and Bartlett, Boston, MA, 1992, pp. 153–178. MR1187341

[CP09] Emmanuel J. Candès and Yaniv Plan, *Tight oracle inequalities for low-rank matrix recovery from a minimal number of noisy random measurements*, IEEE Trans. Inform. Theory **57** (2011), no. 4, 2342–2359, DOI 10.1109/TIT.2011.2111771. MR2809094

[CP10] Emmanuel J. Candès and Yaniv Plan, *A probabilistic and RIPless theory of compressed sensing*, IEEE Trans. Inform. Theory **57** (2011), no. 11, 7235–7254, DOI 10.1109/TIT.2011.2161794. MR2883653

[CR05] E. Candès and J. Romberg. Signal recovery from random projections. In *Proc. SPIE Conference on Computational Imaging III*, volume 5674, pages 76–86. SPIE, 2005.

[CRT06a] Emmanuel J. Candès, Justin Romberg, and Terence Tao, *Robust uncertainty principles: exact signal reconstruction from highly incomplete frequency information*, IEEE Trans. Inform. Theory **52** (2006), no. 2, 489–509, DOI 10.1109/TIT.2005.862083. MR2236170 (2007e:94020)

[CRT06b] Emmanuel J. Candès, Justin K. Romberg, and Terence Tao, *Stable signal recovery from incomplete and inaccurate measurements*, Comm. Pure Appl. Math. **59** (2006), no. 8, 1207–1223, DOI 10.1002/cpa.20124. MR2230846 (2007f:94007)

[CSZ06] Tony F. Chan, Jianhong Shen, and Hao-Min Zhou, *Total variation wavelet inpainting*, J. Math. Imaging Vision **25** (2006), no. 1, 107–125, DOI 10.1007/s10851-006-5257-3. MR2254441 (2007g:94006)

[CT05] Emmanuel J. Candes and Terence Tao, *Decoding by linear programming*, IEEE Trans. Inform. Theory **51** (2005), no. 12, 4203–4215, DOI 10.1109/TIT.2005.858979. MR2243152 (2007b:94313)

[CT06] Emmanuel J. Candes and Terence Tao, *Near-optimal signal recovery from random projections: universal encoding strategies?*, IEEE Trans. Inform. Theory **52** (2006), no. 12, 5406–5425, DOI 10.1109/TIT.2006.885507. MR2300700 (2008c:94009)

[CW08] E. J. Candès and M. Wakin. An introduction to compressive sampling. *IEEE Signal Proc. Mag.*, 25(2):21–30, 2008.

[Dau92] Ingrid Daubechies, *Ten lectures on wavelets*, CBMS-NSF Regional Conference Series in Applied Mathematics, vol. 61, Society for Industrial and Applied Mathematics (SIAM), Philadelphia, PA, 1992. MR1162107 (93e:42045)

[DH01] David L. Donoho and Xiaoming Huo, *Uncertainty principles and ideal atomic decomposition*, IEEE Trans. Inform. Theory **47** (2001), no. 7, 2845–2862, DOI 10.1109/18.959265. MR1872845 (2002k:94012)

[DNW13] Mark A. Davenport, Deanna Needell, and Michael B. Wakin, *Signal space CoSaMP for sparse recovery with redundant dictionaries*, IEEE Trans. Inform. Theory **59** (2013), no. 10, 6820–6829, DOI 10.1109/TIT.2013.2273491. MR3106865

[Don06] David L. Donoho, *Compressed sensing*, IEEE Trans. Inform. Theory **52** (2006), no. 4, 1289–1306, DOI 10.1109/TIT.2006.871582. MR2241189 (2007e:94013)

[DSMB09] W. Dai, M. Sheikh, O. Milenkovic, and R. Baraniuk. Compressive sensing DNA microarrays. *EURASIP journal on bioinformatics and systems biology*, pages 1–12, 2009.

[DSP] Compressive sampling webpage. http://dsp.rice.edu/cs.

[Dut89] *Wavelets*, Inverse Problems and Theoretical Imaging, Springer-Verlag, Berlin, 1989. Time-frequency methods and phase space; Edited by J. M. Combes, A. Grossmann and Ph. Tchamitchian. MR1010895 (90g:42062)

[DW11] Mark A. Davenport and Michael B. Wakin, *Compressive sensing of analog signals using discrete prolate spheroidal sequences*, Appl. Comput. Harmon. Anal. **33** (2012), no. 3, 438–472, DOI 10.1016/j.acha.2012.02.005. MR2950138

[EA06] Michael Elad and Michal Aharon, *Image denoising via sparse and redundant representations over learned dictionaries*, IEEE Trans. Image Process. **15** (2006), no. 12, 3736–3745, DOI 10.1109/TIP.2006.881969. MR2498043

[EFM10] M. Elad, M. Figueiredo, and Y. Ma. On the role of sparse and redundant representations in image processing. *Proceedings of the IEEE - Special Issue on Applications of Sparse Representation and Compressive Sensing*, 98(6):972–982, 2010.

[EJCW09] Y. M. E. J. Candès, X. Li and J. Wright. Robust Principal Component Analysis? *Journal of ACM*, 58(1):1–37, 2009.

[Ela12] M. Elad. Sparse and redundant representation modeling–what next? *IEEE Signal Processing Letters*, 19(12):922–928, 2012.

[FS98] *Gabor analysis and algorithms*, Applied and Numerical Harmonic Analysis, Birkhäuser Boston, Inc., Boston, MA, 1998. Theory and applications; Edited by Hans G. Feichtinger and Thomas Strohmer. MR1601119 (98h:42001)

[F15] S. Foucart. Dictionary-sparse recovery via thresholding-based algorithms. Submitted.

[Gab46] D. Gabor. Theory of communication. *J. Inst. Electr. Eng.*, 93(26):429–457, 1946.

[GE12] Raja Giryes and Michael Elad, *RIP-based near-oracle performance guarantees for SP, CoSaMP, and IHT*, IEEE Trans. Signal Process. **60** (2012), no. 3, 1465–1468, DOI 10.1109/TSP.2011.2174985. MR2859009 (2012m:94114)

[GN13] Raja Giryes and Deanna Needell, *Greedy signal space methods for incoherence and beyond*, Appl. Comput. Harmon. Anal. **39** (2015), no. 1, 1–20, DOI 10.1016/j.acha.2014.07.004. MR3343799

[GNE+12] R. Giryes, S. Nam, M. Elad, R. Gribonval, and M. E. Davies, *Greedy-like algorithms for the cosparse analysis model*, Linear Algebra Appl. **441** (2014), 22–60, DOI 10.1016/j.laa.2013.03.004. MR3134336

[HN07] J. Haupt and R. Nowak. A generalized restricted isometry property. *Univ. of Wisconsin-Madison, Tech. Rep. ECE-07-1*, 2007.

[HS09] Matthew A. Herman and Thomas Strohmer, *High-resolution radar via compressed sensing*, IEEE Trans. Signal Process. **57** (2009), no. 6, 2275–2284, DOI 10.1109/TSP.2009.2014277. MR2641823 (2011a:94028)

[HX13] H. Huang and N. Xiao. Image deblurring based on sparse model with dictionary learning. *Journal of Information & Computational Science*, 10(1):129–137, 2013.

[Jan81] A. Janssen. Gabor representation of generalized functions. *J. Math. Anal. and Applic.*, 83(2):377–394, 1981.

[Jol02] I. T. Jolliffe, *Principal component analysis*, 2nd ed., Springer Series in Statistics, Springer-Verlag, New York, 2002. MR2036084 (2004k:62010)

[Kee03] Stephen L. Keeling, *Total variation based convex filters for medical imaging*, Appl. Math. Comput. **139** (2003), no. 1, 101–119, DOI 10.1016/S0096-3003(02)00171-6. MR1949379 (2003k:92013)

[KK99] George Karypis and Vipin Kumar, *A fast and high quality multilevel scheme for partitioning irregular graphs*, SIAM J. Sci. Comput. **20** (1998), no. 1, 359–392 (electronic), DOI 10.1137/S1064827595287997. MR1639073 (99f:68158)

[KLW+06] S. Kirolos, J. Laska, M. Wakin, M. Duarte, D. Baron, T. Ragheb, Y. Massoud, and R. Baraniuk. Analog-to-information conversion via random demodulation. In *Proc. IEEE Dallas Circuits and Systems Workshop (DCAS)*, pages 71–74. IEEE, 2006.

[KTMJ08] B. Kai Tobias, U. Martin, and F. Jens. Suppression of MRI truncation artifacts using total variation constrained data extrapolation. *Int. J. Biomedical Imaging*, 2008.

[KNW15] Author = F. Krahmer, and D. Needell and R. Ward, Compressive Sensing with Redundant Dictionaries and Structured Measurements. Submitted.

[KW11] Felix Krahmer and Rachel Ward, *New and improved Johnson-Lindenstrauss embeddings via the restricted isometry property*, SIAM J. Math. Anal. **43** (2011), no. 3, 1269–1281, DOI 10.1137/100810447. MR2821584 (2012g:15052)

[LDP07] M. Lustig, D. Donoho, and J. Pauly. Sparse mri: The application of compressed sensing for rapid mr imaging. *Magnetic Resonance in Medicine*, 58(6):1182–1195, 2007.

[LDSP08] M. Lustig, D. Donoho, J. Santos, and J. Pauly. Compressed sensing MRI. *IEEE Sig. Proc. Mag.*, 25(2):72–82, 2008.

[LH07] T. Lin and F. Herrmann. Compressed wavefield extrapolation. *Geophysics*, 72(5):SM77, 2007.

[LLR95] Nathan Linial, Eran London, and Yuri Rabinovich, *The geometry of graphs and some of its algorithmic applications*, Combinatorica **15** (1995), no. 2, 215–245, DOI 10.1007/BF01200757. MR1337355 (96e:05158)

[LW11] Y. Liu and Q. Wan. Total variation minimization based compressive wideband spectrum sensing for cognitive radios. Preprint, 2011.

[Mac67] J. MacQueen, *Some methods for classification and analysis of multivariate observations*, Proc. Fifth Berkeley Sympos. Math. Statist. and Probability (Berkeley, Calif., 1965/66), Univ. California Press, Berkeley, Calif., 1967, pp. Vol. I: Statistics, pp. 281–297. MR0214227 (35 #5078)

[Mal89] S. Mallat. A theory of multiresolution signal decomposition: the wavelet representation. *IEEE T. Pattern Anal.*, 11:674–693, 1989.

[Mal99] S. Mallat. *A Wavelet Tour of Signal Processing.* Academic Press, London, 2nd edition, 1999.

[MBPS09] J. Mairal, F. Bach, J. Ponce, and G. Sapiro. Online dictionary learning for sparse coding. In *Proceedings of the 26th Annual International Conference on Machine Learning*, volume 11, pages 689–696, 2009.

[MBPS10] Julien Mairal, Francis Bach, Jean Ponce, and Guillermo Sapiro, *Online learning for matrix factorization and sparse coding*, J. Mach. Learn. Res. **11** (2010), 19–60. MR2591620 (2011i:62121)

[ME11] Moshe Mishali and Yonina C. Eldar, *Xampling: compressed sensing of analog signals*, Compressed sensing, Cambridge Univ. Press, Cambridge, 2012, pp. 88–147, DOI 10.1017/CBO9780511794308.004. MR2963168

[MPTJ08] Shahar Mendelson, Alain Pajor, and Nicole Tomczak-Jaegermann, *Uniform uncertainty principle for Bernoulli and subgaussian ensembles*, Constr. Approx. **28** (2008), no. 3, 277–289, DOI 10.1007/s00365-007-9005-8. MR2453368 (2009k:46020)

[MSE] J. Mairal, G. Sapiro, and M. Elad. Learning multiscale sparse representations for image and video restoration.

[MSW+10] M. Mohtashemi, H. Smith, D. Walburger, F. Sutton, and J. Diggans. Sparse sensing DNA microarray-based biosensor: Is it feasible? In *IEEE Sensors Applications Symposium (SAS)*, pages 127–130. IEEE, 2010.

[Mut05] S. Muthukrishnan. *Data Streams: Algorithms and Applications.* Now Publishers, Hanover, MA, 2005.

[MYZC08] S. Ma, W. Yin, Y. Zhang, and A. Chakraborty. An efficient algorithm for compressed mr imaging using total variation and wavelets. In *IEEE Conf. Comp. Vision Pattern Recog.*, 2008.

[MZ93] S. Mallat and Z. Zhang. Matching pursuits with time-frequency dictionaries. *IEEE T. Signal Proces.*, 41(12):3397–3415, 1993.

[NDEG11] S. Nam, M. E. Davies, M. Elad, and R. Gribonval, *The cosparse analysis model and algorithms*, Appl. Comput. Harmon. Anal. **34** (2013), no. 1, 30–56, DOI 10.1016/j.acha.2012.03.006. MR2981332

[NJW01] A. Ng, M. Jordan, and Y. Weiss. On spectral clustering: Analysis and an algorithm. In *Advances in Neural Information Processing Systems 14*, pages 849–856, 2001.

[NRWY10] S. Negahban, P. Ravikumar, M. Wainwright, and B. Yu. A unified framework for high-dimensional analysis of M-estimators with decomposable regularizers. *Adv. Neur. In.*, 1348–1356, 2009.

[NT08a] D. Needell and J. A. Tropp. CoSaMP: Iterative signal recovery from incomplete and inaccurate samples. ACM Technical Report 2008-01, California Institute of Technology, Pasadena, July 2008.

[NT08b] D. Needell and J. A. Tropp, *CoSaMP: iterative signal recovery from incomplete and inaccurate samples*, Appl. Comput. Harmon. Anal. **26** (2009), no. 3, 301–321, DOI 10.1016/j.acha.2008.07.002. MR2502366 (2010c:94018)

[NTLC08] B. Nett, J. Tang, S. Leng, and G. Chen. Tomosynthesis via total variation minimization reconstruction and prior image constrained compressed sensing (PICCS) on a C-arm system. In *Proc. Soc. Photo-Optical Instr. Eng.*, volume 6913. NIH Public Access, 2008.

[NV07a] D. Needell and R. Vershynin. Signal recovery from incomplete and inaccurate measurements via Regularized Orthogonal Matching Pursuit. *IEEE J. Sel. Top. Signa.*, 4:310–316, 2010.

[NV07b] D. Needell and R. Vershynin. Uniform uncertainty principle and signal recovery via Regularized Orthogonal Matching Pursuit. *Found. Comput. Math.*, 9(3):317–334, 2007.

[NW12] Deanna Needell and Rachel Ward, *Near-optimal compressed sensing guarantees for total variation minimization*, IEEE Trans. Image Process. **22** (2013), no. 10, 3941–3949, DOI 10.1109/TIP.2013.2264681. MR3105153

[NW13] Deanna Needell and Rachel Ward, *Stable image reconstruction using total variation minimization*, SIAM J. Imaging Sci. **6** (2013), no. 2, 1035–1058, DOI 10.1137/120868281. MR3062581

[OF96] B. A. Olshausen and D. J. Field. Emergence of simple-cell receptive field properties by learning a sparse code for natural images. *Nature*, 381:607–609, 1996.

[OH10] S. Oymak and B. Hassibi. New Null Space Results and Recovery Thresholds for Matrix Rank Minimization. Preprint, 2010.

[OSV03] Stanley Osher, Andrés Solé, and Luminita Vese, *Image decomposition and restoration using total variation minimization and the H^{-1} norm*, Multiscale Model. Simul. **1** (2003), no. 3, 349–370 (electronic), DOI 10.1137/S1540345902416247. MR2030155 (2004k:49004)

[PETM09] M. Protter, M. Elad, H. Takeda, and P. Milanfar. Generalizing the non-local-means to super-resolution reconstruction. *IEEE T. Image Process.*, 16(1):36–51, January 2009.

[PPM] S. Pudlewski, A. Prasanna, and T. Melodia. Compressed-sensing-enabled video streaming for wireless multimedia sensor networks. *IEEE T. Mobile Comput.*, (99):1–1.

[PRT11] Götz E. Pfander, Holger Rauhut, and Joel A. Tropp, *The restricted isometry property for time-frequency structured random matrices*, Probab. Theory Related Fields **156** (2013), no. 3-4, 707–737, DOI 10.1007/s00440-012-0441-4. MR3078284

[Rau08] Holger Rauhut, *On the impossibility of uniform sparse reconstruction using greedy methods*, Sampl. Theory Signal Image Process. **7** (2008), no. 2, 197–215. MR2451767 (2010m:94051)

[RBE10] R. Rubinstein, A. M. Bruckstein, and M. Elad. Dictionaries for sparse representation modeling. In *Proceedings of the IEEE*, volume 98, pages 1045–1057, 2010.

[RFP07] Benjamin Recht, Maryam Fazel, and Pablo A. Parrilo, *Guaranteed minimum-rank solutions of linear matrix equations via nuclear norm minimization*, SIAM Rev. **52** (2010), no. 3, 471–501, DOI 10.1137/070697835. MR2680543 (2012a:90137)

[ROF92] Leonid I. Rudin, Stanley Osher, and Emad Fatemi, *Nonlinear total variation based noise removal algorithms*, Phys. D **60** (1992), no. 1-4, 259–268. Experimental mathematics: computational issues in nonlinear science (Los Alamos, NM, 1991). MR3363401

[Rom08] J. Romberg. Imaging via compressive sampling. *Signal Processing Magazine, IEEE*, 25(2):14–20, 2008.

[RS97] Amos Ron and Zuowei Shen, *Affine systems in $L_2(\mathbf{R}^d)$: the analysis of the analysis operator*, J. Funct. Anal. **148** (1997), no. 2, 408–447, DOI 10.1006/jfan.1996.3079. MR1469348 (99g:42043)

[RS05] J. Rennie and N. Srebro. Fast maximum margin matrix factorization for collaborative prediction. In *Proc. 22nd int. conf. on Machine learning*, pages 713–719. ACM, 2005.

[RV06] M. Rudelson and R. Vershynin. Sparse reconstruction by convex relaxation: Fourier and Gaussian measurements. In *Proc. 40th Annual Conf. on Info. Sciences and Systems*, Princeton, Mar. 2006.

[RV08] Mark Rudelson and Roman Vershynin, *On sparse reconstruction from Fourier and Gaussian measurements*, Comm. Pure Appl. Math. **61** (2008), no. 8, 1025–1045, DOI 10.1002/cpa.20227. MR2417886 (2009e:94034)

[Sch86] R. Schmidt. Multiple emitter location and signal parameter estimation. *Antennas and Propagation, IEEE Transactions on*, 34(3):276–280, 1986.

[SED04] J.-L. Starck, M. Elad, and D. Donoho. Redundant multiscale transforms and their application for morphological component analysis. *Adv. Imag. Elect. Phys.*, 132, 2004.

[SFM07] Jean-Luc Starck, Jalal Fadili, and Fionn Murtagh, *The undecimated wavelet decomposition and its reconstruction*, IEEE Trans. Image Process. **16** (2007), no. 2, 297–309, DOI 10.1109/TIP.2006.887733. MR2462723 (2009j:94052)

[Sin08] Amit Singer, *A remark on global positioning from local distances*, Proc. Natl. Acad. Sci. USA **105** (2008), no. 28, 9507–9511, DOI 10.1073/pnas.0709842104. MR2430205 (2009h:62064)

[SM00] J. Shi and J. Malik. Normalized cuts and image segmentation. *IEEE Transactions on Pattern Analysis and Machine Intelligence*, 22(8):888–905, August 2000.

[Sre04] Nathan Srebro, *Learning with matrix factorizations*, ProQuest LLC, Ann Arbor, MI, 2004. Thesis (Ph.D.)–Massachusetts Institute of Technology. MR2717223

[SY07] Anthony Man-Cho So and Yinyu Ye, *Theory of semidefinite programming for sensor network localization*, Math. Program. **109** (2007), no. 2-3, Ser. B, 367–384, DOI 10.1007/s10107-006-0040-1. MR2295148 (2007j:90097)

[TG07] Joel A. Tropp and Anna C. Gilbert, *Signal recovery from random measurements via orthogonal matching pursuit*, IEEE Trans. Inform. Theory **53** (2007), no. 12, 4655–4666, DOI 10.1109/TIT.2007.909108. MR2446929 (2009h:94042)

[Tro06] Joel A. Tropp, *Just relax: convex programming methods for identifying sparse signals in noise*, IEEE Trans. Inform. Theory **52** (2006), no. 3, 1030–1051, DOI 10.1109/TIT.2005.864420. MR2238069 (2007a:94064)

[TSHM09] G. Tang, R. Shahidi, F. Herrmann, and J. Ma. Higher dimensional blue-noise sampling schemes for curvelet-based seismic data recovery. *Soc. Explor. Geophys.*, 2009.

[TWD+06] J. Tropp, M. Wakin, M. Duarte, D. Baron, and R. Baraniuk. Random filters for compressive sampling and reconstruction. In *Proc. 2006 IEEE Int. Conf. Acoustics, Speech, and Signal Processing*, volume 3. IEEE, 2006.

[Vid11] R. Vidal. Subspace clustering. *IEEE Signal Processing Magazine*, 28(2):52–68, 2011.

[WLD+06] M. Wakin, J. Laska, M. Duarte, D. Baron, S. Sarvotham, D. Takhar, K. Kelly, and R. Baraniuk. An architecture for compressive imaging. In *Image Processing, 2006 IEEE International Conference on*, pages 1273–1276. IEEE, 2006.

[ZCP+09] M. Zhou, H. Chen, J. Paisley, L. Ren, G. Sapiro, and L. Carin. Non-parametric bayesian dictionary learning for sparse image representations. In *Advances in Neural and Information Processing Systems (NIPS)*, 2009.

[ZLW+10] Z. Zhou, X. Li, J. Wright, E. Candes, and Y. Ma. Stable principal component pursuit. In *IEEE International Symposium on Information Theory Proceedings (ISIT)*, pages 1518–1522. IEEE, 2010.

Dept. of Mathematics and Statistics
Current address: San Jose State University
E-mail address: `guangliang.chen@sjsu.edu`

Dept. of Mathematical Sciences
Current address: Claremont McKenna College
E-mail address: `dneedell@cmc.edu`

Index

ℓ_1-analysis, 213
ℓ_1-minimization, 206
p-frame potential, 131
  probabilistic p-frame potential, 136

abc-problem, 47
adjoint operator, 7
algorithm
  Gerchberg-Saxton, 195
  IRLS, 196
  PhaseLift, 194
  Wirtinger flow, 195
analysis operator, 18
  fusion, 38

Balian-Low Theorem, 46
Bessel, 17
  fusion sequence, 37
Bézout determinant, 88
biorthogonal system, 25
bound
  Cramer-Rao Lower Bound, 190
bounded orthogonal matrices, 205

canonical dual frame, 20, 23
coherence, 131
complement property, 47
complete, 2
compressed sensing, 57, 80
  phase transition, 99
compressible, 203
Compressive Sensing, 44
compressive signal processing, 202
concatenations, 212
condition number, 12
cone, 119
conference graph, 74
continuous frames, 108
convex polytope, 119
correlation network, 84
CoSaMP, 207
curvelet frames, 212

data-dependent dictionaries, 217
diagonalizable, 12
diagram vector, 108, 121
dictionary learning, 217
difference set, 72
direct sum, 6
discrete forward GWT, 230
discrete inverse GWT, 231
distance, 2
  matrix-norm induced, 180
  natural metric, 180
distance between two frames, 44
D-RIP, 213
dual frame, 23
  canonical, 20, 23
  local, 37
dyadic cubes, 227

eigensteps, 66
  Gelfand-Tsetlin patterns, 79
  liftings, 89
  map, 89
eigenvalue, 12
eigenvector, 12
ellipsoid, 124
  ellipsoid of minimal volume, 124
  Fritz John ellipsoid, 124
  minimal ellipsoid, 124
equiangular, 40
equiangular frame, 41
equiangular lines, 40
equiangular tight frames, 41
erasure
  erasure channel, 55
erasures, 25
error diffusion, 157

Farkas lemma, 117
Feichtinger Conjecture, 45
fingerprint, 56
finite unit norm tight frames, 79
Fisher Information Matrix, 189
Fourier frame, 40
frame, 17
  bounded, 17
  bounds, 17
  canonical dual, 20, 23, 106
  coefficients, 17
  distance between, 44
  dual, 23
  equal norm, 17
  equiangular, 17, 41
  equiangular tight, 41, 131, 71
  exact, 25
  Fourier, 40
  frame operator, 106
  frame potential, 107
  FUNTF, 107
  fusion, 36
  Gabor, 46
  Gabor frame, 56
  Grassmannian frames, 71
  Grassmanian frames, 131
  harmonic frames, 59
  isomorphic, 22
  local, 37
  local frame bounds, 37
  Mercedes Benz, 18
  Mercedes-Benz frame, 58
  nearly equal norm, 43

nearly Parseval, 43
operator, 19
optimal frame bounds, 17
Parseval, 17
phase retrievable frame, 176
scalable frames, 108
tight, 17, 106
unit norm, 17
unit norm tight frame, 54
unitarily isomorphic, 22
frame force, 61
frame graph, 84
frame homotopy problem, 70, 80
frame operator, 19, 81
fusion, 38
probabilistic frame operator, 127
frame potential, 62, 127
probabilistic frame potential, 129
frame theory, 220
Fritz John, 124
full-spark, 80, 99
full spark, 183
fusion analysis operator, 38
fusion frame, 36
bounds, 37
Parseval, 37
system, 37
tight, 37
fusion frame operator, 38
fusion synthesis operator, 38
future directions, 234

Gabor frame, 46
Gabor frames, 212
geometric multi-resolution analysis, 227
geometric wavelet transforms, 230
GMRA, 227
Gram operator, 106
probabilistic Gram operator, 128
Grammian, 81
Gramian matrix, 35
Gramian operator, 35
greedy algorithm, 206

Hilbert-Schmidt norm, 82
Hilbert space isomorphism, 11

image, 7
image deblurring, 223
image denoising, 221
image inpainting, 223
image processing, 221
image super-resolution, 223
incoherent, 204
inner product, 2, 39
inverse operator, 7
isometry, 9
isomorphic, 11
isomorphic frames, 22

unitarily, 22

kernel, 7
Kirkman equiangular tight frames, 42
Krein conditions, 73
K-SVD, 224

lifting, 192
linear operator, 6
linearly independent, 2, 39
Lipschitz
bi-Lipschitz, 184
bounds, 186
constants, 187
local dual frames, 37
local frame bounds, 37
local frames, 37
low-rank data matrix, 210

majorization, 31
matrix
Fisher Information Matrix, 189
matrix representation, 7
Mercedes Benz Frame, 18
Minimal Moments, 25
minimal scaling, 121
model
AWGN, 189
nonAWGN, 189
modulation, 46
Moore-Penrose inverse, 9
multiscale geometric decomposition, 227
multiscale singular value decomposition, 227

Naimark complement, 73
Naimark's Theorem, 26
norm, 2
operator, 6
p-norm, 177

online dictionary learning, 232
operator
adjoint, 7
analysis, 18
bijective, 7
bounded, 7
fusion anaylsis, 38
fusion synthesis, 38
Gramian, 35
inequality, 19
injective, 7
inverse, 7
invertible, 7
linear, 6
norm, 6
normal, 9
positive, 9
powers of, 15

self-adjoint, 9
surjective, 7
synthesis, 18
trace, 16
unitary, 9
operator inequality, 19
operator norm, 6
optimal frame completion, 70
orthodecomposable, 64, 85
orthogonal, 2
direct sum, 6
complement, 4
subspaces, 4
to a subspace, 4
orthogonal matching pursuit, 206
orthogonally partitionable, 64
orthonormal, 2
basis, 2
orthonormal fusion basis, 37
oversampled DFT, 211

Parseval's Identity, 3
partial isometry, 9
Paulsen Problem, 43
Paulsen problem, 63
phase retrieval, 47
PhaseLift, 194
Platonic solid, 58
Polarization Identity, 3
powers of an operator, 15
principal component analysis, 211
probabilistic p-frame, 137
tight probabilistic p-frame, 137
probabilistic frame, 108, 126
convolution of probabilistic frames, 129
probabilistic canonical dual frame, 128
tight probabilistic frame, 126
product
$\langle\langle\cdot,\cdot\rangle\rangle$, 193
Hilbert-Schmidt, 177
symmetric outer, $[\![\cdot,\cdot]\!]$, 177
projection, 4
nearest point, 4
orthogonal, 4
pulse code modulation, 146
Pythagorean Theorem, 3

quadratic surface, 122
quantization, 145
$\Sigma\Delta$ quantization, 152
memoryless scalar quantization, 146
quantization alphabet, 145

range, 7
rank, 7
rank-nullity theorem, 7
realification, 181
reconstruction
phaseless, 175
redundancy, 24, 37
relative interior, 119
restricted isometry property, 204
Restricted Isometry Property, 44
Riesz basis, 21
robustness, 24

sampling operator, 203
scalable frame, 109, 110
m-scalable, 110
strictly m-scalable, 110
strictly scalable frame, 110
scalar quantizer, 146
scaling problem, 42
Schur–Horn theorem, 68
second moment, 108
signal space CoSaMP, 216
simplex, 27
Smale's 7th problem, 60
space
projective space, 176
real linear, 175
spanning set, 2
sparse, 44, 203
sparse and redundant modeling, 222
sparse coding, 219
sparse decomposition, 57
sparse reconstruction, 206
sparsity, 44, 203
Spectral Tetris, 32, 59
Spectral Theorem, 13
spherical t-design, 74, 134
Steiner system, 72
Stiefel manifold, 81
strongly regular graph, 72
subgaussian matrices, 205
subspace clustering, 220
synthesis operator, 18
fusion, 38

Thomson problem, 60
tight frames, 213
Top Kill, 69
total variation, 208
trace, 16
Trace formula for Parseval frames, 22
translation, 46
transversal intersection, 82
Triangle Inequality, 3

uniform noise model, 147
unitarily isomorphic frames, 22

Wasserstein metric, 109
wavelet frames, 212
Welch bound, 71
window function, 46
worst-case coherence, 57

Published Titles in This Series

73 **Kasso A. Okoudjou, Editor,** Finite Frame Theory, 2016

72 **Van H. Vu, Editor,** Modern Aspects of Random Matrix Theory, 2014

71 **Samson Abramsky and Michael Mislove, Editors,** Mathematical Foundations of Information Flow, 2012

70 **Afra Zomorodian, Editor,** Advances in Applied and Computational Topology, 2012

69 **Karl Sigmund, Editor,** Evolutionary Game Dynamics, 2011

68 **Samuel J. Lomonaco, Jr., Editor,** Quantum Information Science and Its Contributions to Mathematics, 2010

67 **Eitan Tadmor, Jian-Guo Liu, and Athanasios E. Tzavaras, Editors,** Hyperbolic Problems: Theory, Numerics and Applications, 2009

66 **Dorothy Buck and Erica Flapan, Editors,** Applications of Knot Theory, 2009

65 **L. L. Bonilla, A. Carpio, J. M. Vega, and S. Venakides, Editors,** Recent Advances in Nonlinear Partial Differential Equations and Applications, 2007

64 **Reinhard C. Laubenbacher, Editor,** Modeling and Simulation of Biological Networks, 2007

63 **Gestur Ólafsson and Eric Todd Quinto, Editors,** The Radon Transform, Inverse Problems, and Tomography, 2006

62 **Paul Garrett and Daniel Lieman, Editors,** Public-Key Cryptography, 2005

61 **Serkan Hoşten, Jon Lee, and Rekha R. Thomas, Editors,** Trends in Optimization, 2004

60 **Susan G. Williams, Editor,** Symbolic Dynamics and its Applications, 2004

59 **James Sneyd, Editor,** An Introduction to Mathematical Modeling in Physiology, Cell Biology, and Immunology, 2002

58 **Samuel J. Lomonaco, Jr., Editor,** Quantum Computation, 2002

57 **David C. Heath and Glen Swindle, Editors,** Introduction to Mathematical Finance, 1999

56 **Jane Cronin and Robert E. O'Malley, Jr., Editors,** Analyzing Multiscale Phenomena Using Singular Perturbation Methods, 1999

55 **Frederick Hoffman, Editor,** Mathematical Aspects of Artificial Intelligence, 1998

54 **Renato Spigler and Stephanos Venakides, Editors,** Recent Advances in Partial Differential Equations, Venice 1996, 1998

53 **David A. Cox and Bernd Sturmfels, Editors,** Applications of Computational Algebraic Geometry, 1998

52 **V. Mandrekar and P. R.Masani, Editors,** Proceedings of the Norbert Wiener Centenary Congress, 1994, 1997

51 **Louis H. Kauffman, Editor,** The Interface of Knots and Physics, 1996

50 **Robert Calderbank, Editor,** Different Aspects of Coding Theory, 1995

49 **Robert L. Devaney, Editor,** Complex Dynamical Systems: The Mathematics Behind the Mandelbrot and Julia Sets, 1994

48 **Walter Gautschi, Editor,** Mathematics of Computation 1943–1993: A Half-Century of Computational Mathematics, 1995

47 **Ingrid Daubechies, Editor,** Different Perspectives on Wavelets, 1993

46 **Stefan A. Burr, Editor,** The Unreasonable Effectiveness of Number Theory, 1992

45 **De Witt L. Sumners, Editor,** New Scientific Applications of Geometry and Topology, 1992

44 **Béla Bollobás, Editor,** Probabilistic Combinatorics and Its Applications, 1991

43 **Richard K Guy, Editor,** Combinatorial Games, 1991

42 **Carl Pomerance, Editor,** Cryptology and Computational Number Theory, 1991

41 **Roger Brockett, Editor,** Robotics, 1990

40 **Charles R. Johnson, Editor,** Matrix Theory and Applications, 1990

35 **Harry H. Panjer, Editor,** Actuarial Mathematics, 1986

ISBN 978-1-4704-2019-2

9 781470 420192

PSAPM/73